third edition
Basic
Statistical
Analysis

RICHARD C. SPRINTHALL
American International College

PRENTICE HALL, Englewood Cliffs, New Jersey 07632

Library of Congress Cataloging-in-Publication Data

Sprinthall, Richard C.,
    Basic statistical analysis / Richard C. Sprinthall. -- 3rd ed.
        p.   cm.
    Includes bibliographies and index.
    ISBN 0-13-066094-9
    1. Social sciences--Statistical methods. 2. Statistics.
I. Title.
HA29.S858   1990
519.5'0243--dc20   89-32941
                                                            CIP

Editorial/production supervision
  and interior design: Rob DeGeorge
Cover design: Ben Santora
Manufacturing buyer: Peter Havens

 © 1990, 1987, 1982 by Prentice-Hall, Inc.
A Division of Simon & Schuster
Englewood Cliffs, New Jersey 07632

Printed in the United States of America

10 9 8 7 6 5 4 3

ISBN 0-13-066094-9

PRENTICE-HALL INTERNATIONAL (UK) LIMITED, *London*
PRENTICE-HALL OF AUSTRALIA PTY. LIMITED, *Sydney*
PRENTICE-HALL CANADA INC., *Toronto*
PRENTICE-HALL HISPANOAMERICANA, S.A., *Mexico*
PRENTICE-HALL OF INDIA PRIVATE LIMITED, *New Delhi*
PRENTICE-HALL OF JAPAN, INC., *Tokyo*
SIMON & SCHUSTER ASIA PTE. LTD., *Singapore*
EDITORA PRENTICE-HALL DO BRASIL, LTDA., *Rio de Janeiro*

Contents

PART II   INFERENTIAL STATISTICS

Preface

As with any new edition, although significant changes and additions have been incorporated into this third edition of *Basic Statistical Analysis,* the major coverage remains the same. In fact, for readers of past editions, this new version may seem, as Yogi Berra once said, like experiencing "deja vu all over again." Like previous editions, this book is for students in the social sciences who are confronting their first "stats" course.

Significant changes that have been made in this edition include the following:

1. A new chapter (chapter 17) has been written focusing on the use of computers to solve large data-set problems. This chapter is intended to emphasize that some basic knowledge of the procedures and assumptions of general statistics are needed to use the computer effectively as an analytic aid. The centerpiece of this chapter showcases a series of computer print-outs which, for one reason or another, have all produced errors (due to mistakes in entry, machine problems, or even program bugs). The reader's job will be to identify the errors on the basis of the "Logic Checks" presented throughout the text. This will accomplish at least two important objectives: 1) to remind you that the computer is not an infallible genius but a fast idiot who needs a smart leader, and 2) to reinforce the various statistical logic checks by making them "live" in the context of a computer printout.

2. A special section is now included at the end of the text which covers the essentials of the binomial distribution and its relationship to the z distribution. This section includes methods for obtaining exact probabilities for the binomial distribution, as well as the z approximations. The coverage also includes the z test for proportions and the t test for differences between proportions (and its relationship to chi square). Problems are worked out within the exposition of the text and a series of student problems is presented at the end.

3. In chapter 3 the new material includes the semi-interquartile range for establishing variability around the median.

4. Chapter 6 has a new section on combining probabilities.

5. In chapter 8, the sections on the single-sample t and interval estimation have been rewritten in order to show the complimentary nature of parameter estimation and hypothesis testing.

6. The coverage of power has been extended and placed much earlier in the text (in chapter 9).

7. Chapter 14 incorporates more precision in introducing and calculating both partial and multiple correlations.

8. At the end of each chapter new and hopefully interesting problems have been added.

This book is appropriate for students in a variety of fields: psychology, sociology, education, social work, and health-related fields. It is wide-ranging because a standard deviation is, after all, forever a standard deviation, whether it describes IQ scores, hours spent watching television, or the body temperatures of a group of flu-stricken patients in a nursing home. Statistical techniques of data analysis cross all fields of social inquiry: the procedures remain the same but the examples differ. For this reason, examples have been chosen to represent most of the disciplines within the social sciences.

For many students, approaching that first statistics course is a rather scary and mystifying experience. The prospect is laced with a large dose of ambivalence, like a two-edged sword which has "can't live with it" on one side and "can't live without it" on the other. The "can't-live-with-it" feeling often owes its genesis to the remembrance of past traumas, perhaps a ninth grade algebra course in which everything went smoothly "until we got to the word problems." And the "can't-live-without-it" conviction may be even more direct and compelling, since the statistics course is probably required or, at the very least, "strongly recommended."

To alleviate the pangs of anxiety, many students adopt one of two main defenses, both of which revolve around the same core notion that this is a course that is simply not needed. First, there are those students who are convinced they do not need it because the intricacies of statistical proof are irrelevant. To them, reading the conclusions of a research report is enough: "If it's got numbers in it, especially *significant* numbers, then I believe it." Understanding whether or not the logic of the design warrants those conclusions is deemed a waste of time. Students feel that this is like asking them to understand the theory of the internal combustion engine when all they want to do is drive the car. The attitude of these students resembles that of the person who earnestly intones that "they say to starve a cold," without having the vaguest notion as to who "they" are ("they" are always the anonymous voices of supreme and unchallenged authority). This group of students is blithely willing to accept all statistical results. Second, there are those students who are convinced they do not need the course because statistics is all a pack of lies. To them, statistical proof is tantamount to no proof at all and so they smugly and cynically reject all statistical evidence. Their attitude is like that of the person who says "I don't think it works this way," and when questioned as to how it does work, replies "some other way!"

There is, of course, a germ of truth in both arguments. Significant numbers do mean something, although not always what the researcher concludes. Some statistical proofs are like a sleight-of-hand show—now you see it, now you don't. In order to know what the significant results mean and when the statistical charlatan is operating, at least one course in statistics is essential. Newspapers, magazines, and television newscasts constantly bombard us with the results of research studies: drinking coffee causes this, drinking beer causes that, washing clothes in a certain detergent causes something else, and so on.

The avowed goal of this book is to demystify statistics—to state the case for statistical analysis and inference in clear, no-frills language. The student is told specifically what an X is, what a Y is, and whether the twain shall meet. As in law school, the student is presented with rules of evidence and the logic behind those rules. The focus will constantly remain on how statistical techniques *can be used.* It will not be a case of presenting the best method to calculate a standard deviation and then leaving it up to the student to find some use for it. Statistical concepts are imbedded in the hard rock of research methodology. The student will learn at a practical level how to read and do statistical research. The student will be given a guided tour of the most important and practical exhibits in the statistician's showcase: not to create feelings of awe, but to teach.

For most students, it will be easy to read and, at times, perhaps even fun to read. The book assumes *little or no background in mathematics.* The student will not be stunned by finding an elegant, but tricky, derivation on page 3 or by finding that the author suddenly assumes on page 5 that everyone remembers enough calculus to integrate the normal curve equation. The student does not even have to remember arithmetic, let alone calculus.

The use of this book does require, however, that the student own an electronic calculator. Although the calculator need not be expensive, it must at least have a square-root key. Pressing the square-root key is easier and more accurate than looking up and interpolating table values. Therefore, the back of this book is not cluttered with pages of square and square-root tables.

Although the text is designed primarily for a one-semester, beginning course in statistics, enough added material is presented to allow its use by students taking more advanced courses. Chapters 1–13 contain topics usually covered in a one-semester course and if this is what is needed, the course can end with Chapter 13. At this point, the student will have gained enough understanding of statistical reasoning and research methodology to be able to read and comprehend a large part of the research in the social science literature. Because many students must later take courses in experimental psychology or in research methods in sociology and/or education, topics sometimes found in the more advanced courses are also included here in Chapters 14 to 18.

The book is divided into three major units: Descriptive Statistics, Inferential Statistics, and Advanced Topics in Inferential Statistics. How can a book of this size cover so much? Because some topics will not be covered here. First, little consideration will be given to grouped-data problems. Finding class intervals and standard deviations from the frequency data inherently creates some error and also loses track of the individual score. When the amount of data is so large that grouped-data techniques are really needed, statisticians usually turn to computers anyway. Second, the coverage of probability theory will be shortened. Not that probability theory is unimportant; it is absolutely crucial. But the only probability concepts found in this book are those that bear directly on statistical tests of significance. What is practical? How to calculate and understand the logic behind such things as z scores, the t test, ANOVA, chi square, and the Pearson r.

Special features of this book include:

1. *Definitions of key concepts in the glossary.* Brief, but thorough, definitions are conveniently presented. Experience has clearly shown that much of the trauma experienced by students taking their first "stats" course can be traced to confusion over

terminology. Conscientious use of the glossary can alleviate most if not all of this confusion.

2. *A programmed approach to the computation of each statistical test.* Computational procedures are set forth in a step-by-step, programmed format. A student who can follow a recipe or build a simple model plane can do an ANOVA.

3. *Constant stressing of the interaction between statistical tests and research methodology.* Examples are used from the literature of the social sciences to illustrate strategic methodological problems. Statistical analysis, if not carried out in the context of methodology, can degenerate into an empty and sterile pursuit. Two chapters have been specially designed to bridge the analysis-methodology gap: Chapter 11 focuses directly on the essentials of the research enterprise and Chapter 18 presents 26 research simulations which are programmed in such a way as to lead directly to the appropriate statistical test.

4. *A literary style both easy to read and attention-getting.* This book attempts to generate a feeling of excitement and enthusiasm by talking directly to the student and focusing on the student's own life space. Students obviously learn best when their interest is aroused.

5. *A large number of problems and test questions.* John Dewey's "learn by doing" axiom was never more true than in the field of statistics. Over 400 problems and test questions are placed both within and at the end of each chapter. Students need the opportunity to "try their hand" at practice problems and get some immediate feedback as to their progress.

6. *Instructor preference on the standard deviation.* All of the inferential tests using the standard deviation are presented using both the sample standard deviation *and* with the unbiased population estimate of the standard deviation. The instructor is free to choose the technique best suited to the students' needs.

7. *A list of key points and names.* Each chapter contains a list of key points and names, which also appear in the glossary. At the end of each chapter, a convenient, wrap-up summary, is also provided.

8. *Computer Program.* A statistical program (PH-STAT) is available which covers all the statistical tests presented. The program is totally menu driven and can be easily handled by the first-time user.

9. *Computer printouts.* A series of computer printouts, all containing errors of one kind or another, is presented. Students may then use the "logic checks" found throughout the text to identify these errors.

10. *Supplements.* An "Instructor's Manual" containing well over 1500 test items and problems is available.

Putting together a book of this type requires a lot of help. Special thanks must go to the "significant others" in my academic life, the professors who first initiated and then sustained my interest in statistical analysis: Greg Kimble, then of Brown University; Nate (Mac) Maccoby, then of Boston University; and the late P. J. Rulon of Harvard University. Without them, this book would never have happened. Also, I am grateful to the Literary Executor of the late Sir Ronald A. Fisher, F.R.S., to Dr. Frank Yates, F.R.S., and to Longman Group Ltd. London, for permission to reprint Tables III, IV, and VI from their book *Statistical Tables for Biological, Agricultural, and Medical Research,* (6th edition, 1974).

More recently, I must thank my colleagues at American International College: James Brennan for his overall help on topic sequencing; Lee Sirois for his significant role in putting together Chapter 8; Gregory Schmutte for his valuable contributions to the research Chapter, 11; to Gail Furman for her editing of the binomial section; and especially Carol Spafford both for her creative examples in Chapter 10 as well as for her careful reading and editing of the entire manuscript. I also wish to extend a special thanks to Barry Wadsworth at Mt. Holyoke College for his insightful comments on topic inclusions (and exclusions), to Barbara Anderson at Rhode Island College for her help and encouragement during the early stages of the manuscript's development, to Jim Vivian at Boston College for his ideas on sampling theory, to Steve Fisk at Bowdoin College, for his expertise in putting together the computer program, and to Ken Weaver at Emporia State University in Kansas for his careful reading of the entire manuscript. Without his help, many errors of omission and commission would have crept into the text.

At Prentice Hall I received many valuable suggestions from both Susan Willig and Carol Wada. They were truly the patient and motivating "coaches" of the home-office team. I also wish to thank Diane Lange for her meticulous copy editing and Rob DeGeorge for his skill in managing the overall production of the book. I am grateful to the following reviewers for their many helpful and sophisticated comments: Christine E. Bose, SUNY at Albany, Roger B. Frey, University of Maine, and Roger Drake, Western State College of Colorado.

Also, a special note of gratitude must go to both Dianne Sprinthall and Lou Conlin for their illustrations, cartoons, and general help in artistic design. Finally, and most of all, I wish to thank my students who, over the years, taught me how to become a better teacher.

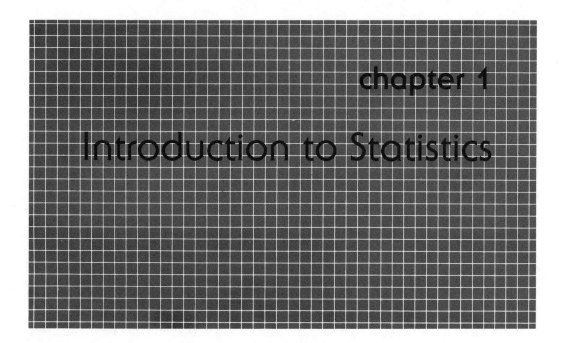

chapter 1
Introduction to Statistics

Rather than beginning by listing all the glorious reasons why you should take a first course in statistics, let us assume that it is probably a required course and that you have to take it anyway. Perhaps you have put it off for quite a while, until there is no choice left but to "bite the bullet" and get it over with. This is not to say that all of you have been dragged, kicking and screaming, into this course; however, as statisticians would put it, the probability is high that this hypothesis is true for some of you.

STUMBLING BLOCKS TO STATISTICS

Let us look at some of the most common objections raised by students when confronted with this seemingly grim situation. Perhaps your feelings of intimidation arise because you know you have a math block. You're still being buffeted by lingering anxieties left over from some math course taken in the perhaps distant past. Or maybe it's that you have read or heard a research report and been totally confused by the seemingly endless and seemingly meaningless stream of jargon. Perhaps you're a person who simply does not trust statistical analysis. If this is the case, you're in good company. Benjamin Disraeli, Queen Victoria's prime minister, once said, "There are three kinds of liars: liars, damned liars, and statisticians." Disraeli obviously agreed with the sentiment expressed by many—that you can prove anything with statistics.

PEANUTS

Many students, frightened by their first statistics course, claim to have a math phobia.
© 1979 United Feature Syndicate, Inc.

Whatever their basis, your doubts about taking this course will probably prove unfounded. You may even, believe it or not, get to like it and voluntarily sign up for a more advanced course.

Math Block.   First, although it is obvious that people do differ in math ability, a case of true math block is extremely rare and difficult to substantiate. It is true that some very fortunate people have a kind of perfect pitch for math. They take to math as gifted musicians take to harmony. (You remember the kid we all hated in high school, the one who completed calculus during his sophomore year and was angry because the school didn't offer any more advanced math courses.) At the other end of the continuum, we find those people who are definitely math phobics. To them, merely drawing a number on the chalkboard evokes strangulating feelings of sheer panic. They avoid any course or any situation (even keeping a checkbook) that deals with those spine-chilling little inscriptions called numbers. If you're one of those who suffers from or borders on this condition, relax— this is not a math course. While numbers are involved and certain arithmetic procedures are required, thanks to the magic of electronics you won't have to do the arithmetic yourself.

Go to your friendly neighborhood discount store and, for less than ten dollars (less than the price of a good slide rule), purchase a small electronic calculator. You don't need a fancy calculator with several memories, but do insist on one with a square root key. Statisticians, as you will see, love to square numbers, add them, and then extract the square root. In fact, you really must get a calculator, for this text assumes that you own one. The back of this book is not cluttered with page after page of square and square root tables. It's not only that such tables are a relatively expensive waste of space, but it is easier for you to push a button than it is to interpolate from a table. Your job is to focus on the logic of statistics. To rephrase the bus ad that says "leave the driving to us," leave the arithmetic to the calculator.

While we're on the topic of arithmetic, a quick word of caution is in order. *Do not* scare yourself by thumbing through later chapters of the book at this point. Some of the procedures and equations found there will seem absolutely harrow-

Don't guess; use your calculator. © 1967 United Feature Syndicate, Inc.

ing to contemplate. When the time comes for you to confront and use these techniques, the mystery and fear will have disappeared. With the background provided by the early chapters, the later chapters will, like a perfect bridge hand, play themselves. If you can follow step-by-step directions well enough to bake a cake or assemble a simple model airplane, you can do any of the procedures in this book and even more important, understand them.

Statistical Jargon.   As to the objection concerning the jargon or technical language, again, relax—it's not nearly as bad as it seems. Too often students are dismayed by the number of technical terms and the seemingly endless statistical lexicon, written in (oh no!) both Greek and English. Too often students are traumatized by the rigors of statistical analysis and its vast and mysterious body of symbols. Social scientists especially seem to make a fetish of the intricacies of significance tests and measurement theory. They seem to spend countless hours debating and fretting over statistical details that often seem trivial to the casual observer. There is also the psychology instructor who gives exam grades back in the form of standard scores. "Never mind about the standard error, or the amount of variance accounted for, did I pass? is the oft heard plea of many a student.

Is the researcher's use of statistical terms simply a case of sound and fury, signifying nothing? Obviously, it is not. The jargon of the trade represents an attempt to be precise in the communication of meaning. This effort is especially important in the social sciences because the concepts being considered are not always as precise as they are in the physical sciences like physics and chemistry. In short, there are some terms and symbols that must be learned. However, you can also get some help. At the end of the book, important terms, concepts, and equations are set down and defined. Faithful use of these glossary items definitely increases the retention of learned material.

Statistical Sleight-of-Hand.   Finally, the objection that the field of statistics resembles a sleight-of-hand show (now you see it, now you don't) is valid only when the research consumer is totally naïve. The conclusions that "figures don't lie, but liars can figure" and that one can prove anything with statistics are only true when the audience doesn't know the rules of the game. To the uninitiated, liars can indeed figure plausibly. But an audience with even a patina of statistical sophistication will not be easily misled by such artful dodgers. Unscrupulous persons will probably always employ faulty statistical interpretations to gain their ends. By the time you finish this course, however, they will not be able to lie to you.

Frankly, some statistical studies, especially correlation studies, have been grossly misinterpreted by certain researchers. This does not mean that all statisticians are charlatans. If statistics, as a result of some trickery and deceit, has a bad name, it is a result of the *misuse* and not the use of statistics. There are some booby traps lying in wait out there, but you'll be guided safely along the right path and each potential pitfall will be carefully pointed out. It should be stressed that most statisticians *do* use the correct statistical tests, and it is almost unheard of that the data have been faked. The traps usually result from the way the conclusions are drawn from the data. Sometimes the conclusions are simply not logically derived. Let's consider a couple of examples.

An often-heard argument against capital punishment is that it simply does not have a deterrent effect. To support this it is stated that years ago in England when convicted pickpockets were publicly hanged, other pickpockets worked the crowd that was there to witness the hanging. Can you see any problems with the logic of this example? The fallacy is that there is no comparison (or, as it will later be called, control) group. What about the number of pockets picked in crowds at a horse race or a carnival? If the frequency of pocket picking was lower at the public hangings, then perhaps capital punishment did have a deterrent effect.

Or statistics show that, say, 50% of a certain country's population are so illiterate they can't read the morning newspaper. Watch it! Perhaps 25% are under age 6, perhaps another 10% are blind, and so on.

In another example, *Parade* magazine headlined the story that "Americans are happier in marriage than their parents were." This news was based on a survey of young married couples, 70% of whom described themselves as being "happily married," while only 51% could say the same about their parents. The question of course is, "Compared to whom?" Had the parents participated in the survey, which they didn't, it might be important to find out how they described their own marital satisfaction, and, more important, the marital success of their offspring. With those data, a meaningful comparison could at least be attempted.[1]

Also in 1985, New York City released figures showing that the uniformed police made only 30% of the subway arrests, while the plainclothes police hit the 70% arrest figure. Does this mean that the plainclothes police work that much harder, or could it possibly be that the uniform is indeed a deterrent?

Or a paint company runs a TV ad showing a house that originally sold for $75,000. The new owner then repaints the house with the brand being advertised and later sells the house for $90,000. The implicit message, of course, is that the new paint job increased the value of the house by $15,000. However, in the absence of a definite time frame, there is no way to factor out the obvious effects of inflation. The house may easily have sold for $90,000, or more, given enough time—with or without its reglossed exterior.

Or consider this one. A researcher with an obvious antifeminist bias concludes that virtually all America's problems can be traced directly to the women's liberation movement. Statistics are paraded before us to prove that as women left the kitchen and entered the fields of psychiatry, criminal justice, politics, real estate, and law, for example, the prison population tripled. The increase in the number of women entering business coincides directly with the number of small business failures. Finally, the increasing number of women's individual bank accounts correlates with the increasing number of felonies. This is the kind of misuse of statistics that causes honest, competent statisticians to blanch and also casts a pall of suspicion over the whole field. If people accept such artful juxtaposing of statistics as proof of a causal relationship, then, indeed, statistics can be used to prove anything. The point, of course, is that just because two events occur simultaneously is no reason at all to conclude that one of these events has in any way caused the other. As we shall see later, the only way we can ferret out a cause-and-effect relationship is through the use of the controlled experiment.

You must always be careful when evaluating groups of different sizes. For example, it might be argued that Alaska is a far healthier area in which to live than New York City is. Why, do you realize that in New York City last year over 70,000 people died, whereas in Ugashik, Alaska, only 22 people died (the other 8 probably moved to Fairbanks)? Only by converting the numbers of deaths to *percentages* of the two total populations could a meaningful comparison be made in a case like this.

Another example of this sort of unequal comparison is when we are told how remarkably safe it is to fly in a privately owned airplane. Do you realize that during 1978 there were over 50,000 fatalities in the United States from auto accidents, but only 1690 fatalities from plane accidents? This is exactly like comparing New York City with Ugashik, Alaska, since there are over 100 million automobiles on the highways but only 187,000 privately owned planes. In fact, if we compare autos and planes using the more realistic yardstick of the rate of fatal accidents per vehicle mile, automobiles are at 2.9 and private planes at 29.8. Therefore, despite the grisly statistics indicating the carnage on United States highways, the fatality rate is 10 times higher among those involved in private aviation.[2]

Unequal comparisons of this type bring to mind the fish store owner who beat all the competition by selling lobster meat for two dollars a pound less. The owner admitted that the lobster meat was stretched out some with shark meat. The best estimate of the proportions was 50–50, one shark to one lobster. Always be alert to the sizes of the groups or the things that are being compared.

Also, you must be aware of the size of the sample group from which the statistical inference is being made. Recently, a well-known company reported the results of a comparative study on the effects of using its brand of toothpaste. "The study was conducted over a period of six months, and millions of brushings later," the company concluded that the other popular fluoride toothpaste was "no better than ours." This is interesting in that the company isn't even claiming victory, but is apparently proud to have merely achieved a tie. Even more interesting, however, is that we are never told *how many subjects* participated in this study. Millions of brushings later could mean four people brushing every minute of every day for the full six months. If so, no wonder they don't have any cavities, they probably haven't any teeth left.

Even percentages can be grotesquely misleading if population size is left unknown. Suppose a listenership survey reports that 33% of the respondents polled said that they listen regularly to radio station WFMF. Perhaps only three people were polled, and one of those, the last one surveyed, mentioned WFMF—at which point the polling ended.

Be extremely skeptical when reading reported corporate profits, especially huge percentage increases. A company can truthfully report a 100% increase in profits and still be having a very poor year. Suppose in the previous year the company earned 1 cent for each dollar spent—a very modest return on capital. This year they earn 2 cents for each dollar. That is a bona fide 100% increase in profits, but it's hardly enough to keep a prudent stockholder happy or the company long in existence.

In his book *How to Lie with Statistics,* Darrell Huff tells us how the president of a flower growers' association loudly proclaimed that certain flowers were now 100% cheaper than they were four months ago. Huff notes that "he didn't mean that florists were now giving them away—but that's what he said."[3]

The National Safety Council is always telling us how dangerous it is to drive our cars near our own homes. We are warned that over 85% of all auto accidents occur within 10 miles of the driver's home. It is alleged that one person became so frightened after hearing about this high percentage of accidents occurring close to home that he moved! The other side to the story is, of course, that more than 95% of all driving is done within 10 miles of home. Where does the danger really lie? It lies in not being alert to statistical misdirection. While statistics do have an aura of precision about them, the interpretations are often anything but precise.

It's All in the Question. You must be especially careful when reviewing the results of all the polling data that constantly bombard us. In the first place, the questions themselves may be loaded—phrased in such a way as to favor a specific answer. For example, a question such as "Do you prefer socialized medicine as practiced in England or free-enterprise medicine as practiced in the United States" might well be slanted. Patriotic respondents might opt for the American variety simply because it is American. Also, the terms "socialized" and "free-enterprise" might bias the results, since some respondents might be negatively disposed to anything socialistic without always realizing what it means in context. When a New York Times—CBS poll asked respondents their opinion on a constitutional amendment "prohibiting abortions," the majority (67%) opposed it. But when the question was reworded, the majority (51%) favored "protecting the life of the unborn child." Finally, poll data may even be reported in such a way as to belie the questions that were actually asked. A sample of Americans was asked,

"Would you be concerned if the family unit were to disintegrate?" Of those polled, 87% said yes. The headlined conclusion in the newspaper was that "87% of Americans are concerned over the disintegration of the family unit." One might just as well ask, "Would you be concerned if the sun were to explode?" The conclusion would then be, "100% of Americans are concerned that the sun will explode."

A BRIEF LOOK AT THE HISTORY OF STATISTICS

The general field of statistics is of fairly recent origin, and its birth and growth were spurred on by some very practical considerations. Although some form of both mathematics and gambling has existed since the earliest days of recorded history, it wasn't until about 300 years ago that any attempt was made to bring the two together.

It is rather curious that it took the human race such a long time to gain any real understanding of the probability concept. Primitive persons used numbers, had a counting system, and were not averse to gambling. Well-formed dice (which must have played true since even now they show almost no bias whatsoever) have been found and dated at least as far back as 3000 B.C. Perhaps early humans were afraid to think about probability, believing that it was the sole province of the gods. Perhaps it was felt that to deny the gods' control over events would be an invitation to personal disaster, either at the hands of the gods or at the hands of the religious authorities. It was probably easier and a good deal safer to speak fatalistically in terms like "the wheel of fortune" and "the throw of the dice," rather than to dare to penetrate the mysteries of the dieties and thereby bring on a charge of impiety.

In Book I of *De Divinatione,* Cicero wrote 50 years before the birth of Christ, "They are entirely fortuitous you say? Come! Come! Do you really mean that? When the four dice produce the venus-throw you may talk of accident; but suppose you made a hundred casts and the venus-throw appeared a hundred times; could you call that accidental?"[4] Implicit in his statement, of course, is that the gods must intervene to cause the occurrence of so improbable an event. In this passage, Cicero is voicing the popular view of his day, but in later writings he indicates his own mistrust of this opinion.

The advent of Christianity didn't do much to advance thinking in this area. Writing in the fifth century A.D., St. Augustine said that "nothing happened by chance, everything being minutely controlled by the will of God. If events appear to occur at random, that is because of the ignorance of man and not in the nature of the events."[5]

Even today, many people prefer not to calculate probabilities, but instead to trust blind luck. All of us have met bold adventurers who grandly dismiss the dangers inherent in their newest sport. "After all, you can get killed in your own driveway," they intone, not caring to be bothered with the blatant probability dif-

Blaise Pascal (1623–1662). (The Bettman Archive.)

ferences between getting hurt by falling on the driveway and getting hurt while hang gliding off the top of Mount Everest.

Pascal and Gossett.   During the seventeenth century, the birth of statistics finally took place. It happened one night in France. The scene was the gaming tables, and the main character was the Chevalier de Mère, a noted gambler of his time. He had been having a disastrous run of losing throws. To find out whether his losses were indeed the product of bad luck or simply of unrealistic expectations, he sought the advice of the great French mathematician and philosopher **Blaise Pascal** (1623–1662). Pascal worked out the probabilities for the various dice throws, and the Chevalier de Mère discovered that he had been making some very bad bets indeed. Thus, the father of probability theory was Pascal. His motive was to help a friend become a winner at the dice table. Although Pascal's motive may seem not to have been overly idealistic, it was extremely practical as far as the Chevalier de Mère was concerned.

Another milestone for statistics occurred at the turn of the century in Ireland at the famous Guinness brewery, now known worldwide for the record books of the same name. In 1906, to produce the best beverage possible, the Guinness Company decided to select a sample of people from Dublin to do a little beer tasting. Since there turned out to be no shortage of individuals willing to participate in this taste test, the question of just how large a sample would be required became financially crucial to the brewery. They turned the problem over to the mathematician **William Sealy Gossett.** In 1908, under the pen name "Student,"

Gossett produced the formula for the standard error of the mean (which specified how large a sample must be, for a given degree of precision, to extrapolate accurately its results to the entire beer-drinking population).

So that's the history—craps and beer—hardly likely to strike terror in the hearts of students new to the field. The point is that the hallmark of statistics is the very practicality that gave rise to its existence in the first place. This field is not an area of mysticism or sterile speculations. It is a no-nonsense area of here-and-now pragmatism. You will not be led upstairs to a dark and dingy garret, with a taper and a crust of bread, to contemplate heavy philosophical issues. Instead, you will, with your trusty calculator in your hand, be brought into the well-lit arena of practicality.

BENEFITS OF A COURSE IN STATISTICS

If, as the Bible says, the truth shall set you free, then learning to understand statistical techniques will go a long way toward providing you with intellectual freedom. Choosing to remain ignorant of statistical concepts may doom you to a life sentence of half-truths. Essentially, the benefits of a course like this are twofold. You will learn to read and understand research reports, and you will learn to produce your own research. As an intelligent research consumer, you'll be able to evaluate statistical reports read at professional conventions or printed in your field's journals. Also, as a student of the social sciences, you will probably be called on at some time to do original research work. This prospect will not seem so overwhelming after you've mastered the tools available in this book. More basic than that, you'll have a far better chance of understanding research items in newspapers or magazines, or on TV. Who should take statistics? Virtually anyone who wishes to be informed.

THE GENERAL FIELD OF STATISTICS

Statistics as a general field consists of two subdivisions: descriptive statistics and predictive or inferential statistics.

Descriptive Statistics

Descriptive statistics involves techniques for describing data in abbreviated, symbolic fashion. It's a sort of shorthand, a series of precise symbols for the description of what could be great quantities of data.

For example, when we are told that the average score on the verbal section of the Scholastic Aptitude Test (SAT) is 435, we are being provided with a description of one characteristic of hundreds of thousands of college-bound high school students. The descriptive tool used in this case is the arithmetic average, or the mean. To arrive at this value, the SAT verbal scores of all the high school students taking the test throughout the country were added together, and then the total was divided by the number of students involved. The resulting mean value of 435 *describes* one characteristic of this huge group of high school students.

Perhaps we would also like to know how wide the range of SAT scores was. To arrive at this value, the difference between the highest and lowest scores is calculated. In the case of the SAT distribution, where the highest score is 800 and the lowest 200, the range is found to be 600.

By a knowledge of this value, our description of the group gains additional refinement. Other important descriptive statistics are the median, the mode, the standard deviation, and the variance. Chapters 2 and 3 will introduce these descriptive techniques.

Inferential Statistics

Inferential statistics involves making predictions of values that are not really known. Suppose we wished to estimate the height of the average American male. Since it would be impossible to line up all the men in the country and actually measure them, we would instead select a small number of men, measure their heights, and then predict the average height for the entire group. In this way, inferential statistics makes use of a small number of observations to predict to, or infer the characteristics of, an entire group.

This process of inference is obviously risky. The small group of observations from which the inference will be made must be representative of the entire group. If not, the predictions are likely to be way off target. A person who takes a small amount of blood for analysis knows that the sample is fairly representative of all the blood in the entire circulatory system. But when a researcher takes a sample of adult males, no one can be absolutely sure that true representation has been achieved. Also, the researcher seldom, if ever, gets the chance to verify the prediction against the real measure of the entire group. One exception, however, is in political forecasting. After pollsters like Gallup, Harris, and Yankelovich make their predictions as to how the population of voters will respond, the actual results are made compellingly (and sometimes embarrassingly) clear on the first Tuesday in November.

Despite the riskiness of the endeavor, statisticians do make predictions with better than chance accuracy (actually, far better than chance) about the characteristics of an entire group, even though only a small portion of the group is actually measured. Inferential statistics is not an infallible method. It does not offer eternal truth or immutable reality carved in stone. As one statistician said, "There is no such thing as eternal truth until the last fact is in on Judgment Day." It does offer a probability model wherein predictions are made and the limits of their accuracy are *known*. As we will see, that really isn't bad.

THEORY OF MEASUREMENT

According to an apostle, "the very hairs of your head are all numbered" (Matt. 10:30). Although that may be true, it's only part of the story. The history of civilization reveals that human beings began counting long before they began writing. Today, we are beset by such a dizzying array of measures, counts, estimates, and averages that modern life is beginning to resemble an ongoing problem in applied mathematics. Our society is so enamored of measures that current descriptions of life on this planet are as often conveyed by numbers as words. Economists keep creating new financial ratios; psychologists, new personality assessments; sociologists, new sociometrics; and political scientists, new demographics—all at a mind-boggling pace. We have, as H. G. Wells put it, "fallen under the spell of numbers." How can we cope? By learning a few fundamental concepts of statistical measurements, we can soften and make more intelligible the noise from our numerical Tower of Babel.

Measurement is essentially the assigning of numbers to observations according to certain rules. When we take our temperature, or step on the bathroom scales, or place a yardstick next to an object, we are measuring, or assigning numbers in a prescribed way. Perhaps every student entering college has had the uncomfortable experience of being measured dozens of times, in everything from height and weight to need for achievement. Obviously, then, scoring an IQ test and reading a tape measure placed around your waist are examples of measurement. But so, too, are counting the number of times a politician's speech mentions the word inflation, or noting the order of finish of the horses at the Kentucky Derby. In each case, a number is being assigned to an observation. People are even measured as to such complex qualities as scientific aptitude and need for affiliation. The basic measurement theory involved, however, is the same as it is for something as precisely physical as being fitted for a new pair of ski boots.

Variables and Constants

Throughout this discussion and the rest of the book, the word "variable" will keep appearing. At the risk of stating the obvious, a **variable** is anything that can be measured and observed to vary. It is any measured quantity that the researcher allows to assume different values. Keep this in mind: anything that is kept at only a single value *cannot be a variable*. Measures of this type are instead called constants. This will become an issue of great importance during our discussion (Chapter 11) of research methods.

SCALES OF MEASUREMENT

The way in which the numbers are assigned to observations determines the scale of measurement being used. Earlier we noted that measurement is based on assigning numbers according to rules. The rule chosen for the assignment process, then, is the key to which measurement scale is being used.

The classification system which follows has become a tradition among statisticians, especially those involved in the social sciences, and is even followed in such major statistical computer programs as SPSS-X. It was first introduced in 1946 by S. S. Stevens.[6] The Stevens system remains as an extremely handy set of rules for determining which statistical test should apply in specific research situations, and it will be used throughout this book. (The student should be aware, however, that it has had its share of critics.)[7]

The Nominal Scale: Categorical Data

Nominal scaling, or simply using numbers to label categories, is the lowest order of measurement. Of all the scales, it contains the least information, since no assumptions need be made concerning the relations among measures.

A **nominal scale** is created by assigning observations into various independent categories and then counting the frequency of occurrence within those categories. In effect, it is "nose-counting" data, such as observing how many persons in a given voting district are registered as Republicans, Independents, or Democrats. Or, it might be categorizing a group of children on the basis of whether or not they exhibit overt aggressive responses during recess, and then noting the frequencies or numbers of children falling within the categories.

The only mathematics involved in nominal scaling is the rule of equality versus nonequality; that is, the same number must be assigned to things that are identical, and different numbers must be assigned to things that differ from each other. The categories are thus independent of each other, or mutually exclusive, which means that if a given observation is placed in category number 1, it *cannot* also be placed in category number 2. In nominal scaling, then, we discover *how many* persons, things, or events have X or Y or Z in common.

With nominal scaling the concept of quantity cannot be expressed, only identity versus nonidentity. If we were to measure people according to gender, for example, by assigning a "1" to females and "0" to males, we aren't, of course, saying that females have more gender than males, or that a classroom of students could possibly have an average gender of .75. Nominal scaling is simply a rule that arbitrarily substitutes, in this case, the number "1" for females and "0" for males.

Nominal scales may also be used in the design of experiments, where the number "1" might be assigned to one group and "0" to the other. The numerical value is, thus, again being used as a substitute for a verbal label.

The Ordinal Scale: Ranked Data

It is often not sufficient to know merely that X or Y is present. As inquiring social scientists we wish to find out how much X or how much Y. The **ordinal scale** answers this need by providing for rank ordering the observations in a given category. Suppose that at a given schoolyard, 60% of the children were nominally categorized as aggressive. We may then examine that category alone and rank the children in it from most to least aggressive.

Mathematically, an ordinal scale must satisfy two rules: the equality/non-equality rule and also the greater-than-or-less-than rule. This means that if two individuals have the same amount of a given trait, they must be assigned the same number. Further, if one individual has more or less of a given trait than another individual, then they must be assigned different numbers. The main thing to remember about ordinal scaling is that it provides information regarding greater than or less than, but it does not tell how much greater or how much less. A good illustration is knowing the order of finishing of a horse race, but not knowing whether the first-place horse won by a nose or by 8 furlongs. The distance between the points on an ordinal scale is unknown.

An example of ordinal scaling from sociology or political science is the ranking of a population on the basis of socioeconomic status, from the upper-upper class, through the middle class, and down to the lower-lower class. In the field of psychology or education, a researcher creates an ordinal scale by ranking a group of individuals on the basis of how much leadership ability each has exhibited in a given situation. Ordinal data are, therefore, rank-ordered data. Ordinal scaling defines only the order of the numbers, not the degrees of difference between them. It tells us that A is greater than B ($A > B$) or that A is less than B ($A < B$). It is a mistake to read any more into it than that, although sometimes it is the kind of mistake that is easy to make. It can be psychologically seductive to assume that a person who ranks, say, fifth, is about the same distance ahead of the one who ranks sixth as the person who ranks first is ahead of the one who ranks second. This psychological tendency must be restrained, however, when evaluating ordinal data. If someone were to tell you that a certain item costs "more than a dollar," you are only being told that the item costs anywhere from $1.01 to infinity.

In short, be careful of ordinal positions, because at times they can be very misleading. You might be told that the *Bugle* is the second largest newspaper in the city, even though in reality it may, in effect, be a one-newspaper city (with second place going to a seventh grader who happens to own a mimeo machine). Or, it could be like the old joke where the freshman is told that he is the second-best-looking guy on the campus. All the others are tied for first.

The Interval Scale: Measurement Data

A still further refinement of scaling occurs when the data are in the form of an interval scale. In an **interval scale,** the assigning of numbers is done in such a way that the intervals between the points on the scale become meaningful. From this kind of scale, we get information not only as to greater than or less than status, but also as to how much greater than or how much less than. Theoretically, the distances between successive points on an interval scale are equal. As a result, inferences made from interval data can be broader and more meaningful than can those made from either nominal or ordinal data. In general, the more information contained in a given score, the more meaningful are any conclusions that are based on that score.

The Fahrenheit (or Celsius) temperature scale provides interval data. The difference between 80°F and 79°F is exactly the same as the difference between 35°F and 34°F. The thermometer, therefore, measures temperature in degrees that are of the same size at any point on the scale.

Psychologists have attempted to standardize IQ tests as interval scales. An IQ score of 105 is considered to be higher than a score of 100 by the same amount that a score of 100 is higher than a score of 95. Although there is some disagreement within the field of psychology on this point (some purists insist that IQ scores can only form an ordinal scale), the vast majority of researchers treat IQ data as interval data. The general opinion seems to be that if it quacks like a duck and walks like a duck, then it is a duck. There is no question, however, about such measures as height, weight, and income—these all form scales of equal intervals. As a matter of fact, weight and height have the added advantage of having an absolute zero value, and thus form an even more sophisticated scale called a ratio scale.

The Ratio Scale

When a scale has an absolute zero (as opposed to an arbitrary zero such as the 0° Fahrenheit or Celsius temperature measure), then valid ratio comparisons among the data can legitimately be made. That is, we can say that someone who is 6 feet tall is twice as tall as someone who is 3 feet tall. In any case, like interval data, data from a **ratio scale** do have equal interval distances between successive scale points, with the *added feature* of an absolute, nonarbitrary zero point.

The social sciences use very few ratio measures. No researcher can define an absolute zero IQ, or zero prejudice, or zero interest in politics. It is, therefore, incorrect to say that someone with an IQ of 100 is twice as intelligent as someone with an IQ of 50. The statistical tests of significance (which are presented later in the book) designed to be used with interval data may also be used with ratio data.

Implications of Scaling

The social scientist's interest in measurement scaling is both acute and profound. *The choice as to which statistical test* can legitimately be used for data analysis rests largely on which scale of measurement has been employed. Further, the inferences that can be drawn from a study cannot, or at least should not, outrun the data being used. It is not correct to employ nominal data and then draw greater-than or less-than conclusions. Neither is it correct to employ ordinal data and then summarize in terms of how much greater or how much less.

Measurement scales are extremely important. Although most of the material in the first several chapters of this book is devoted to interval scaling, you should keep the other types of scales in mind. Later in the book, the ordinal and nominal scales are in the full spotlight. It is very likely that your appreciation of these measurement distinctions will not be fully developed until you get to later chapters and begin testing ordinal and nominal research hypotheses. So, hang in there. There may be some patches of fog on the way up, but the view from the top is worth it.

SUMMARY

It has often been said that one can prove anything with statistics. However, this is only true if the audience is naïve about statistical procedures and terms. The terms used by statisticians are exact and their definitions are important, since they are designed to facilitate very precise communication.

The field of statistics is of fairly recent origin. The laws of probability were not formulated systematically until the seventeenth century. Blaise Pascal is popularly credited with these first formulations. Not until the beginning of the twentieth century were the strategies devised (by W. S. Gossett in 1906) for measuring samples and then using those data to infer population characteristics.

The general field of statistics includes two subdivisions, descriptive statistics and inferential statistics.

Descriptive statistics: Those techniques used for describing large amounts of data in abbreviated, symbolic form.

Inferential statistics: Those techniques used for measuring a sample (subgroup) and then generalizing these measures to the population (the entire group).

Measurement is the assigning of numbers to observations according to certain rules. The rules determine the type of scale of measurement being constructed.

Nominal scale: Using numbers to label categories, sorting observations into these categories, and then noting their frequencies of occurrence.

Ordinal scale: Rank ordering observations to produce an ordered series. Information regarding greater-than or less-than status is contained in ordinal data, but information as to how much greater or less than is not.

Interval scale: Scale in which the distances between successive scale points are assumed to be equal. Interval data do contain information as to how much greater than or how much less than.

Ratio scale: Special interval scale for which an absolute zero can be determined. Ratio data allow for such ratio comparisons as one measure is twice as great as another. All statistical tests in this book that can be used for the analysis of interval data can also be used on ratio data.

[handwritten note in left margin:] most used when making an inference!

KEY TERMS AND NAMES

descriptive statistics	measurement	Pascal, Blaise
Gossett, William Sealy	nominal scale	ratio scale
inferential statistics	ordinal scale	variable
interval scale		

PROBLEMS

1. During the crisis in Vietnam, the number of Americans who died in Vietnam was *lower* (for the same time period) than was the number of Americans who died in the United States. Therefore, one can conclude that it is safer to participate in a war than to remain at home. Criticize this conclusion based on what you know of the two populations involved—Americans who go to war versus those who do not.

2. A certain Swedish auto manufacturer claims that 90% of the cars it has built during the past 15 years are still being driven today. This is true, so say the ads, despite the fact that the roads in Sweden are rougher than are those in the United States. What important piece of information must you have before you can evaluate the truth of this auto-longevity claim?

3. A recent TV ad tried to show the risks involved in not taking a certain antacid tablet daily. The actor, wearing the obligatory white coat and stethoscope, poured a beaker of "stomach acid" onto a napkin, immediately creating a large hole. The actor then menacingly intoned, "If acid can do that to a napkin, think what it can do to your stomach." On the basis of the "evidence" included in the commercial, what can the acid do to your stomach?

4. An oil company grandly proclaims that its profits for the fourth quarter increased by 150% over those in the same period a year ago. On the basis of this statement, and assuming that you have the money, should you immediately rush out and buy the company's stock?

5. A toothpaste company says that a large sample of individuals tested after using its brand of toothpaste had 27% fewer dental caries. Criticize the assumption that using the company's toothpaste reduces the incidence of caries.

6. A statewide analysis of speeding tickets found that state troopers in unmarked cars were giving out 37% more tickets over holiday weekends than were troopers working from cruisers that were clearly visible as police cars. Criticize the suggestion that troopers assigned to unmarked cars were obviously more vigilant and more vigorous in their pursuit of highway justice.

7. A marketing research study reported that a certain brand of dish-washing detergent was found by a test sample to be 35% more effective. What else should the consumer find out before buying the product?

8. Two major automakers, one in the U.S. and the other in Japan, proudly announce from Detroit that they will jointly produce a new car. The announcement further states (patriotically) that *75% of the parts for this car will be produced in the U.S.* Is this necessarily good news for U.S. workers?

9. A research study reported a linkage between learning disabilities and crime. Data from the Brooklyn Family Court (1988) indicated that 40% of the juveniles

who appeared in court were learning disabled. The report further suggested that these data show that juveniles with learning disabilities are likely to engage in anti-social behavior for the following reasons: (1) they are unskilled, (2) they suffer from low self-esteem, and (3) they are easily swayed by others. Criticize these conclusions on the basis of the data being offered.

For problems 10 through 14, indicate which scale of measurement—nominal, ordinal, or interval—is being used.

10. The phone company announces that area code 617 serves 2 million customers.
11. Insurance company statistics indicate that the average weight for adult males in the United States is 168 pounds.
12. Post office records show that 2201 persons have the zip code 01118.
13. The Boston Marathon Committee announces individual names with their order of finishing for the first 300 runners to cross the finish line.
14. Central High School publishes the names and SAT scores for the students selected as National Merit Scholars.

REFERENCES

1. *Parade,* April 28, 1985, p. 14.
2. COLLINS, R. L. (1979). 100 accidents: What went wrong? *Flying.* February, 56–57.
3. HUFF, D. (1954). *How to lie with statistics.* New York: W. W. Norton, p. 109.
4. DAVID, F. N. (1962). *Games, gods and gambling.* New York: Hafner, p. 24.
5. Ibid., p 26.
6. STEVEN, S. S. (1946). On the theory of scales of measurement. *Science,* pp. 103, 677–680.
7. BORGATTA, E. F., AND BOHRNSTEDT, G. W. (1980). Level of measurement once over again. *Sociological methods and research, 9,* pp. 147–160.

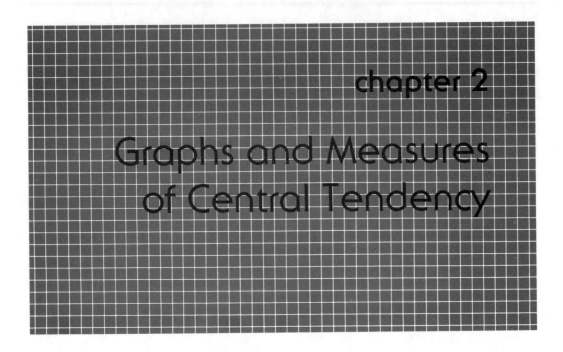

chapter 2
Graphs and Measures of Central Tendency

To enable them to describe large amounts and of course, small amounts of data, statisticians have created a series of tools or abbreviated symbols to give structure and meaning to the apparent chaos of the original measures. Thousands, even millions, of scores can be organized and neatly summarized by the appropriate use of descriptive techniques. These summaries allow for precise communication of whatever story the data have to tell.

During this initial foray into the realm of descriptive statistics, we shall concentrate on two major techniques: graphs and measures of central tendency.

GRAPHS

The process of creating a graph to embody a group of scores always begins with the creation of a distribution.

Distributions

To extract some meaning from original, unorganized data, the researcher begins by putting the measurement into an order. The first step is to form a **distribution** of scores. Distribution simply means the arrangement of any set of scores in order of magnitude. For example, the following IQ scores were recorded for a group of students: 75, 100, 105, 95, 120, 130, 95, 90, 115, 85, 115, 100, 110, 100, and 110.

Table 2.1   Distribution of IQ Scores.

130
120
115
115
110
110
105
100
100
100
95
95
90
85
75

Arranging these scores to form a distribution means listing them sequentially from highest to lowest. Table 2.1 is a distribution of the list of IQ scores.

Frequency Distributions.   A distribution allows the researcher to see general trends more readily than does an unordered set of raw scores. To simplify inspection of the data further, they can be presented as a **frequency distribution.** A frequency distribution is a listing, in order of magnitude, of each score achieved, together with the number of times that score occurred. Table 2.2 is a frequency distribution of the IQ scores. Note that the frequency distribution is more compact than the distribution of all the scores listed separately. The X at the top of the first column stands for the raw score (in this case, IQ) and the f over the second column stands for frequency of occurrence. As we can see, of the 15 peo-

Table 2.2   IQ Scores presented in the form of a frequency distribution.

X (raw score)	f (frequency of occurrence)
130	1
120	1
115	2
110	2
105	1
100	3
95	2
90	1
85	1
75	1

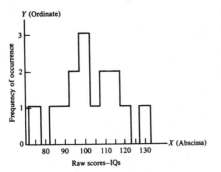

Figure 2.1
A histogram, or bar graph, of IQ scores.

ple taking the test, 2 received scores of 115, 2 received 110, 3 scored 100, 2 scored 95, and 6 achieved unique scores.

In addition to presenting frequency distributions in tabular form, statisticians often present such data in graphic form. A graph has the advantage of being a kind of picture of the data. It is customary to indicate the actual values of the variable (raw scores) on the horizontal line, or X axis, which is called the **abscissa.** The frequency of occurrence is presented on the vertical line, or Y axis, which is called the **ordinate.** It is traditional when graphing data in this way to draw the ordinate about 75% as long as the abscissa. Thus, if the abscissa is 5 inches long, the ordinate should measure about $3^3/4$ inches.

Histograms

Figure 2.1 shows the data that were previously presented in tabular form arranged in a graphic form, called a **histogram,** or bar graph. To construct a histogram, or bar graph, a rectangular bar is drawn above each raw score. The height of the rectangle indicates the frequency of occurrence for that score.

Frequency Polygons

Figure 2.2 shows the same IQ data arranged in another commonly used graphic form called a **frequency polygon.** To construct a frequency polygon, the scores are again displayed on the X axis, and the frequency of occurrence is on the Y axis. However, instead of a rectangular bar, a single point is used to designate the frequency of each score. Adjacent points are then connected by a series of straight lines.

The frequency polygon is especially useful for portraying two or more distributions simultaneously. It allows visual comparisons to be made and gives a quick clue as to whether the distributions are representative of the same population. Figure 2.3 is a frequency polygon that displays the scores of two groups at once. When two distributions are superimposed in this manner, the researcher can glean information about possible differences between the groups. In this case, although the scores of the two groups do overlap (between the scores of 90

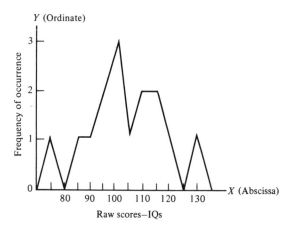

Figure 2.2
A frequency polygon.

and 115), Group B outperformed Group A. Perhaps Group B represents a population different from that of Group A. Perhaps the data show that for the trait being measured (in this case IQ), the difference between these groups is greater than would be predicted to occur by chance. Later in the book, statistical techniques for making that kind of conclusion in a more precise way will be presented.

The Importance of Setting the Base of the Ordinate at Zero

In both the histogram (or its minimum value) and the frequency polygon, the base of the ordinate should ideally be set at zero; if it is not, the graph may tell a very misleading story. For example, suppose we are graphing data from a learning study that show that increasing the number of learning trials increases the amount learned. The number of trials is plotted on the abscissa and the number of correct responses on the ordinate. (See Fig. 2.4.) The graph shows that by the

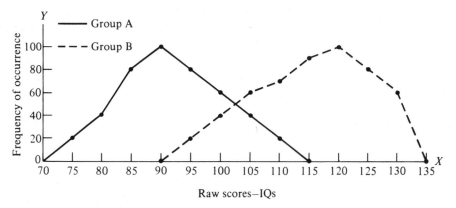

Figure 2.3   A frequency polygon with two distributions.

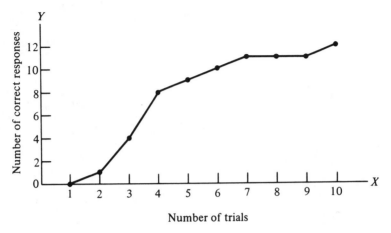

Figure 2.4   A frequency polygon with the ordinate base at zero.

fourth trial the subject made 8 correct responses, and that by the tenth trial the subject made 12 correct responses. These data are typical of the results obtained in many learning studies. A great deal of learning usually occurs during the first few trials, but as the number of trials increases, the rate of increase in learning levels off.

Suppose, however, that a statistician wants to rig the graph to create a false impression of what the data really reveal. The trick is to focus on one small area of the graph. (See Fig. 2.5.) Now the data seem to tell a very different story about how the learning took place. It now appears that no learning took place before the fourth trial and that the great bulk of learning took place between the fourth and tenth trials. We know from the complete graph of the data that this is simply not true. In fact, the majority of the learning actually took place during the first four trials; after the fifth trial, the rate of learning was beginning to top out or level off. This is a blatant instance of how statistics can be used to distort reality if the audience is naïve about statistical methods.

The type of deceit just described is variously known as a "Wow!" "Gee whiz!" or "Oh boy!" graph. Some corporations hire statistical camouflagers to de-

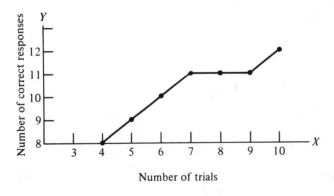

Figure 2.5
A frequency polygon with the ordinate base greater than zero.

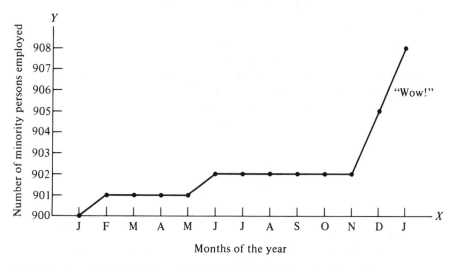

Figure 2.6   Example of a "Wow!" graph. (Note: the ordinate base is greater than zero.)

vise this kind of tinsel for the sales and earnings graphs used in their annual reports. Insignificant sales gains can be portrayed as gigantic financial leaps by adept use of the "Wow!" graph. Another instance of this chicanery was perpetrated by a certain corporation, anxious to create the image of being an equal opportunity employer. It adroitly used a "Wow!" graph (see Fig. 2.6) to prove to the world that it had a liberal and compassionate stance toward hiring minorities.

The graph shows time, as months of the year, on the abscissa and the number of minority persons employed on the ordinate. It seems to show only a modest increase in the number of minority persons employed throughout most of the year, and then ("Wow!") a dramatic increase during the final two months. This increase is, of course, more illusory than real; it reflects the hiring of only six new people, a real increase of less than 1%. If the ordinate is set at zero and the data are graphed again, you need an electron microscope to find the increase. (See Fig. 2.7.)

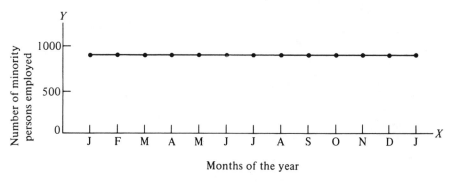

Figure 2.7   Graph of the data in Fig. 2.6 with the ordinate base equal to zero.

Whenever a graph is presented in which the base of the ordinate is not set at zero, be on the alert. The stage has been set for a sleight-of-hand trick. Without question, the most serious error in graphing is using a value other than zero as a base. Where there is no visual relationship to zero, the values are without perspective and thus have no meaning. It's like someone telling you that the temperature outside is 40°, not indicating the scale—Fahrenheit or Celsius—or whether it's 40° above or below zero. You might dress to prevent frostbite and end up suffering a heatstroke.

MEASURES OF CENTRAL TENDENCY

The tools called **measures of central tendency** are designed to give information concerning the average or typical score of a large number of scores. If, for example, we were presented with all the IQ scores of the students at a college, we could utilize the measures of central tendency to give us some description of the typical, or average, intellectual level at that school. There are three methods for obtaining a measure of the central tendency. When used appropriately, each is designed to give the most accurate estimation possible of the typical score. Choosing the appropriate method can sometimes be tricky. The interpretation of the data can vary widely, depending on how the typical score has been found.

Ma'am, do the "N's" justify the means?

The Mean

The most widely, though not always correctly, used measure of central tendency is the **mean,** symbolized as $\overline{X}$. The mean is the arithmetic average of all the scores. It is calculated by adding all the scores together and then dividing by the total number of scores involved. The formula for the mean is as follows:

$$\overline{X} = \frac{\Sigma X}{N}$$

In the formula, $\overline{X}$, of course, stands for the mean. The capital letter X stands for the raw score, or the measure of the trait or concept in question. The Greek capital letter Σ (sigma) is an operational term that indicates the addition of all measures of X. This is usually read as "summation of." Finally, the capital letter N stands for the entire number of observations being dealt with. (It is important that you follow the book's use of capital letters, since lowercase letters often mean something quite different.) Thus, the equation tells us that the mean $(\overline{X})$ is equal to the summation (Σ) of all the raw scores (X) divided by the number of cases (N).

In Table 2.3, the mean of the distribution of IQ scores from Table 2.1 has been calculated. It happens that the mean is an appropriate measure of central

Table 2.3 Calculation of the mean from a distribution of raw scores.

X
130
120
115
115
110
110
105
100
100
100
95
95
90
85
75

$\Sigma X = 1545$

$$\overline{X} = \frac{\Sigma X}{N} = \frac{1545}{15} = 103$$

Table 2.4 Calculation of the mean from a distribution of raw scores where the result is not a whole number.

X
72
71
70
68
68
68
65
63
$\Sigma X = 545$

$$\bar{X} = \frac{\Sigma X}{N} = \frac{545}{8} = 68.125 = 68.13$$

tendency in this case because the distribution is fairly well balanced. Most of the scores occur in the middle range, and there are no extreme scores in either direction. Since the mean is calculated by adding together all of the scores in the distribution, it is not usually influenced by the presence of extreme scores, *unless* the extreme scores are all at one end of the range. The mean is typically a stable measure of central tendency, and, without question, it is the most widely used.

Calculating the Mean. Note in Table 2.3 that the mean is calculated as 103, a whole number. In most situations, however, this won't be the case. A more typical distribution is shown in Table 2.4. The mean in this case is rounded to 68.13.

*Always round to two places.** This requires completing the calculations to three places to the right of the decimal, then rounding the value that is two places to the right of the decimal. If the value in the third place is a 5 or higher, raise the value in the second place by 1. Thus, 68.125 becomes 68.13. But, 68.124 is rounded to 68.12. Remember, whenever you multiply or divide, square or take a square root, you must round your answer to two places. (This is accurate enough for a first course in statistics. When doing a research report for a class or for presentation, three- or four-place accuracy is often required.)

Interpreting the Mean. Interpreting the mean correctly can sometimes be a challenge, especially in situations where either the group or the size of the group changes. For example, the mean IQ of the typical freshman college class is about

---

*Although some statistics texts maintain that a value of 5 in the third place to the right of the decimal should always be rounded to the nearest even number, this book uses the convention of rounding a 5 to the next highest number. This is consistent with the rounding program built into most modern calculators. If your calculator has the fixed decimal feature, simply set it for two places, and the rounding will take place automatically.

115, whereas the mean IQ of the typical senior class is about 5 points higher. Does this indicate that students increase their IQs as they progress through college? No, but since the size of the senior class is almost always smaller than the size of the freshman class, the two populations are not the same. Among those freshmen who never become seniors are a goodly number with low IQs, and their scores are not being averaged in when their class later becomes seniors.

The Mean of Skewed Distributions. In some situations, the use of the mean can lead to an extremely distorted picture of the average value of a distribution of scores. For example, look at the distribution of annual incomes in Table 2.5. Note that one income ($10,000,000.00) is so extremely far above the others that the use of the mean income as a reflection of averageness gives a highly misleading picture of great prosperity for this group of citizens. A distribution like this one, which is unbalanced by an extreme score at or near one end, is said to be a **skewed distribution.**

Figure 2.8 shows what skewed distributions look like in graphic form. In the distribution on the left, most of the scores fall to the right, or at the high end, and there are only a few extremely low scores. This is called a negatively skewed, or skewed to the left, distribution. (The skew is in the direction of the tail-off of scores, not of the majority of scores.) The distribution on the right represents the opposite situation. Here most of the scores fall to the left, or at the low end of the distribution, and only a very few scores are high. This is a positively skewed, or skewed to the right, distribution. Remember, label skewed distributions accord-

Table 2.5 Calculation of the mean of a distribution of annual incomes.

X
$10,000,000.00
20,000.00
20,000.00
19,500.00
19,400.00
19,400.00
19,400.00
19,300.00
19,000.00
18,500.00
18,000.00
18,000.00
17,600.00

$\Sigma X = \$10,228,100.00$

$$\bar{X} = \frac{\Sigma X}{N} = \frac{10,228,100.00}{13} = \$786,776.92$$

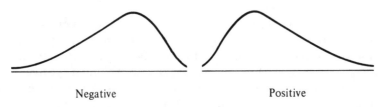

Negative   Positive

Figure 2.8   A graphic illustration of skewed distributions.

ing to the direction of the tail. When the tail goes to the left, the curve is negatively skewed; when it goes to the right, it is positively skewed.

Thanks, Ralph.   A rather dramatic example of how the use of the mean distorts "averageness" in a skewed distribution can be seen in a University of Virginia press release, undoubtedly meant as tongue-in-cheek. During an analysis of how well the members of the college's 1983 graduating class fared in the job market, it was discovered that the highest salaries were earned by graduates of the Department of Rhetoric and Communications Studies, where the beginning average pay was $55,000 a year. It should be pointed out that one student, the 7-foot 4-inch-tall basketball player Ralph Sampson, had a starting salary of well over $1 million. Perhaps the mean height for those same graduates was 6 feet 6 inches.

The Median

The **median** is the exact midpoint of any distribution, or the point that separates the upper half from the lower half. In fact, the median (symbolized as Mdn) is a much more accurate representation of central tendency for a skewed distribution than is the mean. Whereas the mean income of the distribution in Table 2.5 is $786,776.92, the median income is $19,400.00, a much more descriptive reflection of the typical income for this distribution. Since income distributions are almost always skewed toward the high side, you should be on the alert for an inflated figure whenever the mean income is reported. The median gives a better estimation of how the typical wage earner is actually faring.

As a memory trick for recalling the definition of the median, think of the median strip dividing a highway. The same width of road lies both to the left of the median strip and to the right.

Calculating the Median.   To calculate the median, the scores must first be arranged in distribution form, that is, in order of magnitude. Then, count down (or up) through half of the scores. For example, in Table 2.5 there are 13 income scores in the distribution. Therefore, count down 6 scores from the top, and the seventh score is the median (there will be the same number of scores above the median point as there are below it). Whenever a distribution contains an odd number of scores, finding the median is very simple. Also, in such distributions

Table 2.6 Calculation of the median with an
even number of scores.

X
120
118
115
114 $\rbrace$ 114.5 Median
114
112

$\Sigma X = 693$
$\bar{X} = 115.50$ Mean

the median will usually be a score that someone actually received. In the distribution in Table 2.5, someone really did earn the median income of $19,400.

If a distribution is made up of an even number of scores, the procedure is slightly different. Table 2.6 presents a distribution of an even number of scores. The median is then found by determining the score that lies halfway between the two middlemost scores. In this case, the median is 114.5. Don't be disturbed by the fact that in some distributions of even numbers of scores nobody actually received the median score; after all, nobody ever had 2.8 children either.

The Median of Skewed Distributions.   Unlike the mean, the median is not affected by skewed distributions (where there are a few extreme scores in one direction). In Table 2.7, for example, the median score is still found to be 114.5, even with a low score of 6 rather than one of 112 as reported in Table 2.6. The mean of this distribution, on the other hand, plummets to an unrepresentative value of 97.83. The mean is always pulled toward the extreme score in a skewed distribution. When the extreme score is at the high end, the mean is too high to reflect true centrality; when the extreme score is at the low end, the mean is similarly too low.

Table 2.7 Calculation of the median with an
even number of scores and a
skewed distribution.

X
120
118
115
114 $\rbrace$ 114.5 Median
114
6

$\Sigma X = 587$
$\bar{X} = 97.83$ Mean

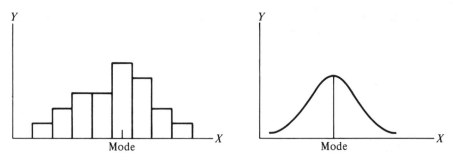

Figure 2.9 The location of the mode in a histogram (left) and a frequency polygon (right).

The Mode

The third, and final, measure of central tendency is called the **mode** and is symbolized as Mo. The mode is the most popular, or most frequently occurring, score in a distribution. In a histogram, the mode is always located beneath the tallest bar; in a frequency polygon, the mode is always found directly below the point where the curve is at its highest. This is because, as was pointed out previously, the Y axis, or the ordinate, represents the frequency of occurrence. (See Fig. 2.9.)

Finding the Mode. When the data are not graphed, just determine which score occurs the most times, and you've got the mode. For example, in the distribution shown in Table 2.8, the score of 103 occurs more often than does any other score. That value is the mode. In Table 2.9 a frequency distribution of the same data is given. Here, to find the mode, all you have to do is to note which score (X) is beside the highest frequency value (f). The mode is a handy tool since it pro-

Table 2.8 Finding the mode of a distribution of raw scores.

X
110
105
105
103 ⎫
103 ⎬ Mode
103 ⎪
103 ⎭
101
101
100
100
98
95

Table 2.9   Finding the mode when a frequency
distribution is given.

X	f
110	1
105	2
103 ← Mode	4
101	2
100	2
98	1
95	1

vides an extremely quick method for obtaining some idea of a distribution's centrality. Just eyeballing the data is usually enough to spot the mode.

Bimodal Distributions.   A distribution having a single mode is referred to
as a **unimodal distribution.** However, some distributions have more than one
mode. When there are two modes, as in Fig. 2.10, the distribution is called bimodal. (When there are more than two modes, it is called multimodal.) Distributions
of this type occur when scores cluster together at several points, or if the group
being measured really represents two or more subgroups.

Assume that the distribution in Fig. 2.10 represents the running times (in
seconds) in the 100-yard dash for a large group of high school seniors. There are
two modes—one at 13 seconds and the other at 18 seconds. Since there are two
scores that both occur with the same high frequency, it is probable that data about
two separate subgroups are being displayed. For example, perhaps the running
times for boys are clustering around one mode and the speeds for girls are clustering around the other.

Interpreting the Bimodal Distributions.   Whenever a distribution does fall
into two distinct clusters of scores, extreme care must be taken with their interpretation. Neither the mean nor the median can justifiably be used, since a bimodal
distribution cannot be adequately described with a single value. A person whose

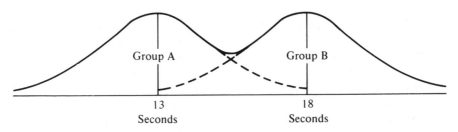

Figure 2.10   Graph of bimodal distribution of running times.

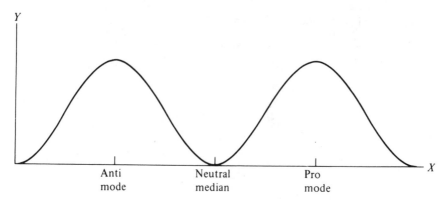

Figure 2.11   Graph of responses to an attitude questionnaire showing two modes.

head is packed in ice and whose feet are sitting in a tub of boiling water cannot be appropriately characterized as being in a tepid condition *on the average.*

Bimodal distributions should not be represented by the use of a single average of the scores. Suppose a group of individuals completed a certain attitude questionnaire. The resulting scores fell in two decidedly different clusters, half the group scoring around a "pro" attitude and the other half around an "anti" attitude. The use of either the mean or the median to report the results would provide a highly misleading interpretation of the group's performance, for in either case the group's attitude would be represented as being neutral. Figure 2.11 shows such a bimodal distribution. The two modes clearly indicate how divided the group really was. Note, too, that while *nobody* scored at the neutral point, using either the mean or the median as a description of centrality would imply that the typical individual in the group was indeed neutral. Thus, when a distribution has more than one mode, the modes themselves, not the mean or the median, should be used to provide an accurate account.

APPROPRIATE USE OF THE MEAN, THE MEDIAN AND THE MODE

Working with Skewed Distributions

The best way to illustrate the comparative applicability of the three measures of central tendency is to look again at a skewed distribution. Figure 2.12 shows an approximation of the income distribution per household in the United States in 1987. Like most income distributions, this one is skewed to the right. This is because the low end has a fixed limit of zero, while the sky is the limit at the high end. Note that the exact midpoint of the distribution, the median, falls at a value of $26,000 a year. This is the figure above which 50% of the incomes fall and below which 50% fall. Because there is a positive skew, the mean indicates a fairly

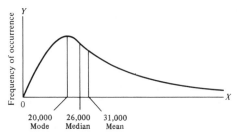

Figure 2.12
Distribution of income per household in the United States.

high average income of $31,000. This value, however, gives a rather distorted pic-
ture of reality, since the mean is being unduly influenced by the few families at the
high end of the curve whose income is in the millions. Finally, the modal income,
which is $20,000 per year, seems to distort reality toward the low side. Although
the mode does represent the most frequently earned income, it is nevertheless
lower than the point separating the income scores into two halves. In the case of a
skewed distribution, then, both the mean and the mode give false, though different, portraits of typicality. The truth lies somewhere in between. Thus, in a positively skewed distribution the order of the three measures of central tendency
from left to right is first the mode, the lowest value; then the median, the midpoint; and finally the mean, the highest value. A negatively skewed distribution
simply reverses this order. (See Fig. 2.13). The point to remember for skewed distributions is that the mean is always located toward the tail end of the curve.

A mnemonic device often used for this purpose is to remember that as
you go up the slope of a skewed curve, the measures of central tendency appear in
alphabetical order, first the mean, then the median, and finally the mode. In using
this memory prod, you must remember that you're going *up* the slope and that
you're going up on the gentle slope, *not* the steep side. Picture yourself at a ski
area, riding the chairlift up the novice slope.

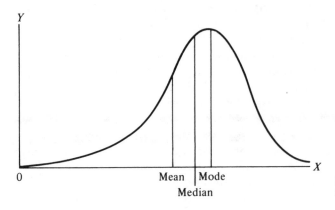

Figure 2.13
A negatively skewed distribution.

Assessing Skewness (Sk)

Although there are techniques for establishing a precise value for the amount of skewness (Sk) present in any unimodal distribution of scores (see the Appendix), a somewhat rough but easy method for spotting Sk is simply to compare the mean and median. Remembering that in skewed distributions the mean always "tips toward the tail," then the mean can be used as a quick indicator of skewness. If the mean lies to the left of the median, the distribution is skewed left. If it lies to the right of the median, the distribution is skewed to the right. Also, the greater the distance between the mean and median, the greater the total amount of Sk. For example, if you are told that the mean price for a new home in the United States is $100,000 and that the median price is $80,000, then you may conclude that the distribution of prices is skewed to the right (which it really is), or has an Sk+. Or, if the median golf score for the members of a certain country club happened to be 95, but the mean was only 85, then the distribution of golf scores would show a negative skew, or Sk−. In general, the greater the discrepancy between the mean and the median, the greater the skewness.

By incorrectly using such descriptive statistics as the measures of central tendency, some extremely interesting bits of con-artistry can be performed. As previously mentioned, if a distribution is skewed to the right, the mode reflects the lowest value of central tendency while the mean portrays the highest. With such a distribution, a researcher could report on a before-and-after study and make it appear that group performance had increased in the "after" condition, even though *every score in the entire group had remained the same.* How? The researcher simply reports the modal score for the results of the "before" condition and the mean score for the results of the "after" condition. Does it matter, then, which measure of central tendency is used? You bet it does!

Effects of the Scale of Measurement Used

Which of the measures of central tendency to use also depends on the scale of measurement on which the data are based. The discussion so far in this chapter has assumed an interval scale, that is, a scale in which the difference between successive scale points is equal. Interval data allow for the calculation of all three measures of central tendency—mean, median, and mode.

Ordinal data, however, cannot be used to calculate the mean. Since ordinal data provide no information regarding the distance between the scale points, calculating an ordinal mean can be extremely misleading. Such a mean gives rise to the assumption that the distances are known. With ordinal data, the median can and should be used. Since the median is the middlemost point in a distribution, it is itself a rank—a rank above which half the scores fall and below which half the scores fall. Recall that before finding the median you had to rank the scores (arrange them in order of magnitude).

Finally, with nominal data, neither the mean nor the median can be used, since each of these measures implies the comparisons of greater than and less than. Nominal data are restricted by the equality–nonequality rule and are in the form of frequency of occurrence within discrete categories. Therefore, the only measure of central tendency permissible for nominal data is the mode, the *most frequently occurring* score. The mode is determined on the basis of frequency of occurrence, which is precisely the kind of data that the nominal scale is designed to handle.

In short, interval data and ratio data allow the use of all three measures of central tendency. If the interval data distribution is unimodal and fairly well balanced, use the mean. If the distribution is skewed to the right or the left, use the median. If the distribution has more than one clustering of score values at different scale positions, use the mode. With ordinal data, since the mean is no longer applicable, use the median. However, if there are clusterings of tied ranks at different ordinal positions, then use the mode. If the data are nominal, there is no choice but to use the mode.

Later in this book, during the discussion of nonparametric statistical techniques, other more subtle distinctions among the measures of central tendency will be presented.

SUMMARY

Descriptive statistics includes those techniques used for describing large amounts of data in convenient and symbolic form.

To get meaning out of the mass of raw data from a study, the statistician arranges the scores in order of magnitude. The arrangement is called a distribution of scores. If the scores are listed with an adjacent column giving the frequency of occurrence for each score, then a frequency distribution has been formed.

Graphs are often used to display data in a more readily comprehensible way. When frequency distributions are graphed, the actual scores are indicated on the horizontal axis (X axis or abscissa), and the frequency of occurrence is presented on the vertical axis (Y axis or ordinate). Histograms are sometimes used. In these figures, a series of rectangles are placed above the scores on the abscissa; the height of each rectangle signifies the frequency. Also, data may be graphed as a frequency polygon, again with scores on the X axis and frequency of occurrence on the Y axis. In this technique, a single point (rather than a rectangle) is used above each score at the appropriate frequency height, and the points are connected by straight lines. In either histogram or frequency polygon, the base of the ordinate should always be set at zero. Otherwise, the graph loses perspective and the values therein have no relational meaning.

To understand how scores are similar, that is, to shed light on the commonality among a set of scores, tools have been designed for finding the average or typical score in a distribution. These techniques are called measures of central tendency. They include the mean, which is the actual arithmetic average; the median, which is the middlemost point in a distribution; and the mode, which is the most frequently occurring point in a distribution. If the data are in interval form, use the mean if the distribution is unimodal and evenly balanced; use the median if the distribution is unimodal and skewed; and use the mode if the distribution is bimodal. With ordinal data, the median must be used for ranked distributions. With nominal data, only the mode is appropriate.

KEY TERMS

abscissa	histogram	mode
distribution	mean	ordinate
frequency distribution	measures of central tendency	skewed distribution
frequency polygon	median	unimodal distribution

PROBLEMS

1. A group of six college students are tested on the Edwards Personality Scale. Their scores on the Need-Achievement subtest are as follows: 14, 12, 9, 10, 12, 12. Find the mean, the median, and the mode.

2. A group of 10 elderly patients in a nursing home are asked how much time they spend watching TV. Their scores (in hours per week) are as follows: 40, 55, 50, 40, 10, 35, 40, 35, 50, 40. Find the mean, the median, and the mode.

3. Immediately following the coverage of a social-science unit, a class of 13 students takes a 20-item quiz on the material presented. Their scores, based on the number of items answered correctly, were as follows: 12, 12, 14, 15, 16, 19, 20, 10, 7, 5, 2, 3, 3.
    a. Find the mean.
    b. Find the median.
    c. Find the mode.

4. A week later the class of 13 students mentioned in problem 3 takes the same test a second time. This time their scores were: 11, 11, 11, 11, 12, 14, 20, 10, 10, 9, 2, 8, 9.
    a. Find the mean.
    b. Find the median.
    c. Find the mode.
    d. Was there a difference in their performance when taking the test a second time?

5. For the set of scores 100, 5, 12, 17, 12, 9, 3, 12,
    a. Find the mean, the median, and the mode.
    b. Which measure of central tendency is the most appropriate, and why?

6. The following is a list of the retirement ages for the workers in a certain production plant: 65, 61, 63, 62, 65, 62, 65, 64, 62.
    a. Find the mean, the median, and the mode.
    b. Which measure of central tendency is the most appropriate, and why?

7. For the following distributions, estimate whether they are skewed, and if so, in which direction.
    a. $\bar{X} = 50$, Mdn = 50, Mo = 50.
    b. $\bar{X} = 110$, Mdn = 100, Mo = 90.
    c. $\bar{X} = 450$, Mdn = 500, Mo = 550.

8. If a distribution is skewed to the right,
    a. which measure of central tendency yields the highest value?
    b. which yields the lowest?

9. If a distribution is skewed to the left,
    a. which measure of central tendency yields the highest value?
    b. which yields the lowest?

10. If the mean is substantially higher than the median, in which direction is the distribution skewed?

11. If the median is substantially higher than the mean, in which direction is the distribution skewed?

12. When the data are ordinal, what is the appropriate measure of their central tendency?

13. When the data are nominal, what is the appropriate measure of their central tendency?

14. The mean number of years of marriage preceding divorce is 7. The median number of years is 6. Most divorces occur, however, either at 3 years of marriage or at 22 years. Which measure of central tendency best describes these data, and why?

True or False—Indicate either T or F for problems 15 through 22.

15. When finding the median, it makes no difference whether one starts counting from the top or the bottom of the distribution of scores.
16. The mean is influenced more than the median by a few extreme scores at one end of the distribution.
17. In a positively skewed distribution, the mean lies to the left of the median.
18. The median is always that point which is arithmetically exactly halfway between the highest and lowest scores in a distribution.
19. If a distribution is perfectly symmetrical, it must have only one mode.
20. The median of a given distribution is 20, the mode is 17, and the mean is 35. The distribution is skewed to the left.
21. When either skewed interval data or ordinal data are used, the median is the best index of central tendency.
22. It makes little or no difference which measure of central tendency is used, since they all "average out" about the same anyway.

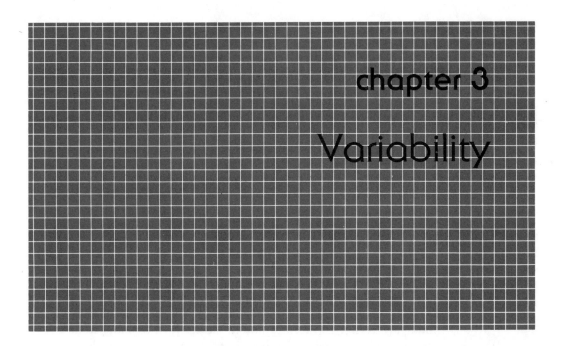

chapter 3
Variability

Just as measures of central tendency give information about the similarity among scores, **measures of variability** give information about how scores differ or vary. Usually when a group of persons, things, or events are measured, the measurements are scattered over a range of values. Sometimes the scatter is large, as in the case of a distribution of the incomes of all wage-earning Americans. Some people earn almost nothing; others earn millions of dollars. Sometimes the scatter is small, as it is among the heights of women in a Las Vegas chorus line. In this case, the shortest and the tallest dancer may differ by only an inch or two. The fact that measures of people vary describes the concept of "individual differences," the theme running through all of the social sciences. The description of data is never complete until some indication of the variability is found.

For example, suppose we are told that the yearly mean temperature in Anchorage, Alaska, is 56°F; in Honolulu, Hawaii, the mean is only 8 degrees higher, 64°F. If this mean temperature comparison were the only information you were given about the two areas, you might be led to believe that the climates are fairly similar. Once some of the variability facts are added, however, you begin to realize how misleading it can be to compare averages alone. In Honolulu, the temperature rarely rises above 84°F or dips below 50°F. In Anchorage, on the other hand, it may reach 100°F during the summer (midnight sun time) and often drops to −40°F during the winter. It is possible that the temperature in Anchorage hits the mean of 56°F only twice a year, once in the spring as the temperatures are going up and once in the fall as they are coming down. Obviously, to interpret the measures of central tendency correctly, we must know something about variability.

MEASURES OF VARIABILITY

There are three major measures of variability: the range, the standard deviation, and the variance.

The Range

One quick and easy way to describe something about the variability of any distribution is to calculate the **range,** symbolized by R. The range is the measurement of the width of the entire distribution and is found simply by figuring the difference between the highest and the lowest scores. The range is always given as a single value. For example, if the highest score in an IQ distribution is 140 and the lowest score is 60, we subtract 60 from 140 and write the range as $R = 80$.

Although the range is certainly a useful device for providing some information about variability, it does have one rather severe limitation—it is based on only two scores, the highest and the lowest. Distributions can have identical means and ranges and yet differ widely in terms of other important measures of variability. Another problem with the range is that it can change with the number of scores in the distribution. That is, as new scores are added to the distribution, there is a distinct possibility that more extreme scores will be among those newly selected. The new scores, of course, can never reduce the range, but the probability is high that they will increase it.

Percentiles, Quartiles, and Deciles.   Percentiles, quartiles, and deciles are descriptive terms used to locate specific points in any distribution. Although they are not variability measures themselves, they are used to create various forms of the just cited range.

Percentiles are an extremely important type of descriptive statistic, since a raw score can be described very precisely by converting it into a percentile (sometimes called a centile). A percentile is defined as that point on a distribution below which a given percentage of scores falls. For example, a score at the 95th percentile (P_{95}) is at the very high end of its distribution, because an enormous number (95%) of scores are below that point. A score at the 5th percentile (P_5), however, is an extremely low score because only 5% of scores are below that point. Since the 50th percentile divides the distribution exactly in half, it is always equal to the median.

Percentiles are also, at times, referred to as **percentile ranks.** A percentile rank of 65 indicates a score at the 65th percentile. When referring to either percentiles or percentile ranks, the actual location on the distribution is considered to be a hypothetical point, one without dimension. The point below which 50% of the scores fall, the 50th percentile, is also the point above which 50% of the scores fall.

There are two possible sources of confusion regarding percentiles. First, if you take a test and score at the 95th percentile, it does not mean that you answered 95% of the items on the test correctly. Instead, it means that 95% of the others who took the test did worse than you. Second, the rank expressed by a percentile is always in reference to the entire group being measured and *compared.* If, on a certain math test, 95% of Harvard Business School students did worse than you, then proudly take a bow. If, however, 95% of the third graders in a metropolitan school district did worse than you, hold your bow—it also means that 5% did better!

Quartiles divide a distribution into quarters. Thus, the 1st quartile, called Q_1, coincides with the 25th percentile, the 2nd quartile, called Q_2, with the 50th percentile (which as we have noted, is also the median), and the 3rd quartile, called Q_3, with the 75th percentile.

Finally, **deciles** divide a distribution into tenths. The 1st decile is equivalent to the 10th percentile, and so on. Note that the 5th decile equals the 2nd quartile equals the 50th percentile equals the median.

The Interquartile Range. The **interquartile range** is the difference between the first and third quartiles. Thus, the interquartile range includes those scores making up the middlemost 50% of the distribution, 25% falling to the right of the median and 25% to its left. (See Fig. 3.1.) Since the interquartile range is not affected by the presence of a few extreme scores, it can be used to advantage with skewed distributions. The interquartile range is always computed with reference to the median, or second quartile, and therefore can be used not only with interval data, but also with ordinal data. Even if the scores are given only in terms of their rank order, the interquartile range can still be used to identify the middlemost 50% of the distribution.

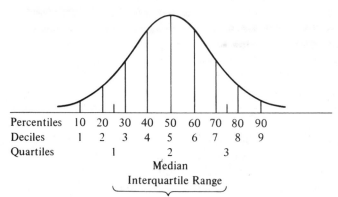

Figure 3.1 Percentiles, deciles, quartiles, and the interquartile range of a distribution.

The Quartile Deviation.   Finally, when the interquartile range is divided in half, it produces what is known as the **quartile deviation,** or as it is sometimes called, the semi-interquartile range. Thus, the quartile deviation, or Q, is equal to $(Q_3 - Q_1)/2$. The quartile deviation gives us a general measure of how much the scores vary around the median. In the next section we introduce the concept of the standard deviation, and it might be well to note here that when the distribution is normal, Q is equal to approximately two-thirds of a standard deviation.

The Interdecile Range.   Statisticians sometimes calculate the **interdecile range,** the difference between the first and ninth deciles. The interdecile range includes the middlemost 80% of the distribution, 40% falling to the right of the median and 40% to its left. This measure of variability is also unaffected by the addition of a few extreme scores and, therefore, can be used with skewed distributions. Like the interquartile range, the interdecile range can be calculated when the scores are ordinal (given in rank-order form).

The Standard Deviation

The **standard deviation** is the absolute heart and soul of the variability concept. It is one term that it is necessary to know and love. You should learn its computation so thoroughly that you can do it with the ease of a surgeon tying knots, virtually blindfolded. Unlike the range, the standard deviation takes into account all the scores in a distribution. The standard deviation is defined as a measure of variability that indicates by how much all of the scores in a distribution typically deviate or vary from the mean. Since the standard deviation is always computed with reference to the mean (never the median or the mode), its calculation demands the use of interval data (or, of course, ratio data).

The standard deviation is, in a sense, the typical deviation in a given distribution. The larger the value of the standard deviation, the more the scores are spread out around the mean; the smaller the value of the standard deviation, the less the scores are spread out around the mean. That is, a distribution with a small standard deviation indicates that the group being measured is homogeneous; their scores are clustered very close to the mean. A distribution with a large standard deviation, on the other hand, indicates a heterogeneous group; their scores are more widely dispersed from the mean. For example, if a distribution of IQ scores has a mean of 100 and a range between a low of 95 and a high of 105, that distribution is homogeneous, and its standard deviation is very small. If another IQ distribution, again with a mean of 100, has scores ranging all the way from 60 to 140, this distribution is more heterogeneous, and its standard deviation is probably larger.

Two methods for calculating the standard deviation are presented here. The first, called the deviation method, is better for understanding the concept of the standard deviation. It is used as a teaching method, because it keeps the real meaning of the term in clear focus. The second, called the computational method, is the one you will actually use after you have some conceptual understanding of the standard deviation. The computational method is easier to use because it has been designed for electronic calculators. Both methods, of course, give the same answer, since the computational method was algebraically derived from the deviation method.

The Deviation Method for Calculating the Standard Deviation. The deviation method is based on the concept of the **deviation score.** We know that any raw score in a distribution is symbolized as X, and the mean of a distribution as $\overline{X}$. Now, using these two values we create the deviation score, or x. The deviation score, x, is equal to the raw score, X, minus the mean, $\overline{X}$.

$$x = X - \overline{X}$$

Because the calculation of the standard deviation, which after all is our goal, must utilize every score in the entire distribution, a deviation score must be found for each and every raw score. The symbol for the standard deviation is SD, and the equation for the deviation method is as follows:

$$SD = \sqrt{\frac{\Sigma x^2}{N}}$$

Although a number of statistics books use the uppercase S symbol for the actual standard deviation of a set of scores, the symbol SD will be used throughout this book. Students new to the field have often had difficulty distinguishing the uppercase S symbol from the lowercase s symbol, which, as will be shown later, represents the estimated standard deviation of a whole population when all the scores are not available.

Example. Take the following distribution of scores: 10, 8, 6, 4, 2. First, calculate the mean. Since the mean, $\overline{X}$, is equal to $(\Sigma X)/N$, start by adding the scores: $10 + 8 + 6 + 4 + 2 = 30$. Then divide this value by N, the number of scores, which in this case is 5. The mean, $\overline{X}$, is therefore 6. Now subtract the mean from each score to get x, the deviation score. Note that all raw scores falling above the mean yield deviation scores with positive signs, whereas all raw scores falling below the mean yield deviation scores with negative signs. (Your calculator will display these negative signs; just pay attention to what the calculator shows.) The raw score of 10 gives a deviation score of $10 - 6 = 4$; 8 gives $8 - 6 = 2$; $6 - 6 = 0$; $4 - 6 = -2$; and $2 - 6 = -4$. The deviation scores for the raw scores of 10, 8, 6, 4, and 2 are, respectively 4, 2, 0, -2, and -4.

The standard deviation is equal to the square root of the sum of all the deviation scores squared, the sum being first divided by the number of scores. Table 3.1 is an illustration of the steps in the procedure for obtaining the standard deviation using this deviation equation.

Table 3.1 Deviation method.

	RAW SCORE X	DEVIATION SCORE X − X̄, or x	x²
	10	(10 − 6) = 4	16
	8	(8 − 6) = 2	4
	6	(6 − 6) = 0	0
	4	(4 − 6) = −2	4
	2	(2 − 6) = −4	16
	N = 5 ΣX = 30		Σx² = 40

$$\bar{X} = \frac{\Sigma X}{N} = \frac{30}{5} = 6.00 \qquad SD = \sqrt{\frac{\Sigma x^2}{N}} = \sqrt{\frac{40}{5}} = \sqrt{8} = 2.828 \text{ or } 2.83$$

The third column in Table 3.1 contains the deviation scores squared, x^2. Plug the first deviation score, 4, into the calculator. Then hit the square (^{2}) button and read, 16. If your calculator does not have a square button, plug in the 4, push the × or times button, and then hit the equals sign. That is, the number, then times, then equals gives the number squared. This trick is no big deal when working with a value like 4, but it can be a real timesaver when working with large values or values that aren't whole numbers.

Next, add the column of squared deviation scores. The total is 40. Divide this by the number of cases, 5, to get 8. Now, simply hit the square root ($\sqrt{\ \ }$) key for the standard deviation, in this case, 2.83. (Remember the rounding rule discussed on page 26, and use it whenever you are multiplying, dividing, squaring, or taking a square root.)

True to its definition, the standard deviation is calculated on the basis of how much *all* of the raw scores in the distribution deviate from the mean of the distribution. In the preceding calculation, it was necessary to obtain the deviation for each raw score to get the crucial Σx^2 value. The relationship between the concept of the standard deviation and its computation is kept clearly evident in this technique.

The Computational Method for Calculating the Standard Deviation. Although it somewhat clouds the concept of the standard deviation, the computational method is much easier and quicker to use with a calculator to do the work. Rather than creating three columns of values to work with, this method sets up only two columns, one of the raw scores and one of the squares of the raw scores. There are no deviation scores to calculate or to square. The equation for the computational method is as follows:

$$SD = \sqrt{\frac{\Sigma X^2}{N} - \bar{X}^2}$$

The standard deviation is equal to the square root of the difference between the sum of raw scores squared, which is divided by the number of cases, and the mean squared. Table 3.2 is an illustration of this computational method for obtaining the standard deviation.

Set up the distribution of raw scores in one column and their squares in another column, and grind the data through the equation. Some calculators even give the ΣX^2 value automatically if the raw scores are entered using the summation ($\Sigma+$) key.

If you happen to be using a more expensive calculator, a "statistician" model with an automatic standard deviation feature, don't fret because you did not get the same answer as that in Table 3.2 (in fact, you got a value of 3.16). This happened because the calculator divides the sum of the squared deviations by $N - 1$ rather than by N. This is a special method which is applied when the sample scores are being used to estimate the standard deviation of the population. It will be further explained in Chapter 8, but for now, just try the following. After entering the raw scores using the $\Sigma+$ key, press the variance (VAR) key and *then hit the square root key*. The standard deviation should be displayed as 2.83, as was previously shown. If your calculator does not have a VAR key, but does have a standard deviation key, plug in the mean as an added score. You will then get 2.83, the actual standard deviation of the distribution of scores being analyzed.

You are urged to use the computational method whenever you work on standard deviations. It is easier, takes less time, and fits perfectly with the capabilities of your calculator. Later, when we cover some of the tests of significance, it will be assumed that this method is being used. Why, then, was the deviation method presented? Simply because it gives a clearer picture (since you must calcu-

Table 3.2 Computational method.

X	X^2
10	100
8	64
6	36
4	16
2	4

$N = 5$ $\qquad \Sigma X = 30 \qquad \Sigma X^2 = 220$

$$\bar{X} = \frac{\Sigma X}{N} = \frac{30}{5} = 6.00$$

$$SD = \sqrt{\frac{\Sigma X^2}{N} - \bar{X}^2}$$

$$SD = \sqrt{\frac{220}{5} - 6.00^2}$$

$$SD = \sqrt{44 - 36}$$

$$SD = \sqrt{8} = 2.828 \text{ or } 2.83$$

Numbers sometimes appear mystifying, but don't let that bother you. Leave the math to your calculator. © 1967 United Feature Syndicate, Inc.

late each deviation score) of the concept of the standard deviation and its reliance on the variability of all the scores around the mean.

Using the Standard Deviation. The standard deviation is an extremely useful descriptive tool. It allows a variability analysis of the data that can often be the key to what the data are communicating. For example, suppose that you are a football coach and that you have two halfbacks with identical gain averages of 5 yards per carry. Back A has a standard deviation of 2 yards, and Back B has a standard deviation of 10 yards. Back A is obviously the more consistent runner—most of the time he gains between 3 and 7 yards, but he almost always gets a gain. When only 2 or 3 yards are needed, Back A is unquestionably the choice. While Back B, having a much larger standard deviation, is more likely to make a longer gain, he, alas, is also more likely to lose substantial yardage. If your team is behind, and it's fourth down, 10 yards to go, and time is running out, Back B is more likely to provide the necessary last-minute heroics.

Or suppose that you are a manufacturer of flashlight batteries. Testing of large numbers of two types of batteries shows that each has the same average life—25 hours. Battery A, however, has a standard deviation of 2 hours, while Battery B has a standard deviation of 10 hours. If you plan to guarantee your battery for 25 hours, you had better decide to manufacture Battery A—there won't be as many consumer complaints or returns. Battery B, however, has more chance of lasting at least 35 hours, but is also more likely to fail after only 15 hours or less.

The standard deviation is indeed a valuable descriptive tool. Its use does, however, require that the researcher have at least interval data.

The Variance

The **variance** is the third major technique for assessing variability. Once you can calculate a standard deviation, you can easily calculate the variance, because the variance, V, is equal to SD^2:

$$V = SD^2$$

$$V = \frac{\Sigma x^2}{N} = \frac{\Sigma X^2}{N} - \bar{X}^2$$

Thus, calculating the variance is precisely the same as calculating the standard deviation, without taking the square root. Table 3.3 illustrates a calculation of the variance using the same distribution of scores as was used previously to calculate the standard deviation. The variance for those scores is, therefore, 8.

Since the variance has such a straightforward mathematical relationship with the standard deviation, it must also be a measure of variability that tells how much all the scores in a distribution vary from the mean. Conceptually, therefore, it is the same as the standard deviation. Heterogeneous distributions with a great deal of spread have relatively large standard deviations and variances; homogeneous distributions with little spread have small standard deviations and variances. It may seem to be redundant to have two such variability measures, one that is simply the square of the other. However, there are situations in which working directly with the variances allows for certain calculations not otherwise possible. We shall see later that one of the most popular statistical tests, the F ratio, takes advantage of this special property of the variance.

No Variability Less Than Zero

Although it may seem to be stating the obvious, all measures of variability must reflect either some variability or else none at all. There can *never be less than zero variability*. The range, the standard deviation, and the variance should never be negative values.

Consider the distribution of values in Table 3.4. Note that, in this rather unlikely situation of all scores being identical, the range, the standard deviation, and the variance must equal zero. When all scores are the same, there simply is no

Table 3.3   Calculation of the variance.

X	X^2
10	100
8	64
6	36
4	16
2	4

$N = 5$ $\qquad$ $\Sigma X = 30$ $\quad$ $\Sigma X^2 = 220$

$$\bar{X} = \frac{\Sigma X}{N} = \frac{30}{5} = 6$$

$$V = \frac{\Sigma X^2}{N} - \bar{X}^2$$

$$V = \frac{220}{5} - 6^2$$

$$V = 44 - 36 = 8$$

Table 3.4 Measures of variability for a distribution of identical scores.

X	X^2
10	100
10	100
10	100
10	100
10	100
10	100
$\Sigma X = 60$	$\Sigma X^2 = 600$

$\bar{X} = 10$

$R = 0$

$SD = \sqrt{\dfrac{600}{6} - 10^2}$

$SD = \sqrt{100 - 100}$

$SD = \sqrt{0} = 0$

$V = SD^2 = 0$

variability. The value of zero is the smallest value any variability measure can ever have. If you ever calculate a negative standard deviation or variance, check over your math and/or the batteries in your calculator.

Grouped Data

Before the days of electronic calculators and computers, statisticians typically used what were called grouped-data techniques. These had the advantages of making large numbers of scores more manageable and making values easier to calculate, but also had the disadvantages of losing track of a given individual's score and losing some degree of accuracy.

Here's how it worked. Assume that we have the following set of 20 scores: 10, 10, 9, 10, 12, 4, 10, 9, 11, 15, 6, 13, 8, 2, 19, 16, 14, 8, 5, 6. Arranging them in distribution form gives the following: X: 19, 16, 15, 14, 13, 12, 11, 10, 10, 10, 10, 9, 9, 8, 8, 6, 6, 5, 4, 2.

Now *group* the data into class intervals, in this case with a width of 3, and select the midpoint of the interval as the score, X. Then identify the number of scores, f (frequency), falling within each interval. Finally, multiply that frequency by the midpoint of the interval, of fX. It's best to use odd-valued intervals (3, 5, 7, etc.), and then center each interval so that the midpoint is a multiple of the width.

Interval	X	f	fX
17–19	18	1	18
14–16	15	3	45
11–13	12	3	36
8–10	9	8	72
5–7	6	3	18
2–4	3	2	6
		$N = \Sigma f = 20$	$\Sigma fX = 195$

Next, add the *fX* column; in this case, the sum equals 195. Then, calculate the mean.

$$\bar{X} = \frac{\Sigma fX}{N} = \frac{195}{20} = 9.75$$

Now, if we had not grouped the data, but had instead simply added the scores, 19, 16, 15, 14, and so on, we would have obtained a value of 197 for ΣX and a mean of 9.85. The reason for the different results is that the grouped-data technique assumes that the mean of the scores within a given interval is equal to the midpoint of that interval. Since this assumption is not always true, the grouped-data technique usually produces results that are slightly in error. In this example, the grouped-data technique underestimated the mean by .10.

The standard deviation is also available from grouped data. Using the values from the preceding example, we calculate it as shown in the following.*

Interval	f	X	fX	XfX or fX²
17–19	1	18	18	324
14–16	3	15	45	675
11–13	3	12	36	432
8–10	8	9	72	648
5–7	3	6	18	108
2–4	2	3	6	18
			$\Sigma fX = 195$	$\Sigma fX^2 = 2205$

$$\bar{X} = \frac{\Sigma fX}{N} = \frac{195}{20} = 9.75$$

$$SD = \sqrt{\frac{\Sigma fX^2}{N} - \bar{X}^2} = \sqrt{\frac{2205}{20} - (9.75)^2}$$

$$= \sqrt{110.25 - 95.06} = \sqrt{15.19}$$

$$= 3.90$$

If we had worked out the standard deviation from these data using the raw score formula $SD = \sqrt{[(\Sigma X^2)/N] - \bar{X}^2}$, we would have obtained a value of 4.11. The grouped-data technique, therefore, underestimated the value of the standard deviation by .21.

*To obtain the values in the last column, multiply the X value by the *fX* value (18 × 18 = 324, 15 × 45 = 675, etc.). *Do not* simply square the *fX* value.

GRAPHS AND VARIABILITY

The concept of variability, or dispersion, can be further clarified through the use of graphs. Look at the two frequency distributions of IQ scores shown in Fig. 3.2. Both distributions have precisely the same mean, 100, and the same range, 60; but they tell dramatically different stories. The distribution in Fig. 3.2a shows a large number of people having both very low and very high IQ scores—about 50 with

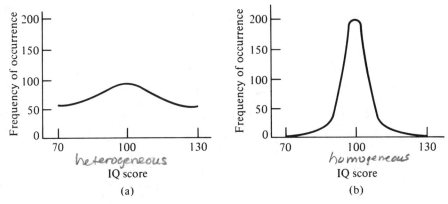

heterogeneous

homogeneous

Figure 3.2 Two frequency distributions of IQ scores.

IQs of 70 and another 50 with IQs of 130. There is a slight bulge in the middle of the distribution. However, it shows that only about 100 persons scored at the mean of 100. (Remember that the height of the curve represents the number of occurrences.) In the distribution in Fig. 3.2b, there are very few people who scored IQs of 70 and very few who scored 130. The great majority (about 200), scored right around the mean, between the IQs of 90 and 110. Therefore, even though both distributions do indeed have the same range and mean, the one in Fig. 3.2b represents a far more homogeneous group than does the one in Fig. 3.2a. The distributions differ in terms of their standard deviations. While Fig. 3.2a has many scores deviating widely from the mean and thus has a large standard deviation, Fig. 3.2b has most of the scores clustering tightly around the mean and thus has a small standard deviation.

Kurtosis

Statisticians use the term **kurtosis** to describe how peaked or how flat a curve is. The distribution in Fig. 3.2a is in the shape called platykurtic. A **platykurtic** curve represents a distribution with a relatively large standard deviation. Figure 3.2b shows a **leptokurtic** curve, which always indicates a relatively small standard deviation. This may sound like extremely pompous and stuffy jargon, but precise terms are necessary to convey precise meanings. Also, you can amaze your friends by working these terms into your conversation at lunch tomorrow.

A curve that is about halfway between platykurtic and leptokurtic extremes is called **mesokurtic.** The importance of the mesokurtic distribution will be discussed at length in the next chapter. It is called the normal curve and is the most important curve we will confront all semester. It is the statistician's "dream" curve, bell shaped and perfectly symmetrical.

We have recently been reminded of two amusing nmnemonics attributed to William Sealy Gossett. Platykurtic curves, like platypuses, are squat with short tails, whereas leptokurtic curves are high with long tails, like kangaroos—noted for their "lepping".[1]

Relationship Between the Standard Deviation and the Range

The relationship between the standard deviation and the range can best be illustrated in the context of kurtosis. First, however, it must be clearly understood that the range always results in a value larger than the value of the standard deviation. Since the range describes the entire width of a distribution, or the difference between the highest and the lowest scores, the value of the range must always be greater than that of any other measure of variability.*

The standard deviation, on the other hand, describes how much all the scores vary from the mean. Therefore, the standard deviation, which takes into account many more than just the top and bottom scores, must result in a value less than the value of the range. The greater the spread of scores around the mean, however, the more closely the value of the standard deviation will approach the value of the range. In a platykurtic distribution, the value of the standard deviation equals a fairly large proportion of the range. The more tightly the scores cluster around the mean, that is, the more leptokurtic a distribution is, the smaller will be the standard deviation relative to the range.

In Table 3.5, the range for distribution A is $20 - 2 = 18$. Because of the large number of both high and low scores, the standard deviation is a relatively high 5.92. Distribution B, with an identical range of 18, but with more of its scores falling close to the mean, has a smaller standard deviation of 3.80.

Is It Reasonable? A couple of very handy checks on the previous calculations, checks that will really prevent gross errors, are based on ensuring that the resulting values are at least reasonable. After all, even when using a calculator or computer, it's always possible to hit the wrong key. Just take a quick time-out and do an "is-it-reasonable?" check.

1. After calculating the mean, look back at the distribution and make sure that the mean's value is within the range of scores. That is, the mean could not possibly be greater than 20 or less than 2. A closer inspection of the data further reveals (for both distributions A and B) that the scores are fairly well evenly balanced, telling us that the mean is somewhere near the middle of the distribution. The calculated value of 10.62, therefore, seems reasonable. Had the distribution been skewed, we would have "guesstimated" the mean to be nearer the skewed end of the distribution rather than centrally located as it was in this case.

*Although the range equals about six times the standard deviation when the distribution is normal, the range must always be larger than the standard deviation. No matter what shape the distribution takes, the range must *always* be *at least* twice the size of the standard deviation, unless $R = 0$.

Table 3.5 Calculating the means and standard deviations of two distributions.

DISTRIBUTION A		DISTRIBUTION B	
X	X^2	X	X^2
20	400	20	400
19	361	14	196
16	256	12	144
15	225	11	121
14	196	11	121
12	144	11	121
12	144	11	121
10	100	10	100
7	49	10	100
5	25	9	81
3	9	9	81
3	9	8	64
2	4	2	4

DISTRIBUTION A:

$N = 13$ $\Sigma X = 138$ $\Sigma X^2 = 1922$

$\bar{X} = \dfrac{\Sigma X}{N} = \dfrac{138}{13} = 10.615 = 10.62$

$SD = \sqrt{\dfrac{1922}{13} - 112.78}$

$SD = \sqrt{147.85 - 112.78} = \sqrt{35.07} = 5.92$

DISTRIBUTION B:

$\Sigma X = 138$ $\Sigma X^2 = 1654$

$\bar{X} = \dfrac{138}{13} = 10.615 = 10.62$

$SD = \sqrt{\dfrac{1654}{13} - 112.78}$

$SD = \sqrt{127.23 - 112.78}$

$SD = \sqrt{14.45} = 3.80$

2. Also, examine the value of the standard deviation. This is a very important check-point, for the standard deviation has certain limits that it cannot exceed. The standard deviation can never be less than zero (it can't be negative) or greater than half the range. Since the previous two distributions each had a range of 18, the highest value the standard deviation could possibly take, no matter how the scores were distributed, would be 18/2, or 9. Had you somehow pressed the wrong button on your calculator and arrived at a standard deviation value of, say, 12.29, you should know you had an illogical answer and you'd better go back to the old drawing board and recheck your work.

Assessing Kurtosis (Ku)—The "1/6" Rule

In the Appendix you will find an equation for establishing the exact value for the kurtosis (Ku) of any unimodal frequency distribution. However, at this point you can simply use the following benchmark for coming up with a quick Ku evaluation. Since the standard deviation of a normal distribution is approximately one-sixth of its range, you may use this "1/6" rule as a kurtosis guidepost. If the standard deviation of a unimodal distribution is less than one-sixth of the range, the distribution tends to be leptokurtic. Similarly, if the standard deviation is more than one-sixth of the range, the distribution is platykurtic. For example, a unimodal distribution with a standard deviation of 50 and a range of 600 is clearly leptokurtic (since the ratio is only 1/12). Or a distribution with a standard deviation of 200 and a range of 600 is platykurtic (since the ratio is 1/3). Thus, the "1/6" rule aids us in clarifying the concept of "relatively large" or "relatively small" standard deviations. Remember, for a quick kurtosis evaluation you don't even have to know the values of the mean and median as you did for assessing skewness. Simply look at the values of the standard deviation and range. Divide the range by 6 and use this value as a kind of "marker" value. This "marker" provides an approximation of what the standard deviation should be if the distribution were normal. Now, compare the actual standard deviation with this marker, and if the actual standard deviation is greater than the marker, the distribution is platykurtic; if smaller, then it's leptokurtic.

SUMMARY

Whereas measures of central tendency define averageness or typicality in a distribution of scores, variability measures describe how scores differ. The three major measures of variability are the range, the standard deviation, and the variance.

1. The range (R) describes the entire width of the distribution and is found by subtracting the lowest score from the highest score. Two variations on the range theme are presented. The interquartile range is the difference between the first quartile (the 25th percentile), and the third quartile (or the 75th percentile). The interdecile range is the difference between the first decile (the 10th percentile) and the 9th decile (the 90th percentile).
2. The standard deviation (SD) describes variability in terms of how much all of the scores vary from the mean of the distribution. Therefore, when the standard deviation is known, precise variability statements can be made about the entire distribution.
3. The variance (V or SD^2) like the standard deviation, measures variability in terms of all of the scores in the distribution. The variance is simply the square of the standard deviation.

In a graph of a unimodal frequency distribution, variability is related to the kurtosis of the curve. Kurtosis is a description of the curvature (peakedness or flatness) of the graph. When a distribution has a small standard deviation (less than $R/6$), its curve has a leptokurtic shape—many scores clustering tightly around the mean. When a distribution has a large standard deviation (greater than $R/6$), its curve has a platykurtic shape—many of the scores spread out away from the mean.

Finally, when a distribution takes on the "ideal" bell shape of the normal curve, the distribution of scores is graphed as mesokurtic. Distributions which either are, or are close to being, mesokurtic in shape enjoy a special relationship between the range and the standard deviation. In this situation the standard deviation equals about one-sixth of the range.

KEY TERMS

deciles	leptokurtic	platykurtic
deviation score	measures of variability	quartiles
interdecile range	mesokurtic	quartile deviation
interquartile range	percentiles	range
kurtosis	percentile rank	standard deviation
		variance

REFERENCES

1. BALANDA, K. P., & MACGILLIVRAY, H. L. (1988). Kurtosis: A critical review. *The American Statistician, 42,* pp. 11–119.

PROBLEMS

1. At Company X, a group of eight blue-collar workers was selected and asked how much of their weekly pay (in dollars) they put into their savings bank account. The following is the list of amounts saved: $12, $11, $10, $20, $1, $9, $10, $10. Calculate the mean, the range, and the standard deviation.

2. At Company Y, another group of eight blue-collar workers was selected and asked how much of their weekly pay they saved. The following is the list of amounts saved: $20, $18, $3, $2, $1, $16, $15, $8. Calculate the mean, the range, and the standard deviation.

3. Despite the facts that the two groups of workers in problems 1 and 2 saved the same mean amount, and that the ranges of amounts saved are identical, in what way do the two distributions still differ? Which group is more homogeneous?

4. A group of seven university students was randomly selected and asked to indicate the number of study hours each put in before taking a major exam. The data are as follows:

STUDENT	HOURS OF STUDY
1	40
2	30
3	35
4	5
5	10
6	15
7	25

For the distribution of study hours, calculate
a. The mean.
b. The range.
c. The standard deviation.

5. The grade-point averages for the seven university students selected above were computed. The data are as follows:

STUDENT	GPA
1	3.75
2	3.00
3	3.25
4	1.75
5	2.00
6	2.25
7	3.00

For the GPA distribution, calculate
   a. The mean.
   b. The range.
   c. The standard deviation.

6. The IQs of all students at a certain prep school were obtained. The highest was 135 and the lowest was 105. The following IQs were identified as to their percentile:

X	PERCENTILE
107	10th
114	25th
118	50th
122	75th
129	90th

   a. What is the range for the distribution?
   b. What is the median?
   c. What is the interquartile range?
   d. What is the interdecile range?

7. Calculate the range and the standard deviation for the following set of scores: 10, 10, 10, 10, 10, 10, 10.

8. Under what conditions can a negative standard deviation be calculated?

9. What is the relationship between the relative size of the standard deviation and the kurtosis of a distribution?

10. For the following distributions, evaluate each for the type of kurtosis.
   a. $\bar{X} = 50$, Mdn $= 50$, Mo $= 50$, $R = 100$, $SD = 2$.
   b. $\bar{X} = 500$, Mdn $= 500$, Mo $= 500$, $R = 600$, $SD = 100$.
   c. $\bar{X} = 100$, Mdn $= 100$, Mo $= 100$, $R = 60$, $SD = 25$.

11. If a distribution is mesokurtic and $R = 120$, find the approximate value of the standard deviation.

12. On what scale of measurement must the data be in order to calculate the standard deviation?

13. If a given mesokurtic distribution has a mean of 100 and a standard deviation of 15, find the approximate value of the range.

14. Using the data in problem 13, calculate the variance.

True or False—Indicate either T or F for 15 through 22.

15. The standard deviation can only be computed with reference to the mean.
16. The distribution with the largest range necessarily has the largest standard deviation.
17. The 9th decile must always coincide with the 90th percentile.
18. The interquartile range includes only the middlemost 25% of the distribution.
19. In any distribution, the median must fall at the 50th percentile.
20. The interdecile range includes only the middlemost 10% of the distribution.
21. The standard deviation can never be greater than the range.
22. The mean is to central tendency as the standard deviation is to variability.

chapter 4

The Normal Curve and z Scores

If the mean and the standard deviation are the heart and soul of descriptive statistics, then the normal curve is its lifeblood. Although there is some dispute among statisticians as to when and by whom the normal curve was first introduced, it is customary to credit its discovery to the great German mathematician **Karl Friedrich Gauss.** Even to this day, many statisticians refer to the normal curve as the Gaussian curve.

As a child, Gauss was one of those "perfect-pitch" mathematical prodigies who put most students and teachers in a state of total terror. It is said that he could add, subtract, multiply, and divide before he could talk. It has never been clear how he communicated this facility to his family and friends. Did he, like the Wonder Horse, stamp out his answers with his foot? We are also told that at the age of 3, when he presumably could talk, he detected a math error on his father's pay envelope. When he was 8, he startled his schoolmaster by adding all the numbers from 1 through 100 in his head. (The teacher had given the class this busy-work assignment to do longhand while he corrected papers. Obviously, little Karl's lightning calculation changed the teacher's plans.)

Gauss is undoubtedly one of the most important figures in the history of statistics. He developed not only the normal curve, but also some other extremely important statistical concepts that will be covered later.

Karl Friedrich Gauss (1777–1855). (Culver Pictures.)

THE NORMAL CURVE

So many distributions of measures in the social sciences conform to the **normal curve** that it is of crucial significance for describing data. The normal curve, as Gauss described it, is actually a theoretical distribution. However, so many distributions of people-related measurements come so close to this ideal that it can be used in generating frequencies and probabilities in a wide variety of situations.

Key Features of the Normal Curve

The normal curve (Fig. 4.1) is first and foremost a unimodal **frequency distribution curve** with scores plotted on the *X* axis and frequency of occurrence on the *Y* axis. However, it does have at least five key features that set it apart from other frequency distribution curves:

1. In a normal curve, most of the scores cluster around the middle of the distribution (where the curve is at its highest). As distance from the middle increases, in either direction, there are fewer and fewer scores (the curve drops down and levels out on both sides).
2. The normal curve is symmetrical. Its two halves are identical mirror images of one another; it is thus perfectly balanced.
3. In a normal curve, all three measures of central tendency—the mean, the median, and the mode—fall at precisely the same point, the exact center or midpoint of the distribution.
4. The normal curve has a constant relationship with the standard deviation. When the abscissa of the normal curve is marked off in units of standard deviation, a

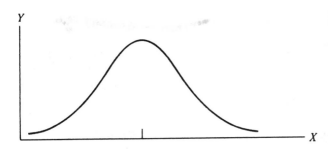

Figure 4.1
The normal curve.

series of *constant* percentage areas under the normal curve are formed. The relationship holds true for all normal curves, meaning that if a certain percentage of scores is found between one and two standard deviation units above the mean, this same percentage is always found in that specific area of any normal curve. Also, because of the symmetry of the curve, that exact percentage is always found in the same part of the lower half of the curve, between one and two standard deviation units below the mean. This constancy of the percentage area under the curve is crucial to an understanding of the normal curve. Once the curve is plotted according to standard deviation units, it is called the standard normal curve.

5. The normal curve is asymptotic to the abscissa. No matter how far out the tails are extended, they will never touch the *X* axis.

The standard normal curve (Fig. 4.2) has a mean of 0 and a standard deviation of 1.00. The standard deviation units have been marked off in unit lengths of 1.00 on the abscissa, and the area under the curve above these units always remains the same. Since, as has been stated, the mean and median of a normal curve (as well as the mode) always fall at exactly the same point on the abscissa under the normal curve, the mean in this case is interchangeable with the median. The importance of this is due to the fact that, since the median always divides any frequency distribution exactly in half, then when the mean and median coincide, the mean also divides this particular distribution in half, with 50% of the scores falling above the mean and 50% below the mean.

Further, the area under the curve between the mean and a point one standard deviation unit above the mean always includes 34.13% of the cases. Because of the symmetry of the curve, 34.13% of the cases also fall in the area under the curve between the mean and a point one standard deviation unit below the mean. Thus, under the standard normal curve, between two points, each located one

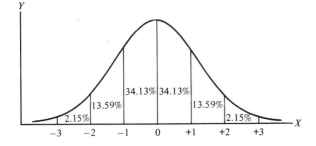

Figure 4.2
The standard normal curve.

standard deviation unit away from the mean, there are always 68.26% (twice 34.13%) of all the cases.

As we go farther away from the mean, between one and two standard deviation units, another 13.59% of the distribution falls on either side. Even though the distance away from the mean has been increased by another full standard deviation unit, only 13.59% of the cases are being added. This is because, as we can see from Fig. 4.2, the curve is getting lower and lower as it goes away from the mean. Thus, although 34.13% fall between the mean and one standard deviation unit away from the mean, only 13.59% are being added as we go another length farther away from the mean, to the area included between one and two standard deviation units. Also, between the mean and a point two full standard deviation units away from the mean, a total of 47.72% (34.13% + 13.59%) of the cases can be found.

Including both halves of the curve, we find that between the two points located two standard deviation units above and below the mean, there will be 95.44% (twice 47.72%) of the distribution. Finally, the area under the curve between two and three standard deviation units away from the mean only holds 2.15% of the cases on each side. There are, then, 49.87% (47.72% + 2.15%) of the cases between the mean and a point three standard deviation units away from the mean. Almost the entire area of the curve, then, that is, 99.74% (twice 49.87%), can be found between points bounded by three standard deviation units above and below the mean.

This may seem confusing. Before throwing in the towel, however, picture for a moment that we are installing a rug in the very small and odd-shaped room shown in Fig. 4.3. We decide to begin measuring from the point marked with an arrow. To both the right and the left from the arrow, we measure 3 feet. The long side of this room therefore measures 6 feet. We mark this distance off in 1-foot intervals, 3 to the right and 3 to the left of the point *X*. Taking great care, we now take measures of the other side of the room. We determine that between 1 foot to the right of *X* and 1 foot to the left of *X*, the area of the room will require roughly two-thirds, or 68%, of our rug. Now, this holds true whether the total area to be covered is 100 square feet, 10 square feet, or 1000 square feet—68% of the rug must fit in that precise area. Between 2 feet to the right and 2 feet to the left of *X*, 95% of the rug will be used. To cover the area between 3 feet to the right and 3

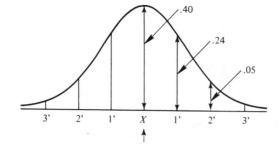

Figure 4.3
Carpeting an odd-shaped room.

feet to the left of the arrow, virtually the entire rug is needed. Again, remember that these percentages hold true for this shape, whether the total area is the size of a gym or the size of a hall closet. (See the box on page 69 for an explanation of how the normal curve is drawn.)

z SCORES

In the preceding section we were, in fact, without naming them as such, discussing z **scores** (also called standard scores). The z **distribution** is a normally distributed set of specially scaled values whose mean is always equal to zero and whose standard deviation must equal 1.00. As seen, the vast majority of the normal curve's total area (99.74%) lies between the z scores of ± 3.00. As we shall soon see, z scores are enormously helpful in the interpretation of raw score performance, since they take into account both the mean of the distribution and the amount of variability, the standard deviation. We can, thus, use these z scores to gain an understanding of an individual's relative performance compared to the performance of the entire group being measured. Also, we can compare an individual's relative performance on two separate normally distributed sets of scores.

The z Score Table

So far the discussion of z scores and the areas under the standard normal curve has assumed that the standard deviation units always come out as nice, whole numbers. We learned, for example, that 34.13% of the cases fall between the mean and a point one standard deviation unit away from the mean. But what happens in the more typical case when the z scores turn out not to be those beautiful, whole numbers?

The z score table (Appendix Table A) should become well worn during this semester. It will be used to obtain the precise percentage of cases falling between any z score and the mean. Notice that the z score values in this table are arrayed in two directions, up and down in a column, and left to right in a row. The column at the far left gives the z scores to one decimal place. The second decimal place of the z score (remember, always round to two places) is given in the top row.

Example. To look up a z score of say, .68, run a finger down the far left column until it reaches .06. Then, using a ruler or the edge of a piece of paper to prevent your eyes from straying off the line, follow across that row until it intersects the column headed by .08. The value there is 25.17. This, like all the values in the table, represents a percentage, the percentage of cases falling between the z score of .68 and the mean. Next, look up the percentage for a z score of 1.65. Run down the far left column until 1.6, then across that row until it intersects the column headed by .05. The value there is 45.05, or 45.05%.

"I don't care if he does have an IQ of
169—I still think he's faking."

Because the normal curve is symmetrical, the z score table gives only the percentages for half the curve. This is all that is necessary since a z score that is a given distance to the right of the mean yields the same percentage as a z score that is the same distance to the left of the mean. That is, positive and negative z scores of a given value give precisely the same percentages, since they deviate the same amount from the mean.

A Fundamental Fact About the z Score Table. It is extremely important that you learn the following: *The z score table gives the percentage of cases falling between a given z score and the mean.* Put the book down for a minute and write that sentence on a piece of paper. Write it several times; repeat it aloud in round, clear tones. Paste it on your bathroom mirror so that it greets you each morning. Memorizing that simple sentence virtually assures success with z score problems by preventing most of the big, fat, egregious errors that so many students seem automatically to commit. Once again now, the z score table gives the percentage of cases falling between a given z score and the mean. It does *not* give a direct readout of the percentage above a z score, or below a z score, or between two z scores (although all of these can be calculated).

Drawing the Curve. When working with z scores, it is extremely helpful to draw the curve each time. It only takes a couple of seconds and is well worth the effort. Drawing the curve and locating the part of the curve the question concerns will give you a much clearer picture of what is being asked. It isn't necessary to waste time or graph paper being compulsive over how the curve looks. Just sketch a quick approximation of the curve, and locate the mean and z scores on the base-line. Remember to place all positive z scores to the right of the mean and negative *z* scores to the left. Also remember that the higher the *z* score, the farther to the right it is located, and the lower the *z* score, the farther to the left it is placed.

Spending a few seconds to draw the curve is crucial. This step is not a crutch for beginners; in fact, it is common practice among topflight statisticians. Although drawing the curve may appear initially to be a trivial waste of time, especially when working with only one *z* score at a time, it is a good habit to get into. A nice picture of the problem right in front of you makes it easier to solve.

The box on page 69 presents the equation for the normal curve, with a couple of examples of ordinates obtained from the *z* scores. Unless you have some math background, it isn't necessary to even look at this equation. Simply take the shape of the curve on faith.

From *z* Score to Percentage of Cases

Case A: To find the percentage of cases between a given *z* score and the mean.
Rule: Look up the *z* score in Table A and take the percentage as a direct readout.
Problems 4a: Find the percentage of cases falling between each of the following *z* scores and the mean

1. 1.25   4.   .10
2. .67   5. − 1.62
3. − 2.58

The Percentage of Cases Below a z Score.   If someone has a *z* score of 1.35 and we want to know the percentage of cases below it, we first look up the *z* score in Table A. The table reveals that 41.15% of the cases occur between the *z* score of 1.35 and the mean. We also know that a *z* score of 1.35 is positive and therefore must fall to the right of the mean. (See Fig. 4.4.) Further, we know that on a normal curve 50% of the cases fall below the mean, so we add 50% to 41.15% and get a total of 91.15%.

If the *z* score happens to be negative, for example, − .75, again we look up the *z* score in Table A, and read the percentage of cases falling between it and the mean, in this case 27.34%. This *z* score is negative and therefore must be below the mean. (See Fig. 4.5.) Since 50% of the cases fall below the mean, and since we have just accounted for 27.34% of that, we subtract 27.34% from 50% and get 22.66%.

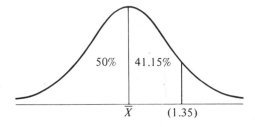

Figure 4.4
Percentage of cases below a *z* score of 1.35.

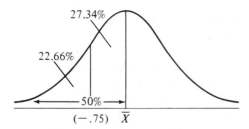

Figure 4.5
Percentage of cases below a z score of −.75.

Case B: To find the percentage of cases falling below a given z score (percentile).
Rule B1: If the z score is positive, look up the percentage and add 50%.
Rule B2: If the z score is negative, look up the percentage and subtract it from 50%.
Problems 4b: Find the percentage of cases falling *below* the following z scores.

1.   2.57   4. 1.61
2. − 1.05   5.   .14
3.   − .79

From z Score to Percentile

The percentile, as defined in Chapter 3, is that point in a distribution of scores below which the given percentage of scores falls. The median, by definition, always occurs precisely at the 50th percentile. Since on the normal curve the median and the mean coincide, then for normal distributions the mean must also fall at the 50th percentile. Since this is true, once the z score is known and the area below the z score found, the percentile can be easily found. Converting z scores into percentiles, then, is a quick and straightforward process.

When converting z scores into percentiles, remember the following points. First, all positive z scores *must* yield percentiles higher than 50. In the case of positive z scores, we always add the percentage from the z score table to 50%. Second, all negative z scores must yield percentiles lower than 50. In this case, we always subtract the percentage from the z score table from 50%. Third, a z score of zero, since it falls right on the mean, must yield a percentile of 50. Finally, it is customary when working with percentiles (and this is the only exception to our round-to-two-places rule) to round to the whole number. Thus, for a z score of 1.35, which falls above 91.15% of all the cases (41.15% between the z and the mean plus 50% below the mean), the percentile rank is stated as 91 or the 91st percentile.

Problems 4c: Find the percentile for each of the following z scores.

1.   .95   4. − 1.65
2.   1.96   5.   .50
3. − .67

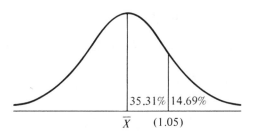

Figure 4.6
Percentage of cases below a *z* score of 1.05.

In later chapters, we will be using a special, timesaving percentile table (Table B) for these kinds of problems. At this point, however, you really should work these out using Table A. The more you use the z score table now, the clearer the later sections on sampling distributions will appear.

The Percentage of Cases Above a z Score.   Many times, the area above the z score must be established. Suppose that we must find the percentage of cases falling *above* a z score of 1.05. First, we use Table A to get the percentage falling between the z of 1.05 and the mean. The table gives us a readout of 35.31%. Since this is a positive z score, it must be placed to the right of the mean. (See Fig. 4.6.) We know that 50% of all cases fall above the mean, so we subtract 35.31% from 50% to get 14.69%.

If the z score is negative, for example, −.85, we look it up in the table and read 30.23% as the percentage of cases falling between the z score and the mean. Since all negative z scores are placed below the mean, 30.23% of cases fall above this z score and below the mean, that is, *between* the z score and the mean. (See Fig. 4.7.) Also, 50% of the cases must fall above the mean. We add 30.23% to 50% and obtain 80.23% as the percentage of cases falling above a z score of −.85.

Case C: To find the percentage of cases above a given z score.
Rule C1: If the z score is positive, look up the percentage and subtract it from 50%.
Rule C2: If the z score is negative, look up the percentage and add it to 50%.
Problems 4d: Find the percentage of cases falling *above* the following z scores.

1. −.15   4. − 2.62
2. 2.00   5.   .09
3. 1.03

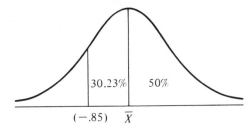

Figure 4.7
Percentage of cases below a *z* score of −.85.

The Percentage of Cases Between z *Scores.* Finding the area under the normal curve between z scores is based essentially on the procedures already covered, except that we work with two z scores at a time. It is important to look up the percentages in the z score table *one at a time.* There are no creative shortcuts!

z *Scores on Opposite Sides of the Mean*

Example. We want the percentage of cases falling between z scores of −1.00 and +.50.

First, as always, we draw the curve and locate the z scores on the X axis. Next, we look up each z score in Table A. For −1.00 we find 34.13%, and for +.50 we find 19.15%. Remember that each of these percentages lies between its own z score and the mean. Now add the two percentages and you discover that 53.28% (that is, 34.13% + 19.15%) of the cases fall between these z scores.

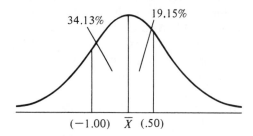

$$(-1.00) \quad \overline{X} \quad (.50)$$

Example. Find the percentage of cases falling between z scores of −2.00 and +1.65. The percentages from Table A are 47.72% and 45.05%, respectively. Adding these percentages, we get a total of 92.77%.

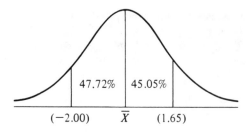

$$(-2.00) \quad \overline{X} \quad (1.65)$$

In each of these examples, the z scores were on opposite sides of the mean, one below the mean and the other above. The rule, then, to find the area between two z scores on opposite sides of the mean is to look up the two percentages and add them together. Add the percentages—*do not add* the z scores.

z Scores on the Same Side of the Mean. Many situations require establishing the area under the normal curve between z scores on the same side of the mean (z scores which have the same signs.)

Example. We wish to determine the percentage of cases between a z score of .50 and a z score of 1.00.

Again, we draw the curve and place the z scores on the X axis, .50 just to the right of the mean and 1.00 a little farther to the right. Look up the two percentages—19.15% for .50 and 34.13% for 1.00. Now, since 34.13% of the cases fall in the area from the mean to the z of 1.00, and since 19.15% of that total fall in the area from the mean to the z of .50, we must *subtract* to obtain 14.98% (34.13% − 19.15%), the percentage of cases between the z scores.

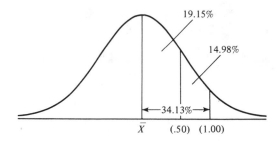

Example. Find the area between the z scores of 1.50 and 1.00. Table A gives 43.32% for 1.50 and 34.13% for 1.00. Subtracting those percentages gives us 9.19% of cases falling between a z of 1.50 and a z of 1.00.

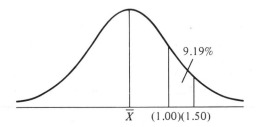

Again, a word of caution on this procedure. You must subtract the percentages, *not* the z scores. If, in the preceding example, we subtracted 1.00 from 1.50 and then looked up the resulting z score of .50 in Table A, our answer would have been 19.15%, instead of the correct answer of 9.19%. The reason for the error is that Table A gives the percentage of cases falling between the z score and the mean, not between two z scores both away from the mean. Remember, the normal curve includes a much greater area near the mean where it is higher than it does where it drops down toward the tails of the distribution.

When working with negative z scores, the procedure is identical. For example, the percentage of cases falling between z scores of −1.50 and −.75 is 15.98% (43.32 − 27.34). (See Fig. 4.8.)

The rule, then, to find the percentage of cases falling between z scores both on the same side of the mean is to look up the two percentages in Table A and subtract the smaller from the larger.

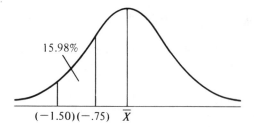

Figure 4.8
Percentage of cases between z scores of −1.50 and −.75.

Case D: To find the percentage of cases *between* two z scores.
Rule D1: If the z scores are on opposite sides of the mean, look up the two percentages and *add* them together.
Rule D2: If the z scores are on the same side of the mean, look up the two percentages and *subtract* the smaller from the larger.
Problems 4e: Find the percentage of cases falling *between* the following pairs of z scores.

1.	− 1.45 and 1.06	4.	− 1.62 and − .17	
2.	− .62 and .85	5.	− 1.65 and − 1.43	
3.	.90 and 1.87			

TRANSLATING RAW SCORES INTO z SCORES

At this point you are fairly familiar with the z score concept. We have found the percentage areas above, below, and between various z scores. However, in all these situations, the z score has been given; it has been handed to you on a silver platter. Using exactly the same procedures, however, we can find the areas under the normal curve for any raw score values. This is actually the more common situation—to find the percentage of cases falling above, below, or between the raw scores themselves. To accomplish this, the values of the mean and the standard deviation for the raw score distribution must be known. Remember, without this knowledge, the raw scores are of little use. If the mean and the standard deviation are known, we can subtract the mean from the raw score, divide by the standard deviation, and we have the z score. This gives us the opportunity to use the characteristics of the normal curve.

The z score equation

$$z = \frac{X - \bar{X}}{SD}$$

defines the z score as a translation of the difference between the raw score and the mean ($\bar{X}$) into units of standard deviation (*SD*). The z score, then, indicates how far the raw score is from the mean, either above it or below it, in these standard deviation units. A normal distribution of any set of raw scores, regardless of the value of the mean or the standard deviation, can be converted into the standard normal

The Equation for the Normal Curve

The equation for the normal curve is as follows:

$$y = \frac{1}{\sqrt{2\pi}} \, e^{-(z^2/2)}$$

where y = the ordinate
  π = 3.14
  e = 2.72 (the base of the Napierian or natural log)
  z = the distance from the mean in units of standard deviation
If $z = 1.00$, then

$$y = \frac{1}{\sqrt{(2)(3.14)}} 2.72^{-(1.00^2/2)}$$

$$y = \frac{1}{\sqrt{6.28}} 2.72^{-(.50)}$$

$$y = \left(\frac{1}{2.51}\right)(.61) = (.40)(.61)$$
$$y = .24$$

Thus, the height of the ordinate when $z = 1.00$ is .24.
Or, if $z = .50$, then

$$y = \frac{1}{\sqrt{(2)(3.14)}} 2.72^{-(.50^2/2)}$$

$$y = \frac{1}{\sqrt{6.28}} 2.72^{-(.13)}$$

$$y = \left(\frac{1}{2.51}\right)(.88) = (.40)(.88)$$

$$y = .35$$

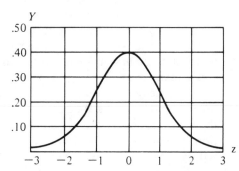

With a z score of .50, the height of the ordinate is .35.

The graph of the normal curve is said to be asymptotic to the abscissa. This means that the curve never touches the X axis, no matter how far it extends from the mean.

Although there are many different normal curves, each with its own mean and standard deviation, there is only one standard normal curve, where the mean is equal to zero and the standard deviation is equal to 1.00. All normal curves can be fitted exactly to the standard normal curve by translating the raw scores into units of standard deviation.

For finding the area between two z scores, calculus could be used to integrate the normal curve equation between the scores. Doing integrations for each normal curve problem is rather spine chilling to contemplate, but fear not. All the integrations have been completed and the results are presented in convenient form in Appendix Table A.

distribution, in which the mean is always equal to zero and the standard deviation is always equal to one.

   Example.   Suppose that the mean on a certain test was found to be 200 with a standard deviation of 22. One person got a raw score of 215 on the test, and we wish to find the z score:

$$z = \frac{X - \bar{X}}{SD} = \frac{215 - 200}{22} = \frac{15}{22} = .68$$

   Let your calculator do this arithmetic for you. Plug in the 215 and hit the minus key; then plug in 200 and hit the equals key. Now, hit the division key, plug in 22 and hit the equals key, and there it is, the z score of .68. This means that the raw score of 215 is exactly .68 *SD* units (68 hundredths of a standard deviation) above the mean—a distance about two-thirds of the way from the mean to the point one full *SD* unit above the mean. A score of 222 falls one full *SD* unit above the mean.

Using z Scores

The z scores are enormously helpful in the interpretation of raw score performances, since they take into account both the mean of the distribution and the amount of variability, its standard deviation. For example, if you get scores of 72 on a history test and 64 on an English test, you do not know, on the face of it, which score reflects the better performance. The mean of the scores on the history test might have been 85 and the mean on the English test only 50; in that case you actually did worse on the history test, even though your raw score was higher.
   The point is that information about the distribution of all the scores must be obtained before individual raw scores can be meaningfully interpreted. This is the reason for using z scores in this situation, because they take the entire distribution into account. The z score allows us to understand individual performance scores relative to all of the scores in the distribution.

Areas of the Normal Curve

As has been shown, if the abscissa of the normal curve is laid out in units of standard deviation (the standard normal curve), the percentage areas of the curve can be found. Suppose that a researcher selects a representative sample of 5000 persons from the adult population of the United States. Each person is given an IQ test, and each of the 5000 resulting IQ scores is duly recorded. The scores are added together and divided by the number of cases, or 5000, in order to find the mean of this distribution, which turns out to be 100. Then, all the IQ scores are squared, the squares added, and the sum divided by *N*. From this value, the square of the mean is subtracted, and finally the square root is taken. The resulting value of the standard deviation turns out to be 15. With the values for the mean and the standard deviation in hand, the normal curve is accessible.

As shown in Fig. 4.9, the mean of 100 is placed at the center of the *X* axis, and the rest of the axis is marked off in IQ points. Since we found that the standard deviation of this IQ distribution is 15, *one standard deviation unit is worth 15 IQ points.* Note that this standard deviation value of 15 was not presented to the researcher as if sprung full grown out of the head of Zeus. It was calculated in the same way as the other standard deviations were in Chapter 3.

Since one standard deviation unit is equal to 15 IQ points, then two standard deviation units must equal 30 IQ points and three must equal 45 points. Each of the standard deviation units can now be converted into their equivalent IQs, ranging from 55 to 145. Also, since we know the percentage areas for standard deviation units in general, we know the percentage areas for the IQs.

Is finding the standard deviation important? It is absolutely crucial. This can best be demonstrated by a quick review of what was known and what is now revealed by having found the standard deviation and the mean of the IQ distribution.

WE KNEW THAT:	*NOW REVEALED IS THAT:*
1. 68.26% of the cases fall between *z* scores of ±1.00.	1. 68.26% of the cases fall between IQs of 85 and 115. (See Fig. 4.10a.)
2. 95.44% of the cases fall between *z* scores of ±2.00.	2. 95.44% of the cases fall between IQs of 70 and 130. (See Fig. 4.10b.)
3. 99.74% of the cases fall between *z* scores of ±3.00.	3. 99.74% of the cases fall between IQs of 55 and 145. (See Fig. 4.10c.)

The conversions of IQs into *z* scores can be shown as follows:

$$z = \frac{X - \bar{X}}{SD}$$

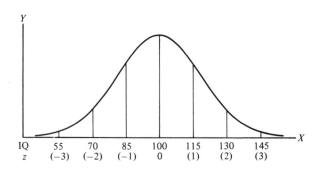

IQ	55	70	85	100	115	130	145
z	(−3)	(−2)	(−1)	0	(1)	(2)	(3)

Figure 4.9
Normal distribution of IQ scores showing mean and standard deviation units.

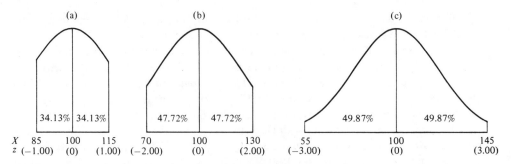

Figure 4.10   Percentage areas of the normal curve of IQ scores.

For an IQ of 115, then,

$$z = \frac{115 - 100}{15} = 1.00$$

And for 85,

$$z = \frac{85 - 100}{15} = -1.00$$

This procedure, then, yields z scores of +2.00 for an IQ of 130, +3.00 for an IQ of 145, −2.00 for an IQ of 70, and −3.00 for an IQ of 55.

z SCORE TRANSLATIONS IN PRACTICE

Let us now review each of the z score cases, only this time placing each case in the context of a normal distribution of *raw scores*.

  Case A: Percentage of Cases Between a Raw Score and the Mean—Example.   A normal distribution of SAT scores has a mean of 440 and a standard deviation of 85. What percentage of scores will fall between an SAT score of 500 and the mean?

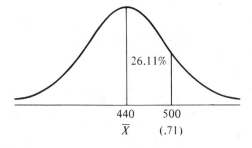

We convert the raw score of 500 into its equivalent *z* score.

$$z = \frac{X - \bar{X}}{SD} = \frac{500 - 440}{85} = \frac{60}{85} = .71$$

Then we draw the curve, placing all values on the abscissa and keeping the *z* score in parentheses to prevent confusing it with the raw score. We look up the *z* score in Table A, and take the percentage as a direct readout; that is, 26.11% of the SAT scores fall between 500 and the mean.

Problems 4f.   On a normal distribution with a mean of 48 and a standard deviation of 6.23, find the percentage of cases between the following *raw scores* and the mean.

1. 52   4. 37
2. 42   5. 40
3. 55

Case B: Percentage of Cases Below a Raw Score—Example.   On a normal distribution of adult male weight scores, with a mean of 170 pounds and a standard deviation of 23.17 pounds, what percentage of weights falls below a weight score of 200 pounds?

We convert the raw score into its equivalent *z* score.

$$z = \frac{200 - 170}{23.17} = \frac{30}{23.17} = 1.29$$

Then we draw the curve, placing all values on the abscissa.

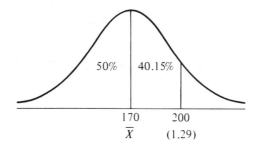

Since the *z* score is positive (above the mean), we look up the percentage in Table A and add it to 50%. That is, 40.15% plus 50% equals 90.15%, which is the percentage of cases below the score of 200. If the *z* score had been negative, then we would have looked up the percentage in Table A and subtracted it from 50%.

Problems 4g. On a normal distribution of raw scores with a mean of 103 and a standard deviation of 17, find the percentage of cases *below* the following raw scores.

1. 115 4. 105
2. 83 5. 76
3. 127

Case C: Percentage of Cases Above a Raw Score—Example. On a normal distribution of systolic blood pressure measures, the mean is 130 and the standard deviation is 9.15. What percentage of systolic measures is above 150?

We convert the raw score into its z score equivalent.

$$z = \frac{150 - 130}{9.15} = \frac{20}{9.15} = 2.19$$

We draw the curve, placing all values on the abscissa (see figure below).

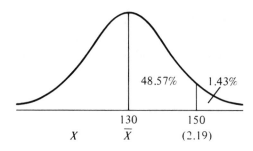

Since the z score is positive (above the mean), we look up the percentage in Table A and subtract it from 50%; that is, 50% − 48.57% equals 1.43%, the percentage of systolic measures above 150. If the z score had been negative, we would have used the Case C2 rule.

Problems 4h. On a normal distribution of raw scores with a mean of 205 and a standard deviation of 60, find the percentage of cases falling above the following raw scores:

1. 290 4. 195
2. 210 5. 250
3. 167

Case D: Percentage of Cases Between Raw Scores—Example. On a normal distribution of adult female height scores, the mean is 65 inches and the standard deviation is 3 inches. Find the percentage of cases falling between heights of 60 inches and 63 inches.

We convert both raw scores into their equivalent z scores.

$$z = \frac{60 - 65}{3} = \frac{-5}{3} = -1.67$$

$$z = \frac{63 - 65}{3} = \frac{-2}{3} = -.67$$

Then we draw the curve, placing all values on the abscissa.

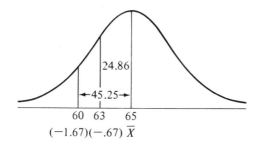

Since both z scores are on the same side of the mean, use Case D2 rule. Look up the percentage in Table A for each z score, and then subtract the smaller from the larger; that is, 45.25% − 24.86% equals 20.39%, the percentage of heights between 60 inches and 63 inches. If the two z scores had been on opposite sides of the mean, we would have looked up the two percentages and added them together.

Problems 4i. A normal distribution of raw scores has a mean of 68 and a standard deviation of 3.06. Find the percentage of cases falling between each of the following pairs of raw scores.

1. 60 and 72   4. 62 and 66
2. 65 and 74   5. 59 and 67
3. 70 and 73

The Normal Curve in Retrospect

Whenever a distribution of values assumes the shape of the normal curve, the z score table, as we have clearly seen, allows us to do some truly remarkable computations. We can find the percentage of cases falling above, below, or between any z score values. This will become a central issue during later discussions of both probability and inferential statistics.

Remember, however, that as remarkable as these percentages are, they only apply when the distribution is normal—not skewed or bimodal, or leptokurtic, or any other shape. If, for example, more than or less than 68% of the scores in a distribution fall in the area within a standard deviation unit of the mean, the distribution simply is not normal, and the z score techniques do not apply. In short, when the distribution is normal, use the z scores. When it isn't, don't!

z Score Percentage Rules

Case A Rule: Percentage of cases between a given z score and the mean ($\bar{X}$). Look up directly in the table (Table A).

Case B Rule: Percentage of cases below a given z score (percentile).
    B1. If z score is positive, look up percentage in Table A and add 50%.
    B2. If z score is negative, look up percentage in Table A and subtract from 50%.

Case C Rule: Percentage of cases above a given z score.
    C1. If z score is positive, look up percentage in Table A and subtract from 50%.
    C2. If z score is negative, look up percentage in Table A and add 50%.

Case D Rule: Percentage between two z scores.
    D1. If z scores are on opposite sides of the mean ($\bar{X}$), look up the two percentages in Table A and *add* them.
    D2. If z scores are on the same side of the mean ($\bar{X}$), look up the two percentages in Table A and *subtract* the smaller from the larger.

SUMMARY

The normal curve is a unimodal frequency distribution, perfectly symmetrical (the mean, the median, and the mode coincide at a single point), and asymptotic to the abscissa (the curve never touches the X axis). The equation for the normal curve provides the curve with a constant relationship to the standard deviation. When the curve is plotted on the basis of standard deviation units, it is called the standard normal curve. Table A in the back of the book contains all the specific percentage areas for the various standard deviation units.

Standard scores, or z scores, result when a raw score is translated into units of standard deviation. The z score specifically defines how far the raw score is from the mean, either above it or below it, in these standard deviation units.

By looking up a z score in Table A, the exact percentage of cases falling between that z score and the mean can be found. A series of rules is outlined for quickly solving the various z score problems in conjunction with Table A. When problems are posed in terms of raw scores, convert the raw score to its equivalent z score and use the procedures outlined earlier.

KEY TERMS AND NAMES

    frequency distribution curve   z distribution
    Gauss, Karl Friedrich   z score
    normal curve

PROBLEMS

Assume a normal distribution for problems 1 through 12.

1. Find the percentage of cases falling between each of the following z scores and the mean.
   a. .70   c. − .42
   b. 1.28   d. − .09

2. Find the percentage of cases falling below each of the following z scores.
   a. 1.39   c.   − .86
   b. .13   d. − 1.96

3. Find the percentage of cases falling above each of the following z scores.
   a.   1.25   c.   1.33
   b. − 1.65   d.   − .19

4. Find the percentage of cases falling between the following pairs of z scores.
   a. − 1.52 and .15   c. − 1.96 and − .25
   b. − 2.50 and 1.65   d.   .59 and 1.59

5. The mean IQ among the members of a certain national fraternity is 118 with a standard deviation of 9.46. What percentage of the members had IQs of 115 or higher?

6. The mean resting pulse rate for a large group of varsity athletes was 72 with a standard deviation of 2.56. What percentage of the group had pulse rates of 70 or lower?

7. The mean grade-point average among freshmen at a certain university is 2.00 with a standard deviation of .60. What percentage of the freshman class had grade-point averages
   a. Below 2.50?   c. Between 1.70 and 2.40?
   b. Above 3.50?   d. Between 2.50 and 3.00?

8. Using the data from problem 7, find the percentile rank for the following grade-point averages.
   a. 3.80   c. 2.10
   b. .50   d. 3.50

9. Road tests of a certain compact car show a mean fuel rating of 30 miles per gallon, with a standard deviation of 2 miles per gallon. What percentage of these autos will achieve results of
   a. More than 35 miles per gallon?   c. Between 25 and 29 miles per gallon?
   b. Less than 27 miles per gallon?   d. Between 32 and 34 miles per gallon?

10. On a normal distribution, at what percentile must
    a. The mean fall?   c. The mode fall?
    b. The median fall?

11. With reference to which measure of central tendency must z scores always be computed?

12. Approximately what percentage of the scores on a normal distribution fall
    a. Below the mean?
    b. Within ±1 SD units from the mean?
    c. Within ±2 SD units from the mean?

True or False—Indicate either T or F for problems 13 through 22.

13. An individual score at the 10th percentile is among the top tenth of the distribution.
14. A negative *z* score always yields a negative percentile.
15. A positive *z* score always yields a percentile rank above 50.
16. A negative *z* score always indicates that the raw score is below the mean.
17. Between *z* scores of ±3.00 under the normal curve can be found almost the entire distribution.
18. A *z* score of zero always indicates that the raw score coincides with the mean.
19. In a normal distribution, it is sometimes possible for the mean to occur above the 50th percentile.
20. Normal distributions must always be unimodal.
21. All unimodal distributions are normal.
22. For a normal distribution, 50% of the *z* scores must be positive.

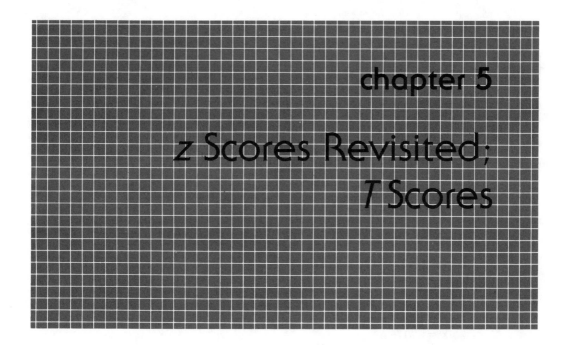

At this point we have learned to solve a great number of z **score** problems. However, in each case the answer has been in percentage terms. That is, the mean and the standard deviation have both been given, and the job was to calculate the frequency of occurrence (percentages) for various areas of the curve. We have consistently solved the z score equation, $z = (X - \bar{X})/SD$, for the unknown value of z. Many z score problems, however, take a different form. Though the z score equation, of course, remains the same, in this chapter it will be shifted around to solve for a variety of other unknowns.

OTHER APPLICATIONS OF THE z SCORE

 From z Score to Raw Score

Just as knowledge of the mean, the standard deviation, and a raw score can be used to generate a z score, so too can knowledge of the z score produce the raw score. Thus, since $z = (X - \bar{X})/SD$, therefore,

$$X = zSD + \bar{X}$$

This important variation on the z score theme provides specific information about an individual's performance, stated in terms of the trait being measured.

PLANE GEOMETRY –

PLANE GEOMETRY IS A BASIC
INTRODUCTORY COURSE INTO PROBLEM SOLVING
USING DEDUCTION AND LOGIC !
IT IS MAINLY FOR THE CURIOUS AND
IS A PREREQUISTE FOR ALL THOSE WISHING TO
TAKE FANCY GEOMETRY !

Plain or fancy, numbers sometimes appear to have lives of their own. (FUNKY WINKERBEAN by Tom Batiuk, © 1981 Field Enterprises, Inc. Courtesy of Field Newspaper Syndicate.)

Example. We know that on a weight distribution with a mean of 150 pounds and a standard deviation of 17 pounds a certain individual had a z score of 1.62. We can utilize this information to specify the measurement of that individual in terms of the units for the trait being measured, in this case, pounds.

$$X = zSD + \bar{X}$$

$$X = (1.62)(17) + 150 = 27.54 + 150$$

$$X = 177.54 \text{ pounds}$$

If the individual scores below the mean (negative z score), the same procedure is used.* Assume the same weight distribution ($\bar{X} = 150$, $SD = 17$), and determine how much an individual weighs who has a z score of $-.65$.

$$X = zSD + \bar{X}$$

$$X = (-.65)(17) + 150 = -11.05 + 150$$

$$X = 138.95 \text{ pounds}$$

Problems 5a: On an IQ distribution (normal), with a mean of 100 and a standard deviation of 15; find the actual IQs (raw scores) for individuals with the following z scores.

1.   .56   4.   − .10
2.   2.73   5.   2.58
3.   − 1.64

From Percentile to z Score

If you know a percentile, you can easily determine the z score, and as we have seen, once you know the z score you can always get the raw score.

*If your calculator cannot handle a negative number placed first in a series of added terms, as in $-11.05 + 150$, then simply change the order of the values and enter the 150 first, 150 $- 11.05$. Since addition is what mathematicians call commutative, it makes no difference in what order you choose to enter the terms.

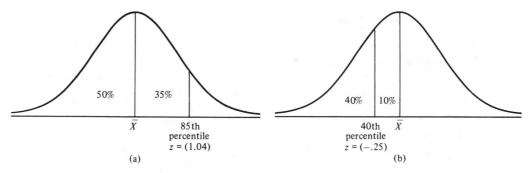

Figure 5.1   Normal curves showing relationship between percentile and percentage of cases.

We learned that a percentile specifies the percentage of scores at or below a given score. We know, therefore, that a percentile of 84 means that 84% of the cases fall at or below that point on the baseline of the curve. The focus in this section, however, is on the percentage of cases falling between the percentile and the mean. Remember that the mean under a normal curve is *always at the 50th percentile,* regardless of its actual value. The mean may be 1.5, or 250, or 10,000, and yet it is always at the 50th percentile under the normal curve. Therefore, as shown in Fig. 5.1a, a percentile of 85 is 35 percentage points away from the mean, in this case, 35% above the mean. Similarly, as shown in Fig. 5.1b, a percentile of 40 is 10 percentage points away from the mean, in this case, 10% below the mean.

Now, since the z score table gives the percentage of cases falling between a given z score and the mean, if we know the percentile we can therefore get the z score. First, determine the percentage of cases between the given percentile and the mean. Then, go right to the heart of the z score table, find the *closest* percentage, and then simply read off the z score from the edges of the table.

THE PERCENTILE TABLE

In the preceding section we explained how any percentile could be converted into a z score by the use of Table A, the main z score table. However, since these particular conversions are so especially important when doing z score (and, later, T score) problems, a separate percentile table (Table B) has been constructed for this book. Next to each percentile can be found its correct z score. Notice that all the percentiles in the first column, from the 1st to the 49th, have negative z scores, which again reinforces the fact that z scores that fall below the mean (those below the 50th percentile) must be given negative signs. In the second column, the z scores are all positive, since all percentiles from the 51st on up are, of course, above the mean percentile of 50.

Problems 5b: Find the z score for each of the following percentiles

1. 95th percentile   4. 55th percentile
2. 10th percentile   5. 75th percentile
3. 39th percentile

From Percentiles to Raw Scores

It is obvious from the preceding that if we have a percentile, we know the *z* score. All it takes is the use of Table B. Also, we have previously seen how the raw score can be determined from the *z* score ($X = zSD + \bar{X}$). Thus, whenever a percentile is known, the raw score can be accurately determined.

Example. A normal distribution has a mean of 250 and a standard deviation of 45. Calculate the raw score for someone who is at the 95th percentile.

First, the *z* score for a percentile of 95 is equal to 1.65. We find this by looking up the 95th percentile in Table B. Now, using the *z* score equation, we solve it for the raw score, *X*,

$$X = zSD + \bar{X}$$

Then plug in all known values:

$$X = (1.65)(45) + 250 = 74.25 + 250$$

$$X = 324.25$$

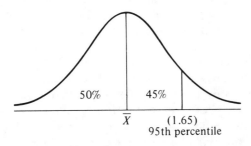

Example. Using the data from the preceding example ($\bar{X} = 250$, $SD = 45$), calculate the raw score for someone who is at the 10th percentile.

Look up the 10th percentile in Table B and get the *z* score of -1.28. Then plug all known values into the equation for *X*.

$$X = zSD + \bar{X}$$

$$X = (-1.28)(45) + 250 = -57.60 + 250$$

$$X = 192.40$$

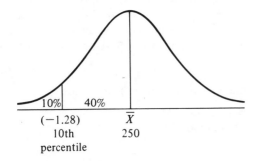

Problems 5c: On a normal distribution with a mean of 1500 and a standard deviation of 90, find the raw score for each of the following percentiles.

1. 45th percentile
2. 55th percentile
3. 15th percentile
4. 79th percentile
5. 5th percentile

From z Score to Standard Deviation

Just as we learned to flip the z score equation around to solve for the raw score, so too, we can solve it for the standard deviation. Since $z = (X - \bar{X})/SD$, then

$$SD = \frac{X - \bar{X}}{z}$$

Therefore, if we are given the raw score, the mean, and the z score (or the percentile from which the z score can be found), we can calculate the standard deviation.

Example. On a normal distribution, the mean is equal to 25. A certain individual had a raw score of 31 and a z score of 1.68. The standard deviation would thus be calculated as follows:

$$SD = \frac{X - \bar{X}}{z} = \frac{31 - 25}{1.68} = \frac{6}{1.68}$$

$SD = 3.57$

Instead of being presented with the z score, we might instead be given the percentile.

Example. On a normal distribution, the mean is 820. A certain individual has a raw score of 800 and is at the 43rd percentile. Find the standard deviation.

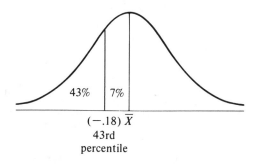

43%   7%

$(-.18)\ \bar{X}$
43rd
percentile

First, we find the z score for a percentile of 43. Looking in Table B, we find that z score to be $-.18$. Using the known values, we calculate the standard deviation:*

$$SD = \frac{X - \bar{X}}{z} = \frac{800 - 820}{-.18} = \frac{-20}{-.18}$$

$$SD = 111.11$$

The Range

Now that you have found the value of the standard deviation, a close approximation of the range may easily be calculated. Since the range is roughly six times the SD (see page 52), the range for the distribution above should be (6) (111.11), or 666.66. Also, knowing the values for both the mean and the range allows for the determination of the highest and lowest scores in the distribution. Since under the normal curve, one-half of the range falls above the mean and one-half below it, simply divide the range by 2, and the resulting value is then added to and subtracted from the mean. Thus, dividing 666.66 by 2 equals 333.33. By adding 333.33 to the mean of 820, the highest score in the distribution is estimated to be 1153.33, and then by subtraction, the lowest score should be approximately 486.67.

Example. The mean on a normal distribution is 500. A certain individual has a raw score of 600 and is at the 85th percentile. Find the standard deviation, the range, and both the highest and lowest scores.

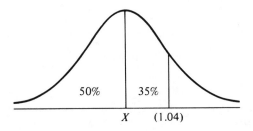

50% | 35%

X (1.04)

From the percentile, the z score is found to be 1.04. Next, plug the known values into the equation for the SD:

$$SD = \frac{X - \bar{X}}{z} = \frac{600 - 500}{1.04} = \frac{100}{1.04}$$

*Dividing a negative value by another negative value always yields a positive value. Also, never forget the fact that all standard deviations *must be positive.* You can have negative z scores and even negative raw scores, but you should never, never get a negative standard deviation.

$SD = 96.15$

$R = (6)(96.15) = 576.90$

$R/2 = 288.45$

Highest score $= 500 + 288.45 = 788.45$

Lowest score $= 500 - 288.45 = 211.55$

Problems 5d: On a normal distribution with a mean of 115, find the standard deviation for the following raw scores.

1. 120 (at the 60th percentile)
2. 100 (at the 30th percentile)
3. 110 (at the 47th percentile)
4. 85 (at the 7th percentile)
5. 140 (at the 83rd percentile)

The ability to determine the standard deviation from a z score can be especially important to the guidance counselor. Perhaps students' scores on a certain standardized test are returned in the form of raw scores and percentiles, but no information regarding the standard deviation is provided. All that is necessary are the values for one student to calculate the standard deviation for the entire distribution.

Example.   On a Verbal SAT distribution with a mean of 440, a certain student received a raw score of 535 and is at the 85th percentile.

The z score for the 85th percentile is found to be 1.04. Then the standard deviation can be calculated.

$$SD = \frac{X - \bar{X}}{z} = \frac{535 - 440}{1.04} = \frac{95}{1.04}$$

$SD = 91.35$

From calculations like those in the preceding example, a guidance counselor can make up a chart showing precisely what raw scores (even those not actually received by the counselor's students) correspond to the various percentiles. Since both the standard deviation and the equation, $X = zSD + \bar{X}$, are available, they can be used to generate the entire distribution.

Example.   With a mean of 440 and a standard deviation of 91.35, a chart can be constructed by finding the z score for each percentile and then solving for X.

For the 90th percentile, the z score is 1.28. Solve for the value of X.

$X = zSD + \bar{X} = (1.28)(91.35) + 440$

$X = 116.93 + 440 = 556.93$

An example of how such a chart could be constructed is the following:

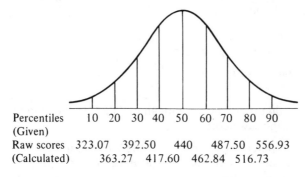

Percentiles (Given)	10	20	30	40	50	60	70	80	90

Raw scores   323.07   392.50   440   487.50   556.93
(Calculated)   363.27   417.60   462.84   516.73

From z Score to the Mean

Since $z = (X - \bar{X})/SD$, it can be shown that

$$\bar{X} = X - zSD$$

Any time a raw score, z score, and the standard deviation are all given, the mean is readily available. Also, since we have shown that a percentile can be quickly and easily converted to a z score, the mean can be found whenever the percentile or the z score is given (assuming the *SD* is known). Also, since the mean, median, and mode all fall at the same point under the normal curve, once the mean is known, so are the median and mode.

Example. A normal distribution has a standard deviation of 15. A raw score of 122 falls at the 90th percentile. Find the mean of the distribution.

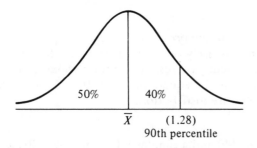

First, we find the z score for the 90th percentile by looking in Table B. The correct z score is 1.28. Using the equation $\bar{X} = X - zSD$, we plug in all known values.

$$\bar{X} = 122 - (1.28)(15) = 122 - 19.20$$

$$\bar{X} = 102.80 \quad \text{(which is also the value for the median and mode)}$$

Next, we consider the same type of problem in a situation where the raw score falls to the left of the mean.

Example. On a normal distribution the standard deviation is 90. A raw score of 400 falls at the 18th percentile. Find the mean.

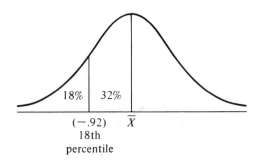

From Table B, we establish the z score for the 18th percentile as $-.92$. Plugging all known values into the equation yields the mean.*

$$\bar{X} = X - zSD = 400 - (-.92)(90)$$

$$\bar{X} = 400 - (-82.80)$$

$$\bar{X} = 400 + 82.80$$

$$\bar{X} = 482.80$$

Problems 5e: Find the mean for each of the following normal distributions.

1. $SD = 10, X = 50$, percentile rank $= 35$
2. $SD = 50, X = 300$, percentile rank $= 76$
3. $SD = 3, X = 15$, percentile rank $= 95$
4. $SD = 23, X = 74$, percentile rank $= 34$
5. $SD = 100, X = 1600$, percentile rank $= 85$

T SCORES

In a great number of testing situations, especially in education and psychology, raw scores are reported in terms of **T scores.** A T score is a converted z score, with the mean always set at 50 and the standard deviation at 10. This ensures that all T scores are positive values (unlike z scores, which are negative whenever the raw score is below the mean).

---

*Note that when subtracting a negative product, simply change the sign to positive; $-(-82.80)$ becomes $+82.80$.

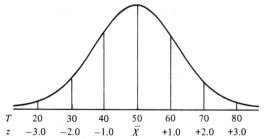

Figure 5.2
Normal curve showing T scores and standard deviation units.

| T | 20 | 30 | 40 | 50 | 60 | 70 | 80 |
| z | −3.0 | −2.0 | −1.0 | $\bar{X}$ | +1.0 | +2.0 | +3.0 |

The *T* score, like the *z* score, is a measure of an individual's performance relative to the mean. The *T* score is the measure of how far a given raw score is from a mean of 50 in standard deviation units of 10. Therefore, a *T* score of 60 is one full standard deviation unit above the mean of 50. A *T* score of 30 is two full standard deviation units below the mean of 50. (See Fig. 5.2.) *T* score values range from 20 to 80—from three standard deviation units below the mean to three standard deviation units above the mean.

Calculating T Scores

Calculating a *T* score is just as easy as finding a raw score when the *z* score, the standard deviation, and the mean are known. To find the raw score, we used the following equation:

$$X = zSD + \bar{X}$$

Since the *T* score assumes a standard deviation of 10 and a mean of 50, these values can be substituted into the equation, so that it becomes

$$T = z(10) + 50$$

With this equation, if the *z* score is known, finding the *T* score becomes automatic. It is not necessary to know the actual mean or standard deviation of the distribution of scores from which the *z* score was derived—just the *z* score itself is sufficient. Any other information about the distribution is superfluous. Information as to the mean and the standard deviation is needed to create the *z* score in the first place; but once *z* is known, so is *T*.

Example. On a normal distribution, an individual has a *z* score of 1.30. Find the *T* score.

$$T = z(10) + 50 = (1.30)(10) + 50 = 13 + 50$$
$$T = 63$$

Or, on a normal distribution, an individual has a z score of $-.50$. Find the *T* score.

$$T = z(10) + 50 = (-.50)(10) + 50 = -5 + 50$$
$$T = 45$$

Applications of T Scores

Comparing Two Raw Scores. The transformation of raw scores to *T* scores can be very helpful, since *T* (like the z score) provides a standard by which performances on different tests can be directly compared.

Example. Suppose that an individual takes two different IQ tests. The mean of both tests is 100. The standard deviation for the first test is 15, and the standard deviation for the second test is 18. On the first test, the individual received an IQ score of 110. On the second test, the IQ score received was 112. Using *T* scores, find out whether the individual did better on the second test.

First, calculate the z score for the raw score on the first test.

$$z = \frac{X - \bar{X}}{SD} = \frac{110 - 100}{15} = \frac{10}{15}$$
$$z = .67$$

Then convert that to a *T* score.

$$T = z(10) + 50 = .67(10) + 50 = 6.70 + 50$$
$$T = 56.70$$

Do the same for the score on the second test.

$$z = \frac{X - \bar{X}}{SD} = \frac{112 - 100}{18} = \frac{12}{18}$$
$$z = .67$$
$$T = z(10) + 50 = (.67)(10) + 50 = 6.70 + 50$$
$$T = 56.70$$

In each case, the *T* score was 56.70. The individual's performance on the two tests was, thus, identical. (This, of course, had to be the case since the z scores were identical.)

Problems 5f: On a normal distribution with a mean of 500 and a standard deviation of 95, find the *T* score for each of the following raw scores.

1. 400   4. 775
2. 610   5. 550
3. 330

From T *to z to Raw Score.*   Sometimes it is necessary to convert *T* scores back into raw scores. A guidance counselor or teacher may be presented with a set of student scores in the form of *T* values. To make this particular conversion, the mean and the standard deviation of the raw score distribution must be known. The procedure involves first translating the *T* score back into its z score equivalent using the following equation.

$$z = \frac{X - \bar{X}}{SD} = \frac{T - 50}{10}$$

Then, the raw score equation, $X = zSD + \bar{X}$, is used to complete the conversion.

Example.   On a normal distribution with a mean of 70 and a standard deviation of 8.50, a certain student reports a *T* score of 65 and wishes to know the equivalent raw score.

Find the z score.

$$z = \frac{T - 50}{10} = \frac{65 - 50}{10} = \frac{15}{10}$$

$$z = 1.50$$

Then, using the known mean and standard deviation values, convert the z score to the raw score.

$$X = zSD + \bar{X} = (1.50)(8.50) + 70 = 12.75 + 70$$

$$X = 82.75$$

For a *T* score of less than 50, a score below the mean, follow the same procedure, being very careful regarding the minus signs.

Example.   Using the data from the previous example ($\bar{X} = 70$, $SD = 8.50$), find the raw score for a student whose *T* score was 37.

$$z = \frac{T - 50}{10} = \frac{37 - 50}{10} = \frac{-13}{10}$$

$$z = -1.30$$

Then solve for X.

$$X = zSD + \bar{X} = (-1.30)(8.50) + 70 = -11.05 + 70$$
$$X = 58.95$$

Problems 5g: The following are from a normal distribution with a mean of 35 and a standard deviation of 4.50. Find the raw score for each **T** score.

1. $T = 43$ 4. $T = 31$
2. $T = 67$ 5. $T = 78$
3. $T = 25$

THE IMPORTANCE OF THE z SCORE

In this and the preceding chapter, we have kept the focus on the z score. We have twisted it, flipped it forward and backward, and studied all its variations, perhaps, you may feel, ad nauseum. The reasons for this scrutiny will become clear as we move into other areas, such as probability and the sampling distributions. This has definitely been time well spent. A thorough understanding of the z score concept will make these other topics far easier to understand. Much of the material to come will rest, like an inverted pyramid, on the perhaps tiny, but stable, base provided by the z score. Over and over again, we will refer to the discussions and examples found in this chapter.

By now perhaps even the skeptics among you are ready to admit that the math involved in doing these problems is not too difficult. If you follow the text step-by-step, the math will *not* get harder. If you and your trusty calculator can handle these z score problems, you can take the rest of the book in stride.

SUMMARY

The z score may be used to obtain the raw score in any normal distribution of scores—by solving the z score equation for X, where the mean and standard deviation of the raw score distribution are known ($X = zSD + \bar{X}$). Further, a percentile score may be converted into a z score and, thus, also used to generate the raw score. All negative z scores yield percentiles less than 50 and translate into raw scores below the mean. Positive z scores, on the other hand, yield percentiles above 50 and raw scores above the mean.

The z score equation may also be used to find the standard deviation, whenever the z score, the raw score, and the mean of the raw score distribution are known. That is, since $z = (X - \bar{X})/SD$, then $SD = (X - \bar{X})/z$. This is especially useful to the psychologist or guidance counselor who works with standardized tests. Simply knowing the raw score and percentile rank (from which the z score can be found) of a single student, along with the mean of the distribution, allows for the calculation of the standard deviation. Also, once the standard deviation is known, the range can be approximated by using the equation $R = 6(SD)$. Knowing the range, in turn, allows for the estimation of both the highest and lowest scores in the distribution. Similarly, if needed, the mean of the raw score distribution can be found whenever the raw score, the z score, and the standard deviation are known.

T scores are z scores which have been converted to a distribution whose mean is set at 50 and whose standard deviation is equal to 10. T scores range in value from 20 to 80, since 20 is three standard deviation units below the mean, and 80 is three standard deviation units above. If the z score is known, the T score can be obtained without having any further information. Finally, if the T score is given, it can be used to find the z score.

KEY TERMS

T score
z score

PROBLEMS

Assume a normal distribution for problems 1 through 13.

1. The employees at a certain manufacturing plant have a mean hourly income of $10.00, with a standard deviation of $2.10. Find the actual hourly wages of the employees who have the following z scores.
   a. 2.50   c. − .97
   b. 1.56   d. − 1.39
2. The mean verbal SAT score at a certain college is 450, with a standard deviation of 87. Find the SAT scores for the students who scored at the following percentiles.
   a. 10th percentile.   c. 5th percentile.
   b. 95th percentile.   d. 45th percentile.
3. The grades of the students in a large statistics class are normally distributed around a mean of 70, with an SD of 5. If the instructor decides to give an "A" grade only to the top 15% of the class, what grade must a student earn to be awarded an A?
4. Suppose the instructor in the class in problem 3 (where the mean was 70 and the SD was 5) decided to give an "A" grade only to the top 10% of the class. What grade would the student now have to earn to be awarded an A?

5. The mean height among adult males is 68 inches. If one individual is 72 inches tall and has a z score of 1.11,
   a. Find the *SD*.
   b. Find the range.
   c. Find the highest score.
   d. Find the lowest score.
6. If a weight distribution for female college students has a mean of 120 pounds, and one individual who weighed 113 pounds was found to be at the 27th percentile,
   a. Find the *SD*.
   b. Find the range.
   c. Find the highest score.
   d. Find the lowest score.
7. A student has a grade-point average of 3.75, which is at the 90th percentile. The standard deviation is .65. Give the values for
   a. The mean.
   b. The median.
   c. The mode.
8. A certain brand of battery is tested for longevity. One battery is found to last for 30 hours and is at the 40th percentile. The standard deviation is 13.52 hours. What is the mean life of the tested batteries?
9. Find the equivalent *T* scores for the following z scores.
   a.   .10   c.   2.03
   b. − 1.42   d.   − .50
10. On a normal distribution of English achievement test scores, with a mean of 510 and a standard deviation of 72, find the *T* score for each of the following raw scores.
    a. 550   c. 610
    b. 400   d. 500
11. On a normal distribution of SAT scores, with a mean of 490 and a standard deviation of 83, find the SAT score for each of the following *T* scores.
    a. $T = 55$ c. $T = 65$
    b. $T = 32$ d. $T = 47$
12. On any normal distribution, the middlemost 68.26% of the cases fall between
    a. Which two z scores?   b. Which two *T* scores?
13. A score that falls at the first quartile (25th percentile) must receive
    a. What z score?   b. What *T* score?

True or False—Indicate either T or F for 14 through 20.

14. If the z score is zero, the *T* score must be 50.
15. The standard deviation of the *T* score distribution must equal the standard deviation of the raw score distribution.
16. A *T* score of less than 50 must yield a negative z score.
17. All *T* scores greater than 50 must yield percentiles that are greater than 50.

18. A normal distribution with a mean of 50 and a standard deviation of 10 must have a range that approximates 60.

19. If the mean of a set of normally distributed values is 11.83, the median is also equal to 11.83.

20. If the mean of a normal distribution is equal to 100, with a range of 50, the highest possible score must be approximately 125.

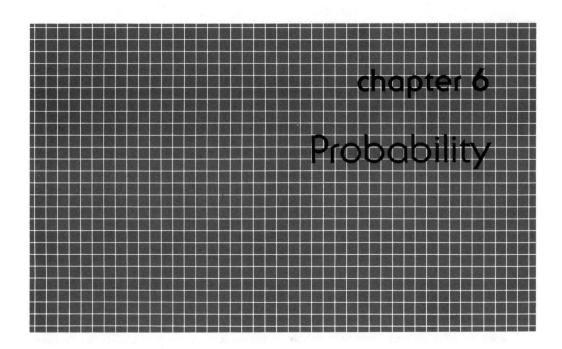

chapter 6

Probability

By the time you arrived at college, you had made literally thousands of probability statements, both consciously and unconsciously. Lurking behind virtually every decision you have ever faced has been an overt or covert probability estimate. This chapter, then, deals with a subject with which you are already thoroughly familiar; this generalization does not apply only to the card players or horse race bettors among you. Every time you cross the street in traffic (can I make it?), decide whether or not to carry an umbrella (will it rain?), choose a space to park your car (is it safe?), or buy one brand of appliance rather than another (will it last?), you are estimating probabilities. Even your selection of a date (can I trust this person?), or a college to attend (will I be accepted?), is affected by the ever present intrusion of probabilities.

Perhaps you and a friend flipped a coin this morning to determine who would buy coffee. You certainly were aware at the time of the probability of your being the one to have to pay. Because of this familiarity with probability, it is therefore fondly hoped that you will know as much about the concept after you finish this chapter as you do now. If asked to specify the exact probability of your having to buy the coffee in the coin-flipping example, you might say "50%" or "fifty-fifty," or perhaps something like "even-up." All these statements are essentially correct, even though they are not stated in the precise terms used by statisticians. Our next step is to take your current knowledge of probability and translate it into statistical terminology.

Probability theory does have a very practical side.
(© Cartoon Features Syndicate.)

"Remember men, it matters not that you win or lose—but, that you beat the point spread!"

THE DEFINITION OF PROBABILITY

Probability is defined as the number of times a specific event, *s,* can occur out of the total possible number of events, *t.* Using the letter *P* for probability, we can say that

$$P = \frac{s}{t}$$

Now, in the case of the coin flip, suppose that you called heads. Since a coin has only two sides, heads is one possible event out of a total of two possible events (heads and tails). The probability of your call being correct, then, is one out of two or

$$P = \frac{1}{2}$$

Statisticians do not express their probability statements in fraction form, however. They divide out the fraction and write the probability in decimal form, as follows:

$$P = \frac{1}{2} = .50$$

Thus, the probability of your being correct when calling heads is .50. This value is a constant for each and every coin flip ever made. Even if you have called it wrong for 20 straight flips, that does not increase, or in any way affect, the probability of your being a winner on the twenty-first flip. Each coin flip is totally independent of every other coin flip. You may remember your long string of losses, *but the coin doesn't remember.* When events are independent of each other, they are just that, independent. They cannot influence or be influenced by one another.

Independent Events

The following story, while undoubtedly apocryphal, nevertheless makes a serious point about the importance of understanding the idea of independent events. It is alleged that in 1960 a political forecaster flipped 1280 pennies, calling heads for Kennedy and tails for Nixon. As "luck" would have it, 640 of the pennies did turn up heads, correctly predicting Kennedy. These were set aside for the next election. In 1964 the forecaster called heads for Johnson and tails for Goldwater. Again, "miraculously," 320 came up heads and Johnson was elected. These were duly set aside. Then, by 1984, our by-now-fearless forecaster had 20 coins that had never been wrong.

In 1984, 10 of these coins correctly called Reagan over Mondale. After the 1984 election, out of the original 1280 coins, 10 coins had correctly predicted every presidential election since Kennedy's win over Nixon. The forecaster was totally convinced that these were 10 "magic" coins. Because of their long and unbroken string of successes, the probability of their predicting the 1988 election "must be 1.00"—after all, they had never been wrong. It is said that 5 of these coins were correct in 1988, turning up heads for Bush and tails for Dukakis. How will our forecaster do in 1992? We don't know. We do know, however, that the probability of any single coin being correct will still be .50.

The Gambler's Fallacy

Gamblers often convince themselves after a long series of losses that the law of averages makes them more certain to be winners. Sometimes after a losing streak, a gambler becomes certain that the probability must swing the other way, to the point, finally, of assuming the next bet is a "sure thing," $P = 1.00$. This is known as the **gambler's fallacy,** and we as researchers and research consumers must be just as aware of it as the inveterate gambler should be.

While it is true that if an unbiased coin is flipped an infinite number of times, heads will turn up 50% of the time, but that does not change the probability for any single coin flip. The so-called law of averages is true only in the long run. This is why casino operators in Atlantic City and Las Vegas have to be winners. The casino is making many more bets than is any individual player. The overall probability in favor of the casino is much more likely, then, to conform to mathematical expectation than is that for the individual player, who, though seemingly involved in an extended series of bets, is really only playing in a short run. As Greg Kimble puts it, "the laws of chance or principles of probability apply, not to single events but to large numbers of them."[1] Also, as John Scarne, an advisor to many casinos, counsels the individual player, "You can beat a race, but not the races."[2] As for the "law of averages," Scarne says, that gamblers

> don't understand that the important word in that phrase is not "law" but "averages." The theory of probability is a mathematical prediction of what may hap-

pen on the average, or in the long run, not a law which states that certain things are inevitable.[3]

People can certainly have hot streaks for a short run but over the long run the more typical probabilities have a way of asserting themselves. For example, you often see a major-league baseball player hitting for an average as high as .450 over the first few weeks of the season but nobody in the history of the game has ever batted .450 for an entire season.

To sum up, then, the so-called laws of chance are really descriptions of what generally happens in the long run, not enforcers of what must happen in the short run. This is why the gambler who drives to Atlantic City in a $60,000 Mercedes often returns home as a passenger in a $400,000 bus.

Probability versus Odds

It is important to distinguish probability from the related concept of **odds.** Probability states the number of times a specific event *will* occur out of the total possible number of events; odds, on the other hand, are usually based on how often an event will not occur. For example, if we toss a normal, six-sided die, the probability of any one of the faces turning up is one out of six.

$$P = \frac{s}{t} = \frac{1}{6} = .17$$

Odds, on the other hand, are typically stated in terms of the chances against a specific event occurring. (In this case, the odds are 5 to 1—there are five sides that will not come up for each single side that will.)

Example. Suppose that we wish to determine the probability of selecting a diamond from a normal deck of playing cards, that is, 52 cards divided into 4 suits—diamonds, clubs, hearts, and spades. The probability of the specific event, a diamond, is, therefore, one out of four (the total number of suits is four).

$$P = \frac{1}{4} = .25$$

The odds against a diamond selection, on the other hand, are 3 to 1, since there are three suits you don't want to select for the single suit you do want to select.

To establish the relationship between probability and odds, we can run through two more conversions. If the probability of an event occurring is 1 out of 20 or .05, the odds against it are 19 to 1. If the probability of an event occurring is 1 out of 100 or .01, the odds against it are 99 to 1.

PROBABILITY AND THE PERCENTAGE AREAS
OF THE NORMAL CURVE

The normal distribution is often referred to as the normal probability curve, since the characteristics of the normal curve, as outlined in Chapter 4, allow its use in making probability statements. The curve, as presented so far, has been set forth in terms of percentage areas. That is, for example, we learned that roughly 68% of the cases fall within ± 1 standard deviation units of the mean, and so on. The z score table (Table A) is outlined in terms of percentage areas. Because of this, all the z score problems given so far have been based on percentages, or frequencies of occurrence. When we did all those problems concerning the area between, above, or below given z scores, our answers were always expressed as frequencies—32% fell here, 14% there, and so on.

There is, however, a direct and intimate relationship between percentage and probability. Specific percentage areas are based on the fact that the total percentage area equals 100%. When expressed in decimal form, probability is based on the fact that the total number of events equals 1.00. Thus, the total number of events in both cases, is, in fact, 100% of all the events. This means that 25% of the possible events is equivalent to a probability of .25 or 25/100. Or, looking at it the other way, if an event has a probability of .25, this means it will occur 25% of the time. As percentages vary from 100 to zero, so do probabilities vary from 1.0 to zero. Thus, if an event is certain to occur, that is, will occur at a frequency of 100%, its probability of occurrence is 1.00. If an event cannot occur, its probability is zero.

Converting Percentages to Probability Statements

The conversion of percentages to probability statements is quick and easy. We simply divide the percentage of cases by 100 and *drop the percentage sign.*

$$P = \frac{\% \text{ of cases}}{100}$$

Thus, if the frequency of occurrence is 90%, then

$$P = \frac{90\%}{100} = .90$$

It's as simple as that. All you need is both ends of a pencil—the lead point to move the decimal two places to the left and the eraser to get rid of the percentage sign. It is crucial to erase the percentage sign. Do *not* write a probability value like .10%. This makes no sense. The problems will be solved *either* in percentage terms or probabilities, so it is not necessary to create new combinations.

Combinations, Permutations, and Statistical Inference

As noted in the preface, this book does not include problems based on combinations and permutations. Although a more thorough understanding of probability may involve these topics, the choice was made to jump you ahead directly into the area of inferential statistics and hypothesis testing, using traditional inferential techniques, such as *t, r, F,* and chi square. You can read the literature in your field without an extensive mathematical grounding in combinations and permutations, but you cannot make sense out of the literature without familiarity with the statistical tests of significance.

To give you at least a nodding acquaintance with the general notions involved, however, consider the following:

Combinations refer to the number of ways a given set of events can be selected or *combined,* without regard to their order or arrangement. Suppose, for example, we have six groups of subjects, and we wish to compare each group with each of the other groups. How many two-group comparisons are required?

$$C_{nr} = \frac{n!}{r!(n-r)!}$$

where C_{nr} = the total number of combinations of *n* events, taken *r* at a time, and ! = factorial, or $n(n-1)(n-2)(n-3) \ldots$. Thus,

$$C_{nr} = \frac{6!}{2!(6-2)!} = \frac{6 \times 5 \times 4 \times 3 \times 2 \times 1}{(2 \times 1)(4 \times 3 \times 2 \times 1)}$$

$$C_{nr} = \frac{720}{(2)(24)} = \frac{720}{48}$$

$$C_{nr} = 15$$

There are a total of 15 ways to compare all the pairs taken from six groups.

A WINNING COMBINATION?

In those states with lotteries, the largest payoff (Big Bucks, Lotto Bucks, Megabucks, etc.) typically results when the drawing is based on a series of consecutive numbers, taken *r* ways. For example, a player might be given a card with the numbers 1 through 36 and be asked to select any 6 different numbers. The total possible combinations, therefore, are

$$C_{nr} = \frac{n!}{r!(n-r)!} = \frac{36!}{6!(36-6)!} = \frac{36!}{(6!)(30!)} = 1,947,792$$

At a dollar a card, then, for an investment of $1,947,792 you could cover every possible combination.

Now (assuming that the logistical problem, no small problem, of buying and filling out almost 2 million cards could be solved), when the jackpot reaches $3 million, $4 million, or even $10 million, is it necessarily a profitable venture to invest this much money to assure a win? Not really, since it is possible that many players could have winning cards, in which case the jackpot would be split. Playing all the combinations does assure you of a win, but in no way does it guarantee the size of the win. Also, any change, even a seemingly slight change, in the total number of items to be selected from, can dramatically alter the chances of winning. In one state, the rules for the lotto-type game were changed from the player selecting 6 numbers out of 44 possible numbers to the player having to select 6 numbers out of 48. This change almost doubled the chance of losing. Using the formula above, 6 out of 44 yields 7,074,468 possible combinations, whereas selecting 6 out of 48 jumps the possible combinations to 12,277,228. But then again, the probability of winning drops to a flat-out zero if the player fails to buy a ticket. You have to be in it to win it!

Permutations refer to the number of *ordered* arrangements that can be formed from the total number of events. For example, assume that an intramural softball league has eight teams. We wish to determine the total possible number of ways the teams could finish:

$$P_n = n!$$

where P_n = the total number of ordered sequences of n events and n = the total number of events. Thus,

$$P_n = 8! = 40,320$$

There are a total of 40,320 possible orders of finish for an eight-team softball league.

z Scores and Probability

All of those z score problems, where we calculated percentage areas, can now be rephrased as probability problems. The only thing that will change is the manner in which we express the answers.

Example.   Let's say that we are given a normal distribution with a mean of 500 and a standard deviation of 87. We are asked to find the percentage of cases falling between the raw scores of 470 and 550. The same problem can be translated into probability form: state the probability that any single score falls somewhere between the raw scores of 470 and 550.

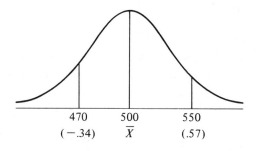

We draw the curve and calculate the two z scores.

$$z_1 = \frac{X - \bar{X}}{SD} = \frac{550 - 500}{87} = .57$$

$$z_2 = \frac{X - \bar{X}}{SD} = \frac{470 - 500}{87} = -.34$$

Then, we look up the percentage for each z score in Table A. Since the two scores fall on either side of the mean, we add the percentages (13.31% + 21.57%) for a total of 34.88% (Rule D1). This is the answer to the original frequency question. To answer the probability question, we simply divide the percentage by 100.

$$P = \frac{34.88}{100} = .348 = .35$$

Example. Another frequency-type problem can be translated to probability. With a mean of 68 and a standard deviation of 3.50, what is the probability that any single score will fall between 70 and 72?

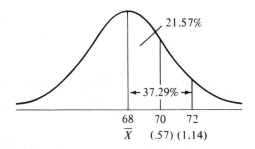

The two z scores are

$$z = \frac{72 - 68}{3.50} = 1.14$$

$$z = \frac{70 - 68}{3.50} = .57$$

From Table A, we find 37.29% for 1.14 and 21.57% for .57. We subtract the percentages (Rule D2) to get 15.72%. Then, it can be found that the probability is as follows:

$$P = \frac{15.72}{100} = .157 = .16$$

Problems 6a: On a normal distribution with a mean of 25 and a standard deviation of 2.42, find:

1. The probability that any single score will fall above a raw score of 27
2. The probability that any single score will fall below a raw score of 30
3. The probability that any single score will fall between raw scores of 20 and 28.
4. The probability that any single score will fall between raw scores of 21 and 24.
5. The probability that any single score will fall between raw scores of 29 and 32.

Combining Probabilities For Independent Events

There are two rules that are of extreme importance when probabilities are combined. We'll call them the "ADD-OR" rule and the "MULT-AND" rule.

1. The ADD-OR Rule:

In order to obtain the probability of one event OR another event occurring, the separate probabilities should be added (hence the label "ADD-OR"). For instance, if you were to flip two coins and wanted the probability of obtaining a head or a tail (which, of course, is a certainty), you would simply add the two separate probabilities:

$$P = .50 + .50 = 1.00$$

If the problem involved playing cards instead of coins, we can just as easily find the "or" probabilities of selecting, say, a king of spades, or a ten of hearts. Since there is only one king of spades and only one ten of hearts in a deck of 52 cards, the probability of selecting one or the other would be

$$P = 1/52 + 1/52 = 2/52 = .0384, \text{ or rounded to } .04$$

The same rule also applies to z scores. Suppose we wished to find the probability of selecting at random someone whose IQ was below 85 or someone whose IQ was above 115. Since the IQ distribution has a mean of 100 and an SD of 15, the two z scores in question would be -1.00 and $+1.00$. Using Table A, we find $50.00\% - 34.13\%$, or 15.87% falling below a z score of -1.00, and, again, $50.00\% - 34.13\%$, or 15.87% falling above the z of $+1.00$. Since this is an "either-or" situation, ADD the percentages $(15.87\% + 15.87\% = 31.74\%)$. To finish the problem, convert the percentage to a probability value (see page 99).

$$P = .3174 \quad \text{or} \quad .32$$

Thus, there are 32 chances out of 100 that the person selected will have an IQ of either below 85 or above 115.

2.   The MULT-AND Rule:

When attempting to find the probability of one event AND another event occurring, we multiply the separate probabilities (thus the label MULT-AND). For example, if we were to flip a coin twice and wanted the probability of getting a head on one toss AND a tail on the second toss, the MULT-AND rule says:

$$.50 \times .50 = .25$$

There are, thus, 25 chances out of 100 that on two coin flips you could obtain both a head and a tail.

Going back to our playing card example, the same rule applies. For example, assume that we wanted to find the probability of selecting an ace and a spade from the deck of cards. We know that there are 4 aces in the deck of 52 cards, so its probability is $4/52 = 1/13$. We also know that there are 13 spades in the deck, yielding a probability of $13/52 = 1/4$. By multiplying, we get

$$P = 1/13 \times 1/4 = 1/52$$

which is the probability of selecting an ace and a spade, or, also in this case, the ace of spades.

We can, of course use the same rule on the z distribution. For example, to find the probability of selecting someone whose IQ was below 85 and on the next draw someone whose IQ was above 115, you multiply the separate probabilities:

$$.1587 \times .1587 = .025, \text{ or rounded, } = .03$$

There are, therefore, only 3 chances out of 100 that one could randomly select someone whose IQ was below 85 and on the next draw someone whose IQ was above 115.

INCLUSION AND EXCLUSION

Before leaving the area of descriptive statistics, one final type of z score problem must be confronted. The problems in this section, called inclusion and exclusion problems, are absolutely critical to prepare you for the next unit on inferential statistics.

We have learned that the normal curve is perfectly symmetrical. A knowledge of one side of the curve gives, by deduction, full knowledge of the other side. Also, the majority of scores fall around the middle of the normal distribution, and extreme scores, both extremely high scores and extremely low scores, fall way out in the tails of the curve. These facts, then, tell us that the middlemost scores in the distribution are those scores clustering around the mean, half on one side and half on the other. Therefore, if we look for the middlemost 30% of the scores, they are found to fall within 15% of the area on either side of the mean, as shown in Fig. 6.1.

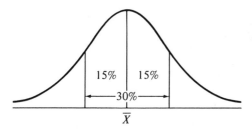

Figure 6.1
The middlemost 30% of the scores on a normal curve.

Similarly, if we have to locate the area containing the most extreme 30% of all scores, it is found by establishing the area containing the highest 15% of the scores *and* the area containing the lowest 15%. (See Fig. 6.2.) The words "most extreme" always refer to those scores farthest from the mean.

Inclusion Area

The **inclusion area** always contains the middlemost scores in the normal distribution, those surrounding the mean. Inclusion problems ask for two scores that *include* the middlemost 30%, or 40%, or whatever, of the scores in a normal distribution.

Assume a normal distribution with a mean of 65 and a standard deviation of 6.25. Which two raw scores include the middlemost 95% of the scores?

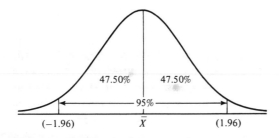

First, we draw the curve and distribute the middlemost 95% evenly around the mean, 47.50% on either side. To get the z score, we look up 47.50% in Table A and find a value of ±1.96. Thus, we have discovered that the middlemost

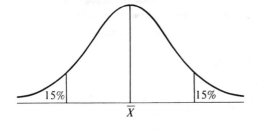

Figure 6.2
The most extreme 30% of the scores on a normal curve.

95% is included between z scores of $+1.96$ and -1.96. Since the question asks for raw scores, we must solve the z score equation for X.

$$X = zSD + \bar{X}$$
$$X = (\pm 1.96)(6.25) + 65 = 12.25 + 65$$
$$X = +12.25 + 65 = 77.25$$
$$X = -12.25 + 65 = 52.75$$

The two raw scores, thus, that include the middlemost 95% of this distribution are 77.25 and 52.75. Inclusion problems, such as these, will become extremely important in later chapters, during the discussions of confidence intervals.

Rule for Inclusion Problems. The rule for solving inclusion problems is to take the percentage given in the problem and divide it by 2. Then, look up the resulting percentage in Table A and locate the z scores. Finally, if necessary, convert the z scores to raw scores.

Problems 6b: On a normal distribution with a mean of 42 and a standard deviation of 6.27, find the two raw scores that include the middlemost

1.	10%	4.	90%
2.	30%	5.	99%
3.	50%		

Exclusion Area

The **exclusion area** is made up of the tails of the distribution and is thus involved in specifying those scores lying *farthest* from the mean. Because of the curve's symmetry, we again divide the percentage in two, placing half at the bottom tail and half at the top. Exclusion problems ask for two scores that exclude some percentage of the most extreme scores of the distribution. An exclusion problem is simply a reverse way of looking at an inclusion problem. If we find the two scores that include the middlemost 60% of the scores, then we have also found the two scores that exclude the most extreme 40% of the scores. We must think in these terms, because the z score table indicates the percentage of cases between the z score and the mean, that is, those cases lying *closest to the mean.*

Remember, exclusion problems are asking you to find those scores that exclude a percentage that deviates the most from the mean. To find the most extreme 10% of the scores, we do not identify the top 10%. Extreme scores are from both the cream of the crop and the bottom of the barrel. The extreme 10%, then, is made up of both the lowest 5% as well as the highest 5%.

Example. The mean of a normal distribution is 15 and its standard deviation is 1.95. Find the two scores that exclude the most extreme 10% of scores.

We draw the curve, splitting the extreme 10% into the highest 5% and the lowest 5%. These points are found to be 45% away from the mean. By looking up 45% in Table A, we get z scores of ± 1.65 (remember, with tie percentages we take the higher of the two z scores).

Solving the z score equation for X, we get

$$X = zSD + \bar{X}$$

$$X = (\pm 1.65)(1.95) + 15 = \pm 3.22 + 15$$

$$X_1 = 3.22 + 15 = 18.22$$

$$X_2 = -3.22 + 15 = 11.78$$

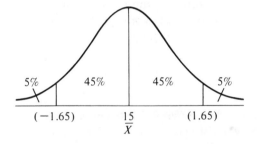

Thus, the raw scores 18.22 and 11.78 exclude the most extreme 10% of the entire distribution. (They also include the middlemost 90%.) Exclusion problems such as these will take on even more importance when we later, in Chapter 8, get to the topic of the alpha error.

Rule for Exclusion Problems.   The rule for solving exclusion problems is to take the percentage given in the problem, divide it by 2, and subtract that result from 50%. Then, look up the resulting percentage in Table A and locate the z scores. Finally, if necessary, convert the z scores to raw scores.

Problems 6c: On a normal distribution with a mean of 146 and a standard deviation of 38, find the two raw scores that exclude the most extreme

1. 10%   4. 22%
2.   5%   5.   1%
3. 15%

A REMINDER ABOUT LOGIC

If you have been able to do the problems presented throughout this chapter, you should have no difficulty with the problems at the end. You must think each one out carefully before plunging into the calculations. Remember, this course has more to do with logic than with math. When appropriate, always draw the curve and plot the values on the abscissa. This small step can prevent big logical errors. For example, if the problem states that someone has a score of 140 and is at the 25th percentile, it is illogical to come up with a mean of less than 140 for the distribution. Or, if a problem asks for a probability value for an event occurring between scores that are both to the right of the mean, it is illogical to come up with a value of .50 or higher. With the curve drawn out and staring back at you, you cannot make these kinds of logical errors.

If you can do the problems so far, you may now revel in the luxury of knowing that you have a working knowledge of probability—at least of probability as it relates to the normal curve. For the next few chapters, that is all you need to know about probability. Estimating population means, deciding whether or not to reject the chance hypothesis, establishing the alpha error—all these depend on a working knowledge of the probability concept. Once you have mastered this chapter, you are ready for inferential statistics.

SUMMARY

Probability, P, is equal to the number of times a specific event can occur out of the total possible number of events, s/t. Statisticians typically divide out this fraction and express probability in decimal form. Thus, a probability of one out of two, 1/2, is written as .50.

Whenever events are independent of each other, as in a succession of dice throws or coin flips, the result of any one of these events is not conditional on the result of any of the preceding events. Assuming that independent events are conditional is known as the gambler's fallacy.

Whereas probability specifies the outcome in terms of the number of times an event will occur out of the total possible number of events, odds are stated in terms of how often an event will not occur as compared to how often it will. Thus, if the probability of an occurrence is one out of two or .50, the odds against it are 1 to 1.

The percentage areas of the normal curve can be used to obtain probability values. If an event occurs at a frequency of 50%, the probability that the event will occur is 50 out of 100, or .50. To obtain probability values from the percentages given in Table A, simply move the decimal two places to the left (divide by 100) and erase the percentage sign.

All the z score problems previously solved in frequency (percentage) terms can be quickly and easily translated into probability statements. Thus, if 68% of the cases fall between z scores of ± 1.00, then the probability that a single case will occur in that same area is .68.

When probabilities are combined, two rules are needed: (1) to get the probability of one event or another on successive occasions, add the separate probabilities, and (2) to get the probability of one event and another on successive occasions, multiply the separate probabilities.

Inclusion problems ask for the middlemost area of the normal curve. Since the curve is symmetrical, the middlemost scores are those that cluster evenly around the mean. The middlemost 50% of the scores, thus, are those that fall within 25% to the left of the mean and 25% to the right of the mean.

Exclusion problems ask for the area in the tails of the normal distribution. Again, due to symmetry, the extreme scores divide evenly between the two tails of the distribution. The extreme 5% of the curve, thus, divides evenly between the highest 2½% and the lowest 2½%.

Inclusion and exclusion problems provide the logic behind future statistical calculations, such as confidence intervals and the alpha error.

KEY TERMS

exclusion area   odds
gambler's fallacy   probability
inclusion area

REFERENCES

1. Kimble, G. A. (1978). *How to use (and misuse) statistics.* Englewood Cliffs, NJ: Prentice-Hall, p. 80.

2. Scarne, J. (1961). *Scarne's complete guide to gambling.* New York: Simon & Schuster, p. 67.
3. Ibid, p. 25.

PROBLEMS

1. A certain event occurs one time out of six due to chance.
    a. What is the probability of that event occurring?
    b. What are the odds against that event occurring?
2. A given event occurs at a frequency of 65%. What is the probability that the event will occur?
3. Assume that a deck of 52 cards is well shuffled.
    a. What is the probability of selecting a single card, for example, the ace of hearts?
    b. What are the odds against selecting that single card?
    c. If that card were selected and then replaced in the deck, and the deck shuffled, what is the probability of selecting it a second time?

Assume a normal distribution for problems 4 through 17.

4. A certain guided missile travels an average distance of 1500 miles with a standard deviation of 10 miles. What is the probability of any single shot traveling
    a. More than 1525 miles?

   b.   Less than 1495 miles?
   c.   Between 1497 and 1508 miles?
5. The mean weight among women in a certain sorority is 115 pounds with a standard deviation of 13.20 pounds. What is the probability of selecting a single individual who weighs
   a.   Between 100 and 112 pounds?
   b.   Between 120 and 125 pounds?
6. On the WISC-R, the mean on the arithmetic subtest is 10.00, with a standard deviation of 3.00. What is the probability of selecting at random two persons who scored either above 11.00 or below 8.00?
7. Using the data from problem 6, what is the probability of selecting at random two persons, one of whom scored above 11.00 and another who scored below 8.00?
8. The middlemost 70% of a normal distribution is included between which two z scores?
9. The most extreme 8% of a normal distribution is excluded by which two z scores?
10. On a normal distribution with a mean of 25.62 and a standard deviation of 6.72, find the two raw scores that include the middlemost
   a.   15%   c.   60%
   b.   28%   d.   95%
11. Using the data from problem 8, ($\bar{X}$ = 25.62, SD = 6.72), find the two raw scores that exclude the most extreme
   a.   60%   c.   10%
   b.   26%   d.   5%
12. A group of young women trying out for the college softball team is asked to throw the softball as far as possible. The mean distance for the group is 155 feet with a standard deviation of 14.82 feet. What is the probability of selecting at random one woman who can throw the ball
   a.   172 feet or more?
   b.   160 feet or less?
   c.   Between 145 and 165 feet?
   d.   Between which two distances are the middlemost 80% of the throws?
   e.   Which two distances exclude the most extreme 12% of the throws?
13. The freshmen basketball coach's office has a doorway which is 77 inches high. The coach will recruit only those student-athletes who have to duck to enter the office. Of all the freshmen who wish to play basketball, the mean height is 75 inches with a standard deviation of 2.82 inches. What is the probability that any single individual in the group will be recruited?
14. Among the shoppers at a certain women's store, dress sizes are normally distributed around a mean of 12.06. The standard deviation is 3.03. What is the probability of a customer entering the store who needs a size 18 or larger?
15. A certain school district has a population of 500 high school seniors, 275 girls and 225 boys. On a single draw, what is the probability of selecting
   a.   A boy?
   b.   A girl?
   c.   A boy or a girl?
16. Using the data from problem 15, what is the probability, on successive draws, of selecting both a boy and girl?

17. In tossing a single coin and a single die, what is the probability of getting
    a. A head on the coin and a three on the die?
    b. A tail on the coin or a three on the die?
    c. Heads or tails on the coin and a six on the die?

Fill in the blanks in problems 18 through 22.

18. When an event cannot occur, its probability value is _____.
19. When an event must occur, its probability value is _____.
20. The middlemost 50% of the scores in a normal distribution fall in such a way that _____% of them lie to the left of the mean.
21. On a normal distribution, _____% of the scores must lie above the mean.
22. The ratio *s/t* (specific event over total number of events) defines what important statistical concept?

True or False—Indicate either T or F for problems 23 through 30.

23. If a given event occurs at a frequency of 50%, its probability of occurrence is .50.
24. Probability values can never exceed 1.00.
25. If a coin is flipped 10 times and, by chance, heads turn up 10 times, the probability of obtaining heads on the next flip is .90.
26. A negative *z* score must yield a negative probability value.
27. The more frequently an event occurs, the higher is its probability of occurrence.
28. To find the probability of selecting one event or another, the separate probabilities should be added.
29. To find the probability of selecting one event and another, the separate probabilities should be multiplied.
30. Probabilities may take on negative values only when the events are occurring less than 50% of the time.

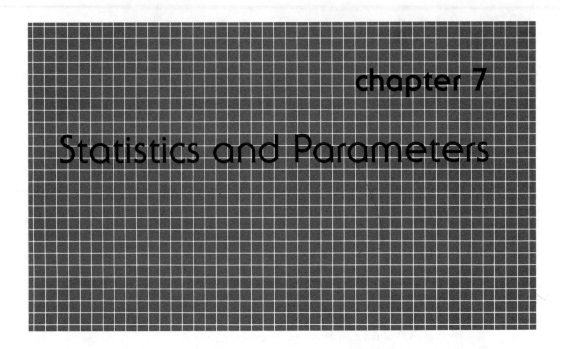

chapter 7
Statistics and Parameters

All of us have, at one time or another, been warned about jumping to conclusions, about forming hasty generalizations. "One swallow doesn't make a summer," as the old adage has it. The statistician puts this message differently, "Never generalize from an N of one." Or, as someone once said, "There are no totally valid generalizations—including this one!" All of this is good advice. It is best to be alert to the dangers of the "glittering generality," which sometimes seems to have a validity of its own, even though it is based on scanty, or even fallacious, evidence. Hasty conclusions and unfounded generalizations are indeed dangerous—they often are used to produce "the big lie."

However, some generalizations are safer than others. For example, you are playing cards and betting against the chance that your opponent will draw an ace from the deck. You will undoubtedly feel more comfortable about your bet if you are holding two aces in your hand than if you are holding none. Furthermore, you will be positively oozing confidence if you happen to be holding three aces. (Of course, if you have all four aces, nothing is left to chance and the probability of your winning the bet is a perfect 1.00.)

In the game of predictive, or inferential, statistics, the statistician rarely, if ever, holds all four aces. Therefore, every inference must be couched in terms of probability. Inferential statistics are not designed, and could not be, to yield eternal and unchanging truth. Inferential statistics, however, do offer a probability model. When making a prediction, the statistician knows beforehand what the probability of success is going to be. This does give the statistician a tremendous edge over the casual observer.

GENERALIZING FROM THE FEW TO THE MANY

The main thrust of inferential statistics is based on measuring the few and generalizing to the many. That is, observations are made of a small segment of a group, and then, from these observations, the characteristics of the entire group are inferred.

For many years, advertisers and entertainers asked, "Will it play in Peoria?" The question reflected their tacit belief that if an ad campaign or show biz production were successful in Peoria, Illinois, it would be successful throughout the United States. Peoria was seen as a kind of "magic" town, whose inhabitants echoed precisely the attitudes and opinions of the larger body of Americans in general. The importance of the question for us lies not in its acceptance that the citizens of Peoria truly reflect the American ethos, but in its implication of a long-held belief that by measuring the few, a picture of the many will clearly emerge.

KEY CONCEPTS OF INFERENTIAL STATISTICS

Population

A **population,** or universe, is defined as an entire group of persons, things, or events having at least one trait in common. The population concept is indeed arbitrary, since its extent is based on whatever the researcher chooses to call the common trait. We can talk of the population of blondes, or redheads, or registered voters, or undergraduates taking their first course in statistics, or whatever we choose as the common trait. Furthermore, since the definition specifies *at least* one trait in common, the number of shared traits can be multiplied at will. It is possible to speak of a population of blonde, left-handed 21-year-old, male college students majoring in anthropology and having exactly three younger siblings.

Adding too many common traits, however, may severely limit the eventual size of the population—even to the point of making it difficult to find anyone at all who possesses all of the required traits. As an instance of this, consider the following list of traits specified to qualify for a certain scholarship: "Must be of Italian extraction; live within a radius of 50 miles from Boston; be a practicing Baptist; be attending a college in North Dakota; be a teetotaler; and be planning to major in paleontology." We read such a list in awe and wonder if anyone ever qualifies. Obviously, the more traits being added, the more you limit the designated population, but as your population becomes increasingly limited, so too does the group to which your findings can rightfully be extrapolated.

* Do not define traits which are too general

You can't study the population of 6-year-old females living in Greenwich Village, New York, and then generalize these findings to include 50-year-old, male, rug merchants working the streets of Teheran. Nor should you do a study on albino rats learning their way through a maze and then quickly extrapolate these findings to cover the population of U.S. college sophomores.

A population may be either *finite*, if there is an exact upper limit to the number of qualifying units it contains, or *infinite*, if there is no numerical limit as to its size.

Parameter

A **parameter** is any measure obtained by having measured an entire population. If we gave IQ tests to every single college student in the country, added the IQ scores, and divided by the total number of students, the resulting mean IQ score would be a parameter for that *specific* population, that is, the population of college students. Furthermore, we could then compute the median, mode, range, and standard deviation for that vast distribution of IQ scores. Each of the resulting values would be a parameter.

Obviously, when a population is of any considerable size, parameters are hard to get. It would be difficult, if not impossible, to measure the IQ of every American college student. Even if it were possible, the time and expense involved would be enormous. Remember, to obtain an official parameter, every student would have to be found and given the IQ test. Because they are often so difficult to obtain, most parameters are estimated, or inferred. There are very few situations where true parameters are actually obtainable. As was pointed out in Chapter 1, an exception to this is in the area of political polling, where the results do become vividly clear, but not until after the election. One final point is that if a parameter is known, there is no need at all for inferential statistics—simply because there's nothing left to predict.

Sample

A **sample** is a *smaller* number of observations taken from the *total* number making up the population. That is, if Omega College has a total student population of 6000, and we select 100, or 10, or even 2 students from that population, only the students so selected form our sample. Even if we select 5999 students, we still only have a sample. Not until we included the last student could we say we had measured the total population. Obviously, for purposes of predicting the population parameters at Omega College, a sample of 5999 is far more accurate than is a sample of 10. After all, with 5999 students already measured, how much could that last student change the values of the mean or the standard deviation?

Therefore, *other things being equal* (which they rarely are), the larger the sample, the more accurate the parameter prediction. Be very careful of that statement, since "other things being equal" is a powerful disclaimer. Later, we will see that where other things weren't equal, some positively huge samples created unbelievably inaccurate predictions. So far, we have simply defined the sample; we have yet to address the important issue of what constitutes a good or a bad sample.

Statistic

A **statistic** is any measure obtained by having measured a sample. If we select 10 Omega College students, measure their heights, and calculate the mean height, the resulting value is a statistic. Since samples are usually so much more accessible than populations, statistics are far easier to gather than are parameters.

This, then, is the whole point: we select the sample, measure the sample, calculate the statistics, and, from these, infer the population parameters. In short, inferential statistics is the technique of predicting unknown parameters on the basis of known statistics—the act of generalizing to the many after having observed the few.

THE TECHNIQUES OF SAMPLING

To make accurate predictions, the sample used should be a **representative sample** of the population. That is, the sample should contain the same elements, should have the same overall coloration from the various subtle trait shadings, as the population does. As we mentioned previously, the nurse who takes a blood sample for analysis is almost positive that the sample is representative of the entire blood supply in the circulatory system. On the other hand, the researcher who takes a sample of registered voters from a given precinct cannot be absolutely certain that accurate representation is involved, or even that everyone in the sample will actually vote on election day. The goal for both the nurse and the political forecaster, however, is the same—to select a sample truly reflective of the characteristics of the population.

A good, representative sample provides the researcher with a miniature mirror with which to view the entire population. There are two basic techniques for achieving representative samples: random sampling and stratified or quota sampling.

Random Sampling

Random sampling is probably one of the media's most misused and abused terms. Newspapers claim to have selected a **random sample** of their readers; TV stations claimed to have interviewed a random sample of city residents; and so on. The misapplication results from assuming that random is synonymous with haphazard—which it is not.

Random sampling demands that each member of the entire population has an equal chance of being included. And the other side of that coin specifies that no members of the population may be systematically excluded. Thus, if you're trying to get a random sample of the population of students at your college, you can't simply select from those who are free enough in the afternoon to meet you at the psychology laboratory at 3 P.M. This would exclude all those students who work in the afternoon, or who have their own labs to meet, or who are members of athletic teams. Nor can you create a random sample by selecting every *n*th person entering the cafeteria at lunch time. Again, some students eat off campus, or have classes during the noon hour, or have cut classes that day in order to provide themselves with an instant, minivacation—the reasons are endless. The point is that, unless the entire population is available for selection, the sample cannot be random. To obtain a random sample of your college's population, you'd have to go to the registrar's office and get a list of the names of the entire student body. Then clip out each name individually, and place all the separate names in a receptacle from which you can, preferably blindfolded, select your sample. Then, *you go out and find each person so selected.* This is extremely important, for a sample can never be random if the subjects are allowed to select themselves. For example, you might obtain the list of entering students, the population of incoming freshman, and, correctly enough so far, select from the population of names, say, 50 freshmen. You then place requests in each of their mailboxes to fill out an "attitude-toward-college" questionnaire, and, finally 30 of those questionnaires are dutifully returned to you. Random sample? Absolutely not, since you allowed the subjects, in effect, to select themselves on the basis of which ones felt duty bound enough to return those questionnaires. Those subjects who did comply may have differed (and probably did differ) systematically on a whole host of other traits from the subjects who simply ignored your request. This technique is really no more random than simply placing the questionnaire in the college newspaper and requesting that it be clipped, filled out, and returned to you. The students who exert the effort to cooperate with you almost surely differ in important ways from those who do not.

Perhaps the most blatant example of the self-selected sample is when a TV station asks its audience to respond to some seemingly momentous question by using the telephone to dial the 50-cent 900 number. Clearly, only the most zealous viewers will then bestir themselves to get up, go to the phone, and be willing to pay for the privilege of casting a vote.

It was a true random selection that the U.S. Selective Service System used to tap young men for the draft. All the days of the year were put in a huge fishbowl and a certain number of dates pulled out. If your birthday fell on one of the dates selected, you were called! Whether you went or not is another story, but since everyone's birth date was included in the fishbowl, the *selection* was random. (Also, note that the government did not wait for you to go to them; they came to get you.)

Along with the Selective Service System, another extremely efficient government operation is the Internal Revenue Service. When the IRS asks its big computer in West Virginia to spit out 2% of the income tax returns for further investigation, the selection is again, strictly random, since a copy of every taxpay-

The IRS doesn't wait for the sample to select itself. (George James/NYT Pictures.)

er's return is coded into the computer. Note that the IRS also does not wait for the sample to select itself. It chooses you!

Compare these examples with a TV station's attempt to select a random sample by interviewing every tenth person who happens to be walking down the main street on a Tuesday afternoon. Is every resident downtown that day, that hour, strolling along the main thoroughfare? Or, a comparable example is if the college newspaper interviews every tenth student entering the library on a Thursday afternoon. This cannot possibly be a random sample. Some students might be in class, or at work, or starting early on their round of weekend parties, or back in their rooms doing statistics problems. As long as any students are excluded from the selection process, the sample cannot, by definition, be random.

Stratified, or Quota Sampling

Another major technique for selecting a representative sample is known as **stratified** or **quota sampling.** To obtain a sample, the researcher must know beforehand what some of the major population characteristics are and, then, deliberately select a sample that shares these characteristics in the same proportions. For instance, if 35% of a student population are sophomores and, of those, 60% are majoring in business, then a quota sample of the population must reflect those same percentages.

Some political forecasters utilize this sampling technique. They first decide what characteristics are important in determining voting behavior, such as religion, socioeconomic status, age, and so on. Then they consult the federal government's census reports to discover the percentage of individuals falling within each category. The sample is selected in such a way that the same percentages are present in it. When all this is done carefully and with real precision, a very few thousand respondents can be used to predict the balloting of over 100 million voters.

Sampling Error

Whenever a sample is selected, it must be assumed that the sample measures will *not* precisely match those that would be obtained if the entire population were measured. Any other assumption would be foolhardy indeed. To distinguish it from the sample mean, $\bar{X}$, the mean of the population is designated by the Greek letter mu, μ. **Sampling error** is, then, the difference between the sample value and the population value.

$$\text{sampling error} = \bar{X} - \mu$$

Note that this is a normal, expected deviation. We expect that the sample mean might deviate from the population mean. Sampling error is *not a mistake!* Sampling error is conceptually different from the error committed by a shortstop booting a grounder. It is an expected amount of deviation. Also, sampling error should be random; it can go either way. The sample mean is just as often below the population mean as it is above it. If we took the means of 100 random samples from a given population, then 50% of the resulting sampling errors should be positive (the sample mean overestimating the population mean) and the other 50% should be negative. Applying the relationship between percentage and probability, as defined in Chapter 6, we can say here that the probability that a given sampling error will have a plus sign is the same as the probability that it will have a minus sign, that is, $P = .50$.

Bias

Whenever the sample differs systematically from the population at large, we say that **bias** has occurred. Since researchers typically deal in averages, bias is technically defined as a constant difference, in one direction, between the mean of the sample and the mean of the population. For example, suppose that the mean verbal SAT score at your college is 500, and yet you select samples for your particular study only from those students who, because of poor performance on the English placement test, have been assigned to remedial English courses. It is likely that the average verbal SAT scores among your samples are *consistently lower* than the mean of the population as a whole. This biased selection would be especially devastating if your research involved anything in the way of reading comprehension, or IQ, or vocabulary measures. Your results would almost certainly underestimate the potential performance of the population at your college.

Bias occurs when most of the sampling error loads up on one side, so that the sample means are consistently either over- or underestimating the population mean. Bias is a *constant sampling error in one direction.* Where there is bias, no longer does the probability that $\overline{X}$ is higher than μ equal .50. The probability now might be as great as .90 or even 1.00 (or, on the other side, might be as low as .10 or even zero).

Assume that John Doe, a major in pseudoscience, is engaged in a research project in which he has to estimate the height of the average American college student. He wants a sample with an N of 15. Late in the afternoon, he runs over to the gym, having heard there were some players out there on the gym floor throwing a ball into a basket. They obligingly stop for a few minutes, long enough for John to measure each of their heights. The sample mean turns out to be 81 inches. Undaunted, our hero then repeats this procedure at a number of colleges, always going to the gym in the afternoon and always coming up with mean heights of 77 inches or more. Since the parameter mean height for male college students is actually about 69 inches, the sample means John obtained were constantly above the population mean. This is bias "writ large." This story may seem a little con-

Well that proves it, the average height of male college students is 6'9".

trived. For a true history of the effects of bias, we can turn to the saga of political forecasting of presidential elections.

Political Polling

The first known political poll took place in 1824. The candidates for president that year were John Quincy Adams and Andrew Jackson. A Pennsylvania newspaper attempted to predict the election's outcome by sending out reporters to interview the "man on the street." The data were tallied and the forecast made—Jackson to win. Jackson, however, lost. The science of political forecasting did not get off to an auspicious start.

The **Literary Digest** *Poll.*   Throughout the rest of the 1800s other newspapers in other cities began to get involved in political polling, and the results were decidedly mixed. In 1916, an important magazine, *Literary Digest,* tried its hand at political forecasting. Over the next several presidential elections, the magazine enjoyed a fair degree of success. The technique was simple: send out sample ballots to as large a group of voters as possible, and then sit back and hope that the people would be interested enough to return the ballots. The measured sample had to separate itself from the selected sample by choosing whether or not to mail in the ballots. Now, a mail-in poll can be accurate, by luck, but the *Digest's* luck ran out.

The candidates during the Depression year 1936 were Alf Landon, Republican, and Franklin D. Roosevelt, Democrat. Beginning in 1895, the *Literary Digest* had compiled a list of prospective subscribers, and by 1936, there were over 20 million names on the list. The names, however, were of people who offered the best potential market for the magazine and its advertisers, or middle- and upper-income people. To "correct" for any possible subscriber bias, the magazine also collected a list of names selected from various telephone books throughout the country and also from names provided by state registries of motor vehicles. This list became known as the famous "tel-auto public." In all, the *Digest* sent out more than 10 million ballots and got back 2,376,523 (almost 25%).[1] Said the *Literary Digest,*

> Like the outriders of a great army, the first ballots in the great 1936 Presidential Campaign march into the open this week to be marshalled, checked and counted. Next week more states will join the parade and there will be thousands of additional votes. Ballots are pouring out by tens of thousands and will continue to flood the country for several weeks. They are pouring back too, in an ever increasing stream. The homing ballots coming from every state, city, and hamlet should write the answer to the question, "Roosevelt or Landon?" long before Election Day.[2]

The *Literary Digest* wrote its answer—Landon in a landslide. The final prediction gave Landon 57% of the vote to a mere 40% for Roosevelt. The results of the election were the other way around. Roosevelt received 27 million votes (over 60%) to Landon's 16 million votes. Following the election, the magazine's next issue had a blushing, pink cover with the cutesy caption "Is our face red!" The humiliation, however, must have been too great to bear, since the *Digest* soon went out of business.

What went wrong? Bias, a constant sampling error in one direction, produced a sample in which Republicans were markedly overrepresented.* Also, as is the case with all mail-in polls, the sample was allowed to select itself. Not everyone who received one returned a ballot, but those who did probably had more fervor about their choice than did the population at large. Perhaps the major finding from the whole fiasco was that in 1936 Democrats did not have phones, did not drive cars, and certainly did not read or respond to the *Literary Digest*.

The Gallup Poll. In 1936, a new statistical star arose. His name was George Gallup, and he headed the American Institute of Public Opinion, headquartered in Princeton, New Jersey. Gallup correctly predicted Roosevelt's win over Landon, even though the margin of victory was off by a full 7 percentage points. In the next two elections Gallup also predicted the winner, Roosevelt over Willkie in 1940 and Roosevelt over Dewey in 1944. In both instances, Gallup reduced his margin of error, to 3% in 1940 and to a phenomenally low 0.5% in 1944.

Gallup, unlike the *Literary Digest*, did not allow the sample to select itself; he went to the sample. He stressed representative sampling, not necessarily large sampling. Using the quota technique described earlier in this chapter, he chose fewer than 3000 voters for inclusion in his sample; compare this with the enormous sample of over 2 million voters used by the *Literary Digest*. Some of the population parameters used by Gallup were geographic location, community size, socioeconomic level, sex, and age.

In 1948 the Republicans, as they had in 1944, again chose Thomas E. Dewey, and the Democrats chose Harry S. Truman. Because of Gallup's three previous successes, his poll was now highly regarded, and his predictions were awaited with great interest. During the last week of September, a full six weeks before the election, Gallup went to press and predicted a win for Thomas Dewey. His poll was considered infallible, and one Chicago newspaper even printed a headline proclaiming Dewey's victory before the final vote was completely counted.

Harry Truman, of course, won, and statistical analysis learned its second hard lesson in a dozen years. We now know that Gallup made two crucial mistakes. First, he stopped polling too soon. Six weeks is a long time, and voters do

*It is now assumed that telephone owners are representative of the voting population. One poll in 1988, conducted by the CBS network, was based on a random sample of fewer than 1500 telephone owners throughout the entire country. The margin of error was considerably less than 3%.

Some newspapers believed more in the polls than in the election results. Harry Truman believed the results. (UPI Photo.)

change their minds during a political campaign. If the election had been held when Gallup took his final count, Dewey might indeed have won. Second, Gallup assumed that the "undecideds" were really undecided. All those individuals who claimed not yet to have made a decision were placed in one pile and then allocated in equal shares to the two candidates. We now know that many persons who report indecisiveness are in reality, perhaps only at an unconscious level, already leaning toward one of the candidates. Questions such as "Who are your parents voting for?" or "your friends?" or "How did you vote last time?" may often peel away a voter's veneer of indecision.

In 1952 Gallup picked Eisenhower, although he badly underestimated the margin of victory. Perhaps by then, having been burned twice when predicting Republican wins, the pollsters no longer trusted the G.O.P. and therefore underplayed the data.

In 1960, however, Gallup became a superhero by predicting Kennedy's razor-thin victory over Nixon. Gallup's forecast even included the precise margin of Kennedy's win, which was 0.5%. Polling had come of age. Since 1960 most of the major political forecasters have done a creditable job, and none use a sample much larger than 3000 voters. This is truly remarkable, considering that the voting population is over 100 million.

The Year of the Hedge. In 1980, many of the forecasters began hedging their predictions as November neared. These statisticians, who during the summer and early fall had seen a huge Reagan lead slip away, began having "Harry Truman" nightmares. Could it happen again? Would Jimmy Carter retain the presidency and be photographed the day after gleefully exhibiting a "Reagan Won" headline? Perhaps the memory of past miscalls clouded their vision, for many of the major forecasters simply refused to call the winner in 1980, especially since it looked like a Republican winner. One exception was Lou Harris, whose final prediction was Reagan with 47% and Carter with 42%. (The actual election results showed Reagan with 51% and Carter with 41%.)

There now seems to be no question that the Truman–Dewey phenomenon had resurfaced in 1980. Many voters either remained undecided until late in the campaign or changed their minds during the final weekend. One poll that was right on the money was NBC's exit poll of voters in the east. NBC declared Reagan the winner at 8:15 P.M. eastern standard time, even though West Coast and Rocky Mountain voters still had from one to three hours in which to cast their ballots. There was some feeling at that time that persons living west of the Mississippi had been disenfranchised.

In any case, the exit poll is obviously very accurate, but, other than providing political analysts with important, though after-the-fact, background information on who voted and why, it is an extremely late predictor. In fact, the three major TV networks have since promised that they will no longer use exit polls to give away the outcome of an election while people are still voting.

The pollsters have continued their glittering string of successes, both in 1984 and 1988. In fact, the Gallup poll (which still uses in-person interviews rather than the telephone) predicted Reagan to be the 1984 winner by 18 percentage points. The final count was just that, 58% for Reagan and 40% for Mondale. In 1988, the final Gallup poll of Nov. 5, just 3 days before the election, predicted that Bush would defeat Dukakis by amassing a total of 53% of the popular vote. Bush actually received 54%.

Sources of Polling Bias

Over the years it has become obvious that the pollsters' mistakes, when they made them, lay in getting nonrepresentative samples. Their samples, in fact, were biased toward the Republicans. This is understandable, considering that Republicans are typically more affluent and, therefore, easier for the pollsters to get at. Republicans are more likely to live in the suburbs, or in apartment houses with elevators, or, generally, in places where the pollsters themselves feel safe. High-crime areas are more apt to trigger that ultimate pollster cop-out of going to the nearest saloon and personally filling out the ballots. This practice, known in the trade as "curb-stoning," is definitely frowned on.

When polling household members via the telephone, it is important that the pollster keep calling back until the person who was chosen gets to the phone. It has been found that when a telephone rings in a U.S. household it will be answered by women more than 70% of the time.

Even the U.S. Census Bureau recently admitted that the 1970 count underreported the number of blacks and Hispanics by a whopping 7.7%. Even worse, among black males of ages 25 to 44, the undercount was an incredible 18.2%.[3] The 1980 census has been charged with similar minority undercounts, a matter of grave concern to local government officials in several northeastern

Polling results may depend on how the respondent interprets the question. (By permission of Johnny Hart and Field Enterprises, Inc.)

states. Federal funding to cities and the reapportioning of the U.S. House of Representatives and state legislatures depend on the census figures.

There are other cases of serious problems at the U.S. Census Bureau. Back in 1920, for example, the census claimed to have counted 2.2 million American-born white children under the age of 10. Ten years later, however, the census counted almost 150,000 more than that among the same group, now aged 10 through 19. Since people are not born 10 years old, something was seriously wrong with either the 1920 figures, the 1930 figures, or both.

Recently, Shere Hite published the results of her survey of U.S. women.[4] Among the more provocative findings was Hite's assertion that 70% of America's women, married at least five years, were involved in extramarital affairs. Now, although that result may have been true of the sample, there is no way of knowing whether this sample accurately reflected the population of American women, since the sample was totally self-selected. Hite mailed out 100,000 questionnaires to women's groups, subscribers to certain women's magazines, and other groups and received only 4500 returns, or less than 5%. Could it not be that the 5% who did take the time and trouble to respond differed systematically from the 95% who did not?

As powerful as the inferential statistical tests are—and they are—nothing can compensate for faulty data. There's an old computer adage, which says "G.I.G.O.," or garbage in, garbage out. Shaky data can only result in shaky conclusions.

SAMPLING DISTRIBUTIONS

Each distribution presented so far has been a redistribution of individual scores. Every point on the abscissa has always represented a measure of *individual* performance. When we turn to **sampling distributions,** this is no longer the case.

This, then, requires a dramatic shift in thinking. In sampling distributions, each point on the abscissa represents a measure of a group's performance, typically the arithmetic average, or mean performance, for the group. Everything else remains the same. The ordinate, or Y axis, still represents frequency of occurrence, and the normal curve remains precisely as it was for all of those z score problems. The spotlight, however, now turns to the mean performance of a sample.

The Mean of the Distribution of Means

Assume that we have a large fishbowl containing slips of paper on which are written the names of all the students at Omega College—in all, 6000 students identified on 6000 slips of paper. Thus, $N_p = 6000$, or the number of individuals making up the entire population. We shake the bowl and, blindfolded, draw out 30 names. The group of 30, so selected, forms a sample (any number less than the population taken from the population). Even more important (since all 6000 names had an equal chance of selection), this group forms a random sample. Next, we search the campus, find the 30 students selected, and give each an IQ test. When we add the scores and divide by 30 ($N_s = 30$, the size of the sample group), the resulting value equals the mean performance for that sample group. It turns out to be a mean of 118 ($\bar{X}_1 = 118$). Now we can't expect a sample mean of 118 to be identical to the population mean. The concept of sampling error tells us that we must expect the sample mean to deviate from the population mean.

Is this an end in itself, or just a means to an end?

Next we go back to the fishbowl, draw out 30 more names, find the students, give the IQ tests, and again calculate the mean. This time it turns out to equal 121 ($\overline{X}_2 = 121$). We continue this process, selecting and measuring random sample after random sample until there are no more names left in the bowl. Since we started with 6000 names and drew them out 30 at a time, we select our last sample on our 200th draw.*

We thereby create a long list of sample means, 200 in all. Since each value is based on having measured a sample, it is, in fact, a list of statistics.

$$\overline{X}_1 = 118$$
$$\overline{X}_2 = 121$$
$$\vdots$$
$$\overline{X}_{200} = 112$$

We now add these mean values, divide by the number of samples, and calculate a kind of supermean, the overall mean of the distribution of sample means, or $\overline{X}_{\bar{x}}$.

$$\overline{X}_{\bar{x}} = \frac{\Sigma \overline{X}}{N}$$

This value is, and must be, a parameter. Although it was created by measuring successive *samples*, it required measuring *all* the samples, and therefore the entire population, to get it. That is, the mean of the distribution of means, $\overline{X}_{\bar{x}}$, is the same value that would have been obtained by measuring each of the 6000 subjects separately, adding their IQs, and dividing by the total N. The mean of the distribution of sample means is equal to the parameter mean μ.

$$\overline{X}_{\bar{x}} = \mu$$

Therefore, if we actually followed the procedure of *selecting successive samples* from a given population until the population was exhausted, we would, in fact, have the parameter μ, and there would be nothing left to predict. Why, then, do we go through such a seemingly empty academic exercise? Because later we will be predicting parameters, and it is crucial that you have a clear idea of exactly what values are then being inferred.

We now have a measure of central tendency in the mean of the sampling distribution of means, and we must next get a measure of variability.

---

*Technically, to create a sampling distribution, we should replace each sample taken out and make an infinite number of selections. But since the concept of infinite selection is hard to grasp (and impossible to accomplish), the example is designed to give a more vivid picture of the reality of a sampling distribution. (See the box on page 127.)

Infinite versus Finite Sampling

The two sampling distributions, the distribution of means and the distribution of differences, are based on the concept that an infinite number of sample measures are selected, with replacement, from an infinite population. Because the selection process is never ending, it is obviously assumed that all possible samples will be selected. Therefore, any measures based on these selections will be equivalent to the population parameters. Because the theoretical conceptualization of infinite sampling, replacement, and populations can be rather overwhelming, this text for purposes of illustration discusses populations of fixed size (finite) and sampling accomplished without replacement. The important message that this should convey is that combining sample measures does not equal population parameters until every last measure, from every last sample, has been counted. For sample measures to be the equivalent of population measures, all possible samples must be drawn until the population is exhausted.

Although sampling distributions are, in fact, theoretical distributions based on infinite random selection with replacement from an infinite population, for practical purposes of illustration, we treat the population as though it were finite.

The noted psychologist and statistician Janet T. Spence acknowledges that although the theoretical distribution of random sample means is indeed based upon the assumption of drawing an infinite number of samples from an infinite population, "it is quite satisfactory for illustrative purposes to go through the statistical logic with a finite population and a finite number of samples.[5] Also, if the sample is large enough, or, indeed, if the population is infinite, the difference between sampling with or without replacement becomes negligible.

"I'm beginning to understand eternity, but infinity is still beyond me." (© 1973 by Sidney Harris/American Scientist.)

The Standard Deviation of the Distribution of Means

When distinguishing between samples and populations, different symbols are used for their means, $\bar{X}$ for the sample and μ for the population. We make a similar distinction regarding variability. The true standard deviation of all the sample scores is designated, as we have seen, as *SD*, whereas the actual standard deviation of the entire population is always assigned the lowercase Greek letter sigma, σ.

The calculations involved in obtaining either standard deviation are identical, since each is producing a true (not an estimated) value. The difference between them is simply a reflection of what each value is describing. The *SD*, when used in this fashion, becomes a statistic (a sample measure), and σ, because it is a population measure, is a parameter. We can, therefore, calculate the standard deviation of the distribution of means just as we did when working with distributions of individual scores. That is, we can treat the means as though they are raw scores and then use the same standard deviation equation.

$$\sigma_{\bar{x}} = \sqrt{\frac{\Sigma \bar{X}^2}{N} - \bar{X}_{\bar{x}}^2}$$

The standard deviation of the distribution of means is equal to the square root of the difference between the sum of the squared means divided by the number of means and the mean of the means squared. (Know what we mean?) Just think about it. You may never have to do it, but you should at least think through the concept so you will know what could be done. Again, to obtain this standard deviation, all the sample means in the entire population had to be included. This value, like the mean of the means, is a parameter (since the whole population, though taken in small groups, had to be measured to get it). If we obtained the actual standard deviation of the distribution of means with the method just described, we would again have achieved a parameter and would have nothing left to predict.

Comparing the Two Standard Deviations. In comparing the standard deviation of the distribution of sample means with the standard deviation of the distribution of raw scores, we find a dramatic difference. The standard deviation of means is much smaller. This is because the standard deviation of means, like all standard deviations, is measuring variability, and the distribution of means has far less variability than does the distribution of individual scores. When shifting from the distribution of individual scores to the distribution of sample means, *extreme scores are lost through the process of averaging.* When we selected our first sample of 30 students from the population of Omega College students, one or two of those subjects might have had IQs as high as 140. But when these were averaged in with the other IQs in the sample (perhaps one student was measured at only 100), the *mean* for the sample turned out to be 118. The distribution of sample means will be narrower than the distribution of individual scores from which the samples were selected. All the measures of variability will be similarly more restricted, meaning a smaller range as well as a smaller standard deviation.[6]

The Standard Error of the Mean

The standard deviation of the distribution of means is providing a measure of sampling error variability. We expect sample means to deviate from the population mean, but the precise amount of this deviation is unknown until the standard deviation is calculated. This particular standard deviation is, thus, called the **standard error of the mean,** since it expresses the amount of variability being displayed between the various sample means and the true, population mean. It is a measure of sample variability, and although it certainly could be calculated by using the method just shown (that is, by treating each sample mean as though it were a raw score and then using the standard deviation equation), it may also be found in a less cumbersome fashion.

Had we not wanted to continue selecting random sample after random sample, we could also have calculated the standard error of the mean on the basis of the entire population's distribution of raw scores. With the population of X's, and a sample of sufficient size, the equation becomes

$$\sigma_{\bar{x}} = \frac{\sigma_x}{\sqrt{N}}$$

That is, to get the parameter standard error of the mean, we could divide the standard deviation of the population's distribution of raw scores by the square root of the number of scores in the sample.

The Central Limit Theorem

That the sampling distribution of means may approach normality is a crucial consideration. If it were not the case, we could no longer utilize Table A. Perhaps you wondered why that particular curve is normal, or perhaps you just accepted it, took it on faith. It is true, and the **central limit theorem** describes this very important fact. The theorem states that when successive random samples are taken from a single population, the means of these samples assume the shape of the normal curve, regardless of whether or not the distribution of individual scores is normal. Even if the distribution of individual scores is skewed to the left, or to the right, the sampling distribution of means approaches normality, and, thus, the z score table can be used.

The central limit theorem applies only when all possible samples have been selected from a single population. Also, the distribution of sample means more closely approaches normality as the sample sizes become larger. To understand this point, assume for a moment that your samples are made up of only one person each. In this farfetched case, the distribution of sample means is identical

---

Galton and the Concept of Error

When reading about or using the term "standard error," it must be kept firmly in mind that to a statistician, the word "error" equals deviation and does not imply a mistake. Hence, a standard error always refers to a standard deviation, or to a value designating the amount of variation of the measures around the mean. It was originally called "error" by the German mathematician Karl Friedrich Gauss, the father of the normal curve. In the early nineteenth century, Gauss had used the term "error" to describe the variations in measurement of "true" physical quantities, such as the measurement errors made by astronomers in determining the true position of a star at a given moment in time. The idea behind the current use of the concept of the "standard error" was largely formulated during the latter half of the nineteenth century by Sir Francis Galton.

Galton noted that a wide variety of human measures, both physical and psychological, conformed graphically to the Gaussian, bell-shaped curve. Galton used this curve, not for differentiating true values from false ones, but as a method for evaluating population data on the basis of its members' variation from the population mean. Galton saw with steady clarity the importance of the normal curve in evaluating measures of human traits. Thus, to Galton, the term "error" came to mean deviation from the average, not falsity or inaccuracy. The greater the error among any set of measures, then, the greater their deviation from the mean, *and the lower their frequency of occurrence.* Said Galton, "there is scarcely anything so apt to impress the imagination as the wonderful form of cosmic order expressed by the Law of Frequency of Error. The law would have been personified by the Greeks and deified, if they had known of it.[7]

---

to the distribution of individual scores. Then, if the distribution of individual scores is skewed, so too is the sampling distribution. If each sample is sufficiently large, however (typically around 30 cases), then we can rest assured that our sampling distribution will be approximately normal.

The Importance of the Sampling Distribution Parameters

Why are the parameters μ and $\sigma_{\bar{x}}$ important? Because the sampling distribution of means can approach a *normal distribution,* which allows us to use the percentages on the normal curve from the z score table. This puts us in the happy position of being able to make probability statements concerning the distribution of means, just as we previously did with the distribution of individual scores. We have already done problems like the following.

Example. On a normal distribution with a mean of 118 and a standard deviation of 10, what is the probability of selecting someone, at random, who scored 125 or above?

$$z = \frac{X - \bar{X}}{SD} = \frac{125 - 118}{10} = \frac{7}{10} = .70$$

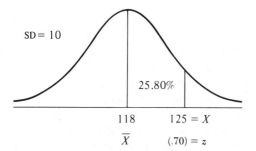

From Table A, z of .70 yields 25.80% falling between the z score and $\bar{X}$. Since 50% fall above the mean, the percentage above the z score of .70 is $50\% - 25.80\% = 24.20\%$.

$$P = \frac{\text{percentage}}{100} = \frac{24.20}{100} = .24$$

Example.   On a sampling distribution of means with a known population mean of 118 (μ) and a known standard error of the mean of 10, what is the probability of selecting a single sample whose mean ($\bar{X}$) would be 125 or above?

Since in the two examples, the values of X, $\bar{X}$, and SD are the same as $\bar{X}$, μ, and $\sigma_{\bar{x}}$, then the same probability value will result:

$$z = \frac{\bar{X} - \mu}{\sigma_{\bar{x}}} = \frac{125 - 118}{10} = .70$$

$P = .24$

Example.   The sampling distribution of means yields an overall mean weight for professional football players of 230 pounds, with a known standard error of the mean of 10 pounds. What is the probability of selecting a sample of football players whose mean weight would be 200 pounds or less?

$$z = \frac{\bar{X} - \mu}{\sigma_{\bar{x}}} = \frac{200 - 230}{10} = -3.00$$

The z score, -3.00, shows a percentage of 49.87 falling between it and the mean, leaving only 0.13% (far less than 1%) falling below that point.

$P = .0013$

Thus, there is almost no chance of finding a random sample of players whose mean weight is 200 pounds or less. The probability of selecting an individual player weighing less than 200 pounds is undoubtedly higher, but selecting a whole group of players, a random sample, that averages out in that weight category is extremely remote. This again illustrates the important point that variation among sample means is considerably less than it is among individual measures. It also illustrates the importance of random sampling when utilizing the z score table. Certainly a group of football players could be culled out, perhaps a group of wide receivers, whose mean weight would be less than 200, but that would be a biased, nonrepresentative sample of the population at large.

These examples demonstrate two important points to remember. First and foremost, you can do them. If you can work out a probability statement for a distribution of individual scores, you can just as easily do the same thing for a sampling distribution. Second, the reason for needing those two key parameters, the mean and the standard deviation of the distribution of means, becomes clear. With them, the probability values built into the normal curve are again available to us.

BACK TO z

The theory underlying parameter estimates rests solidly on the firm basis of z score probability. For example, income distributions, as stated earlier, are typically skewed to the right, not only in general, nationally across industries, but even within a single company (where the salaries of unskilled laborers are averaged in with those of the corporate president and vice-president). However, within a specific industry, and within that industry's specific job classification, worker incomes can be normally distributed. The Bureau of Labor Statistics of the U.S. Department of Labor compiles annual wage information for various metropolitan areas on earnings for standard workweeks and years. Suppose we learn that among the skilled, "component assemblers" in the computer industry, wages for that particular job classification are normally distributed around a mean of $24,000 per year, with a standard deviation of $2740. Next, we select a random sample of 30 female workers and calculate a sample mean ($\overline{X}$) of $23,000, which management is proud to proclaim is "only $1000 less than the national mean for all workers in that category." Thus, we now have four values: the mean of the population, the standard deviation of the population, the mean of the sample, and the size of the sample.

We can now calculate the standard error of the mean, since it has already been shown that

$$\sigma_{\bar{x}} = \frac{\sigma_x}{\sqrt{N}} = \frac{2740}{\sqrt{30}} = \frac{2740}{5.48} = 500$$

Because at this point we did know the population parameters and also because the distribution is known to approach normality, we can use our old friend the z equation for the distribution of means,

$$z = \frac{\overline{X} - \mu}{\sigma_{\bar{x}}} = \frac{23,000 - 24,000}{500} = -2.00$$

and assess the probability of obtaining a difference of $1000 or more in a distribution whose population mean is known to be 24,000.

Looking up the z score of -2.00, we find 47.72% of the cases falling between that point and the mean. We see, then, that the total area included by z scores of ± 2.00 is equal to 47.72% + 47.72%, or 95.44%. The area, thus, excluded by those z scores contains only 4.56% of all the sample means. This produces a probability value of less than .05 of finding a difference in either direction which is this large or larger. We say "in either direction" because, if the distribution of means is indeed normal, we could just as easily have selected a random sample mean of 25,000, or a value that would be 1000 greater than μ. If a sample is chosen randomly, its mean has the same probability of being less than μ as it has of being greater than μ. The probability is therefore extremely slight that the female workers are part of the same population whose mean is known to be $24,000.

Statisticians demand more precise terminology than this. To say that the probability is "extremely slight" is just too vague and subjective. What is needed is a clear line of demarcation, where on this side you have a sample which represents the population and on that side you don't. That line has been drawn, and, by convention, it is set at a probability value of .05 or less. Thus, if a sample mean is in an area of the curve whose frequency is 5% or less, then we conclude that it could not have come from a population whose mean was known to be $24,000—or, that the two means, sample and population, differ by too much to expect them to be part of the same normal distribution of sample means. The extreme 5% of the curve's area, remember, is that area that is excluded by z scores of ± 1.96.

Any z value, then, that falls beyond the ± 1.96 values leads to a conclusion that the mean is not part of the known population distribution. Now, this decision may not always be correct. There are, after all, 5% of the sample means which really do fall way out there under the tails of the curve.

The odds, however, still favor our decision. Although it's possible that a sample mean could deviate from the true mean by z's as large as ± 1.96, the chances are extremely remote that our particular sample mean could have been so far away from the true mean. In fact, these chances are so remote that we probably should reject the possibility. Using this type of logic, we are going to be right far more often than wrong. In fact, this limits our chances of being wrong to an exact probability value of only .05 or less. Later in this chapter and in the next, we shall see that on some occasions a researcher may wish to limit the probability of being wrong to an even more stringent .01 or less.

The z Test

When the parameter mean and standard deviation are known, as in the preceding example, we can use the **z test** to determine the probability of whether a given sample is representative of the known population. In this way we are drawing an inference about the population being represented by that sample. We knew that the overall population mean was 24,000, or that $\mu = 24,000$. We also knew that the sample mean was 23,000, and, on that basis, guessed that μ_1 might equal 23,000. As will be shown in the next chapter, this predicted value of μ_1 is called a *point estimate*. We are hypothesizing that μ_1 is the mean of the population of female workers, since that was the population we sampled from. We can, therefore, test the *chance hypothesis* (called the *null hypothesis*) that $\mu_1 = \mu$. The null hypothesis, symbolized as H_0, is stating that the mean of the population of female workers is equal to the mean of the population of workers in general. Thus, the null hypothesis is suggesting that despite the blatant fact that there is a difference of $1000 between the population mean being estimated by the sample and the known population mean of workers in general, this difference is only due to chance or random sampling error—that in reality there is no difference. The alternative hypothesis is that $\mu_1 \neq \mu$, or that the population mean of female workers is not equal to the population mean in general.

Again, look at the z:

$$z = \frac{\overline{X} - \mu}{\sigma_{\bar{x}}} = \frac{23,000 - 24,000}{500} = -2.00$$

Since the z test is yielding a value that, as we have seen, could have occurred less than 5 times in a 100 on the basis of chance, we say that we *reject* the null hypothesis and conclude that the alternative hypothesis is true—that the female worker's mean *is different from* the mean of the population at large. (In the next chapter we will learn to call this a *significant difference*.) In summary, then,

1. The null hypothesis $H_0: \mu_1 = \mu$ $\begin{cases} \text{where } \mu_1 = \text{hypothesized} \\ \text{population mean and } \mu = \\ \text{known population mean} \end{cases}$
2. The alternative hypothesis $H_a: \mu_1 \neq \mu$

3. Critical value of z at .05 $z_{.05} = \pm 1.96$
4. Calculated value of z test $z = -2.00$ reject H_0

With the z test, the null hypothesis is rejected whenever the calculated value of z is greater than ± 1.96, since its probability of occurring by chance is .05 or less. At the more stringent .01 level of probability, the critical value of z needed to reject the null hypothesis is 2.58 (the z scores that exclude the extreme 1% of the distribution).

From z to t: A Look Ahead

The concepts introduced in the previous section, such as the null hypothesis, the alternative hypothesis, critical values, and so on, will be more fully covered in the next chapter. The z test, although a perfectly valid statistical test, is not as widely used in the behavioral sciences as are some of the tests that follow. After all, how often are we really going to know the parameter values for the population's mean and standard deviation? However, the z test is still a conceptually important test for you to understand. It helps to reinforce your existing knowledge of z scores and also to lay the groundwork for that supercrucial test to be covered in the next two chapters, the *t* test.

SOME WORDS OF ENCOURAGEMENT

This chapter has taken you a long way. Some sections, as in the beginning of the chapter, read as quickly and easily as a novel. Some sections you may have to read and reread. If that happens, you should not be discouraged. Rome wasn't built in a day. We will be reading about and using these terms throughout the rest of the book. If you feel that you have yet to conquer this material completely, there is still plenty of time left. But do give it your best shot now. Memorize some of the definitions: sample, population, random sample, statistic, parameter, and standard error of the mean. The definitions are exact; they must be, because communication among researchers must be extremely precise. Do the problems now. Putting them off may make future problems look like a fun-house maze—easy to enter but almost impossible to get through.

SUMMARY

Statistical inference is essentially the act of generalizing from the few to the many. A sample is selected, presumed to be representative of the population, and then measured. These measures are called statistics. From these resulting statistics, inferences are made regarding the characteristics of the population, whose measures are called parameters. The inferences are basically educated guesses, based on a probability model, as to what the parameter values should be or should not be.

Two sampling techniques by statisticians to ensure that samples are representative of the population are (1) random—where every observation in the population has an equal chance of selection—and (2) stratified—where samples are selected so that their trait compositions reflect the same percentages as exist in the population.

Regardless of how the sample is selected, the sample mean, $\bar{X}$, must be assumed to differ from the population mean, μ. This difference, $\bar{X} - \mu$, is called sampling error. Under conditions of representative sampling, it is assumed to be random. When it is not random, when a constant error occurs in one direction, the resulting deviation is called bias. Biased sample values are poor predictors of population parameters.

When successive random samples are selected from a single population and the means of each sample obtained, the resulting distribution is called the sampling distribution of means.

The standard deviation of the sampling distribution of means is called, technically, the standard error of the mean, a value that can be found by measuring the entire population, either individually, or as members of sample groups. Since the shape of the sampling distribution of means is assumed to approximate normality (central limit theorem), z scores may be used to make probability statements regarding where specific means might fall. Finally, when the population parameters, mean and standard error of the mean, are known, the z test may be used to assess whether a particular hypothesized population mean could be part of the same sampling distribution of means whose parameter mean is known. The null, or chance, hypothesis states that the means represent the same population, while the alternative hypothesis states that they represent different populations.

Inferential statistical techniques do not provide a magic formula for unveiling ultimate truth. They do provide a probability model for making better than chance predictions, in fact, in most cases, far better.

KEY TERMS

bias	representative sample	standard error of the mean
central limit theorem	sample	statistic
parameter	sampling distributions	stratified or quota sampling
population	sampling error	z test
random sample		

REFERENCES

1. GALLUP, G., & RAE, S. F. (1968). *The pulse of democracy.* New York: Greenwood Press, p. 42.
2. Ibid., p. 42.
3. JACKSON, B. Statistical squabble. *The Wall Street Journal,* December 9, 1980, p. 24.
4. HITE, S. (1987). *Women and love, a cultural revolution in progress.* New York: Knopf.
5. SPENCE, J. T., UNDERWOOD, B. J., DUNCAN, C. P., &

COTTON, J. W. (1976). *Elementary statistics* (3rd ed.). Englewood Cliffs, NJ: Prentice Hall, p. 101.
6. ASOK, C. (1980). A note on the comparison between sample mean and mean based on distinct units in sampling with replacement. *The American Statistician, 34,* p. 159.
7. KELVES, D. J. Annals of eugenics. *The New Yorker,* October 8, 1984, p. 60.

PROBLEMS

1. On a certain standardized math test, the distribution of sample means has a known parameter mean of 50 and a known standard error of the mean of 4. What is the probability of selecting a single random sample whose mean could be 52 or greater?
2. The distribution of sample means on a standardized reading test yields a known population mean of 75 and a standard error of the mean of 5. What is the

probability of selecting at random a single sample whose mean could be 70 or lower?

3. The standard deviation of the population of individual IQ scores is 15. A given random sample of 40 subjects produces a mean of 105.
   a. Calculate the standard error of the mean.
   b. Test the hypothesis, via the z test, that the sample mean of 105 is still representative of the population whose mean is known to be equal to 100.

4. The standard deviation of the population of individual pulse rates is 5.00 beats per minute. A given random sample of 100 subjects yields a mean pulse rate of 73 beats per minute.
   a. Calculate the standard error of the mean.
   b. Test the hypothesis, via the z test, that the sample mean of 73 could be representative of a population whose mean is known to be equal to 76.

Fill in the blanks in problems 5 through 17.

5. Statistic is to sample as _____ is to population.
6. The total number of observations sharing at least one trait in common is called the _____.
7. That branch of statistics where estimates of the characteristics of the entire group are made on the basis of having measured a smaller group is called _____.
8. Any measurement made on the entire population is called a _____.
9. Any measurement made on a sample is called a _____.
10. The difference between the sample mean and the population mean ($\bar{X} - \mu$) is called _____.
11. A method of sampling in which every observation in the entire population has an equal chance of being selected is called _____.
12. Every point on the abscissa of a sampling distribution of means represents a _____.
13. A nonrepresentative sample is often due to _____, which is constant sampling error in one direction.
14. If successive random samples are taken from the population, the standard deviation of the resulting distribution is called _____.
15. The central limit theorem states that the shape of the entire distribution of sample means will tend to approximate _____.
16. There are two major assumptions of the central limit theorem: _____ and _____.
17. When samples are selected randomly from a single population,
    a. The probability of selecting a sample whose mean is higher than the population mean is _____.
    b. The probability of selecting a sample whose mean is lower than the population mean is _____.

True or False—Indicate either T or F for problems 18 through 22.

18. Every sample measure is assumed to contain bias.
19. Every sample measure is assumed to contain sampling error.
20. If sample sizes are large, bias is eliminated.

21. The null hypothesis for the z test states that the hypothesized population mean is different from the known population mean.
22. To calculate the z test, the population's parameter values must be known.

For each of the terms in problems 23 through 28, indicate whether it is a statistic or a parameter.

23. Sample mean ($\bar{X}$).
24. Standard error of the mean ($\sigma_{\bar{x}}$).
25. Population mean (μ).
26. Standard deviation of the entire sampling distribution of means.
27. Standard deviation of the sample.
28. The range of sample scores.

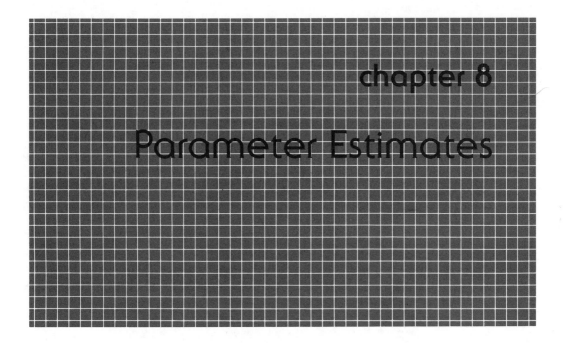

chapter 8

Parameter Estimates

As we have now seen, knowing the mean and the standard error of the sampling distribution of means is of critical importance. It allows us to use the z score table, which in turn permits probability statements regarding specific samples. We have also seen, however, that to obtain the mean and the standard deviation of this important distribution, we must select every last sample and measure each individual in all the samples, in short, measure the entire population. Whenever this is done, the resulting values are the parameters. When the parameters are known, there is nothing left to predict—no need for inferential statistics.

The job of the statistician, then, is to estimate the parameters, to predict their values *without measuring the entire population.* This is done by measuring a sample, calculating the resulting values called statistics, and using the statistics for inferring the parameters.

ESTIMATING THE POPULATION STANDARD DEVIATION

Up to this point, whenever we had to calculate a standard deviation, either from a set of sample scores, *SD*, or population scores, σ, there was really no problem. We simply squared the scores, added the squares, divided the sum by *N*, subtracted the square of the mean, and extracted the square root.

$$SD \text{ or } \sigma = \sqrt{\frac{\Sigma X^2}{N} - \bar{X}^2}$$

This equation, you may recall from Chapter 3, was derived from the deviation equation.

$$SD \text{ or } \sigma = \sqrt{\frac{\Sigma x^2}{N}}$$

where $x = X - \bar{X}$.

The result, in either case, was the true value of the standard deviation for that particular distribution of scores. However, when attempting to estimate the standard deviation of a population, you are simply not in possession of all the scores (if you were, there would be nothing to estimate).

In fact, in obtaining an **estimated standard deviation** not only do you not have all the scores, you don't even have the exact value of μ, the population mean. This does present a problem. Why? Because when calculating the standard deviation, had you used any value other than the mean, the result would have taken on a somewhat higher value. For example, in the following distribution, the deviation method will produce a standard deviation of 2.83:

X	x	x^2
10	4	16
8	2	4
6	0	0
4	-2	4
2	-4	16
$\Sigma X = 30$		$\Sigma x^2 = 40$

$$SD = \sqrt{\frac{\Sigma x^2}{N}} = \sqrt{\frac{40}{5}} = \sqrt{8} = 2.83$$

Suppose, however, that instead of correctly subtracting the mean of 6 each time, you had incorrectly chosen a different value, say, 5, or any other number you may choose:

X	x	x^2
10	5	25
8	3	9
6	1	1
4	-1	1
2	-3	9
		$\Sigma x^2 = 45$

$$SD = \sqrt{\frac{45}{5}} = \sqrt{9} = 3.00$$

The result is showing a higher value, a value which is indicating more variability than this sample of five scores actually possesses. The problem, therefore, is that when estimating a population standard deviation on the basis of a sample mean, the resulting predicted value is in all likelihood too low, since the population mean is almost certainly a value different from the mean of the sample. Because of sampling error, we expect the sample mean to deviate somewhat from the mean of the population. The conclusion, then, is that the standard deviation of the set of sample scores systematically underestimates the standard deviation of the population and is therefore called a biased estimator. To correct for this flaw, statisticians have worked out a special equation for estimating the population standard deviation, called s, the **unbiased estimator:**

$$s = \sqrt{\frac{\Sigma x^2}{N - 1}}$$

By reducing the denominator by a value of 1, the resulting value of the standard deviation is slightly increased. Using the previous distribution, we now get

X	x	x^2
10	4	16
8	2	4
6	0	0
4	−2	4
2	−4	16
$\Sigma X = 30$		$\Sigma x^2 = 40$

$$s = \sqrt{\frac{\Sigma x^2}{N - 1}} = \sqrt{\frac{40}{4}} = \sqrt{10} = 3.16$$

Since, as was pointed out in Chapter 3, the deviation method can become unwieldy when there are a large number of scores and/or the mean is something other than a whole number, the following computational formula has been derived:

$$s = \sqrt{\frac{\Sigma X^2 - (\Sigma X)^2/N}{N - 1}}$$

※ USE FOR
Sample SD.

Before plunging into this equation, it is well to remember that ΣX^2 and $(\Sigma X)^2$ are defining two separate operations. [Whereas ΣX^2 is asking you to square the scores, and then add the squares, $(\Sigma X)^2$ is asking that you add the scores and then square the sum.

X	X^2
10	100
8	64
6	36
4	16
2	4
30	220

$$s = \sqrt{\frac{\Sigma X^2 - (\Sigma X)^2/N}{N-1}} = \sqrt{\frac{220 - 30^2/5}{5-1}} = \sqrt{\frac{220 - 900/5}{4}} = \sqrt{\frac{220 - 180}{4}}$$

$$= \sqrt{\frac{40}{4}} = \sqrt{10} = 3.16$$

Thus, whereas the sample standard deviation for these same data had yielded a value of 2.83, the unbiased estimated standard deviation produces the slightly higher value of 3.16. As the sample size increases, however, the effect of the $N - 1$ factor diminishes markedly. With 50 or 60 scores, the two techniques produce very similar answers. This is, after all, as it should be, since the greater the number of random sample scores, the greater the likelihood of the sample mean coming ever closer to the population mean. This would obviously lead us to expect less of a discrepancy between the sample and population standard deviations.

One note of caution is now in order. When checking the calculated value of the estimated population standard deviation, we must reexamine our old rule (Chapter 3) that it never exceed a value of half the range. An estimated standard deviation may, when taken from very small samples, be slightly higher than half the range of the sample scores. After all, the fact that the variability in the population is likely to be larger than the variability among the scores found in tiny samples is the very reason for using this estimating technique in the first place. Thus, although the estimated standard deviation of the population scores may, at times, be somewhat greater than one-half the range of *sample scores* (among extremely small samples, as much as 72% of the sample range), it is still never greater than one-half the true range existing in the population.

ESTIMATING THE STANDARD ERROR OF THE MEAN

The **estimated standard error of the mean** is a statistic that allows us to predict what the standard error of the distribution of means would be if we had measured the entire population. Since it is a statistic, the estimated standard error of the mean can be calculated on the basis of the information contained in the sample, namely, its variability and its size.

Using the Estimated Standard Deviation

The estimated standard error of the mean is equal to the estimated population standard deviation divided by the square root of the sample size. (The letter $s_{\bar{x}}$ is used here for the estimated standard error of the mean so as not to confuse it with the Greek letter $\sigma_{\bar{x}}$, which represents the parameter.) Thus,

$$s_{\bar{x}} = \frac{s}{\sqrt{N}}$$

Example. A random sample of 10 factory workers ($N_s = 10$) is selected and measured in terms of their hourly pay. The amount each earned ranged from $10.00 an hour to $4.00 an hour.

X	X²
10	100
9	81
8	64
7	49
7	49
7	49
7	49
6	36
5	25
4	16

$\Sigma X = 70 \qquad \Sigma X^2 = 518$

$$\bar{X} = \frac{\Sigma X}{N} = \frac{70}{10} = 7.00$$

$$s = \sqrt{\frac{\Sigma X^2 - (\Sigma X)^2/N}{N-1}} = \sqrt{\frac{518 - 70^2/10}{10-1}}$$

$$s = \sqrt{\frac{518 - 4900/10}{9}} = \sqrt{\frac{518 - 490}{9}} = \sqrt{\frac{28}{9}}$$

$$s = \sqrt{3.11} = 1.76$$

$$s_x = \frac{s}{\sqrt{N}} = \frac{1.76}{\sqrt{10}} = \frac{1.76}{3.16} = .56$$

Using the Sample Standard Deviation: An Alternative Route

If your instructor prefers the use of *SD*, the actual standard deviation of the sample, the equation for the estimated standard error of the mean becomes the following:

$$s_{\bar{x}} = \frac{SD}{\sqrt{N-1}}$$

Example. The wage distribution problem in the preceding example is then calculated as follows for $N = 10$:

X	X^2
10	100
9	81
8	64
7	49
7	49
7	49
7	49
6	36
5	25
4	16
$\Sigma X = 70$	$\Sigma X^2 = 518$

$$\bar{X} = \frac{\Sigma X}{N} = \frac{70}{10} = 7.00$$

$$SD = \sqrt{\frac{\Sigma X^2}{N} - \bar{X}^2} = \sqrt{\frac{518}{10} - 7.00^2}$$

$$SD = \sqrt{51.80 - 49.00} = \sqrt{2.80} = 1.67$$

$$s_{\bar{x}} = \frac{SD}{\sqrt{N-1}} = \frac{1.67}{\sqrt{10-1}} = \frac{1.67}{\sqrt{9}} = \frac{1.67}{3} = .556$$

$$s_x = .56$$

Even though the actual standard deviation of the sample is a biased estimator, in this equation the correction is made by subtracting the value of 1 in the denominator of $s_{\bar{x}}$, allowing the estimated standard error of the mean to remain unbiased.

Either way, regardless of the alternative chosen, we obtain the same value for the estimated standard error of the mean.

The Sample and Estimated Standard Deviations

Statisticians traditionally use the English letters S, SD, or s for the standard deviation and $\bar{X}$ (sometimes M) for the mean when referring to sample measures (statistics), but use the Greek letters σ (standard deviation) and μ (mean) to represent population measures (parameters). Mathematically, $\bar{X}$ and μ create no problem, since they are calculated the same way: adding up sample scores and dividing by N yields $\bar{X}$; adding up all the scores in a population and dividing by N yields μ. With the standard deviation, however, the situation becomes somewhat sticky. In Chapter 3, the deviation technique was presented for calculating the standard deviation, and it was shown that SD could be found by taking the square root of the sum of the deviation scores squared, divided by N.

$$SD = \sqrt{\frac{\Sigma x^2}{N}}$$

This is the actual standard deviation of the sample. We also saw that this equation was algebraically the same as the raw score computational method. Thus,

$$SD = \sqrt{\frac{\Sigma x^2}{N}} = \sqrt{\frac{\Sigma X^2}{N} - \bar{X}^2}$$

These equations can also be used for obtaining σ, the standard deviation of the population (if all the scores in the population are available).

$$\sigma = \sqrt{\frac{\Sigma x^2}{N}} = \sqrt{\frac{\Sigma X^2}{N} - \mu^2}$$

Again, there is no problem so far, but when using the sample scores to estimate the standard deviation of the population, a slight correction is made to produce an unbiased estimate. The estimated population standard deviation, s, is found by taking the square root of the sum of the deviation scores squared, divided by $N - 1$. This method produces a slightly higher value (although with large samples, the difference is negligible).

$$s = \sqrt{\frac{\Sigma x^2}{N - 1}} = \sqrt{\frac{\Sigma X^2 - (\Sigma X)^2/N}{N - 1}}$$

When using s to obtain the estimated standard error of the mean, the equation becomes

$$s_{\bar{x}} = \frac{s}{\sqrt{N}}$$

Or, when using SD, where the 1 has not yet been subtracted, use the following:

$$s_{\bar{x}} = \frac{SD}{\sqrt{N - 1}}$$

In either case, the values obtained are identical, since $SD/\sqrt{N - 1}$ is algebraically identical to $s/\sqrt{N}$.

If you wish to make conversions, use

$$s = SD\sqrt{\frac{N}{N - 1}}$$

and

$$SD = s\sqrt{\frac{N - 1}{N}}$$

In a recent discussion of the two versions of the standard deviation, it was suggested that SD has the intuitive appeal of being the square root of an average taken on n quantities and because of this is used in a number of Statistics text books.[1] The use of s, however, is justified in hypothesis testing, since each of the quantities being averaged is the difference between a data point and the mean and thereby reduces the degrees of freedom by 1. Also s^2 can be used immediately as an unbiased estimator of the population variance.

Estimating the Population Mean

With the estimated standard error of the mean safely in hand, we can now turn to the problem of inferring the population mean, or the true mean of the entire distribution of means.

POINT ESTIMATES

To predict where the true mean of the population, μ, might be, a point estimate may be made, where a single sample value is used to estimate the parameter. A random sample is selected and measured, and the sample value can be used to infer a population value. Thus, the single inferred or predicted population value is called the **point estimate.** Due to sampling error, however, even this "best estimate" can never fully assure us of the true parameter value. Using the point estimate as a focal point, however, a range of possible mean values can be established within which it is assumed that μ is contained a certain percentage of the time. The goal is to bracket μ within a specific interval of high and low sample means. A probability value can then be calculated which indicates the degree of confidence we might have that μ has really been contained within the interval. This range of mean values is called the confidence interval, and its high and low values are called the confidence limits.

There are times when researchers wish to estimate the actual point value of the population mean, μ. To do this, a random sample is selected, and the sample mean, $\overline{X}$, is assumed to represent the population mean, μ. Thus, the researcher is hypothesizing a population parameter on the basis of a measured sample. Because of sampling error, this hypothesis may, of course, not be valid. After all, sample means do deviate from parameter means. But, because it is known (central limit theorem) that the distribution of sample means approaches normality, then most sample means are going to be fairly close to the population mean. As has been shown over and over again, under any normal curve, far more cases are going to lie near the center than out under the tails. Although the point estimate is a guess, it is, nevertheless, not a wild guess but an educated guess. Inferring a parameter value from a sample measure is not just a shot in the dark. It may be a shot which doesn't always hit the bull's eye but at least we can be sure that it is landing somewhere on the target. Why? Because the *sample mean must be one of the possible mean values lying along the entire sampling distribution of means.* Hence, a point estimate of μ is a *hypothesized parameter.* This, in fact, was precisely the same logic used in the last chapter when a parameter mean for the population of female workers was inferred on the basis of a measured sample of those workers. However, in that problem we used z scores, where the parameter mean of the population of workers was known, and the shape of the distribution was also known to be perfectly normal.

What about the real world, where the population parameters are almost never known, and why does this matter? Because when selecting a sample, there is no absolute assurance of obtaining that beautiful, bell-shaped Gaussian curve, even though the population of sample means may indeed be normal. It must by now be fairly obvious that the smaller the sample, the less the chance of obtaining normality. It was exactly for that real world of population inferences that William Sealy Gossett, whom we met in Chapter 1, constructed the t ratio. We now turn to the t distribution, a family of distributions each of which deviates from normality as a function of its sample size.

THE t RATIO

Degrees of Freedom and Sample Size

For the mean, **degrees of freedom** are based on how many values are free to vary once the mean and the sample size are set. Let's say that we have five values ($N = 5$) and a mean of 3:

$$\frac{X}{\begin{array}{c} 5 \\ 4 \\ 3 \\ 2 \\ ? \end{array}}$$

$$\Sigma X = 15 = (N)(\bar{X}) = (5)(3)$$

Once the number of values and the mean are known, then ΣX is fixed, since $(N)(\bar{X}) = \Sigma X$.

Thus, with an N set at 5 and ΣX set at 15, that last value is not free to vary. That last value must equal 1 (? = 1). So with five values, four of the five values are free to vary, but the last one is not, indicating that this distribution has $N - 1 = 4$ degrees of freedom. The larger the sample, naturally, the more the degrees of freedom and, assuming random selection, the closer the distribution gets to normality. The t ratio, unlike the z score, allows us to use smaller samples and compensate for their possible lack of normality via these degrees of freedom.

Table 8.1 Two-tail t (selected values).

df	.05	.01
1	12.706	63.657
10	2.228	3.169
30	2.042	2.750
∞	1.960	2.576 = z

The Two-Tail *t* Table

The two-tail *t* table, Appendix Table C, presents the critical values of *t* for probability values of both .05 and .01, each set for the various degrees of freedom. At the very bottom of the df (degrees of freedom) column, you'll see the sign for infinity ∞. An infinite number of df identifies a situation in which the sample size is also infinite and a distribution which assumes normality. Under such conditions, the *t* distribution approaches the standard normal *z* distribution. Therefore, the values in the .05 and .01 columns are, respectively, 1.96 and 2.58. These, you may recall, are the very same values that exclude the extreme 5% and 1% of the normal curve. This is illustrated in Table 8.1.

This means that, by chance, 95% of all *t* ratios will fall somewhere between ±1.96 and that only 5% of all *t* ratios will be more extreme than ±1.96. Note, also, that as the df are reduced, as sample sizes get smaller, the *t* ratios needed to exclude those extreme percentages of the curve get increasingly higher. For example, at 10 df we need a *t* ratio of at least ±2.23 to exclude the extreme 5% and ±3.17 to exclude the extreme 1%.

The Single-Sample *t* Ratio

Parameter Values Unknown.   Now, although the procedures that follow will *seem* very much like those used in the *z* test, there is one overwhelming difference. When the *z* was used, in Chapter 7, to compare female salaries with the population of salaries in general, we knew for certain what the population mean was for the workers at large and also the real standard deviation of that distribution. In the next example, we will know neither of these important parameters.

Example.   A researcher theorizes that the population mean among college students taking the new Social Conformity Test is a "neutral" 100. (Scores higher than 100 represent more conformity than average, and those lower indicate less conformity.) A random sample of 30 students was selected and the following results were recorded:

Sample mean:	$\bar{X} = 103$
True standard deviation of the sample:	$SD = 10.83$
Estimated standard deviation of the population:	$s = 11$
Sample size:	$N = 30$

From these values, we calculate the estimated standard error of the mean.

$$s_{\bar{x}} = \frac{s}{\sqrt{N}} = \frac{11}{\sqrt{30}} = \frac{11}{5.48} = 2.01 \quad \text{or} \quad s_{\bar{x}} = \frac{SD}{\sqrt{N-1}} = \frac{10.83}{\sqrt{29}} = \frac{10.83}{5.39} = 2.01$$

This value can now be used in the denominator of that special kind of *z* score which must be used whenever the population mean is being estimated rather than known for certain.

$$t = \frac{\bar{X} - \mu}{s_{\bar{x}}} = \frac{103 - 100}{2.01} = 1.49$$

The t ratio tells us specifically how far the sample mean deviates from the population mean in units of standard errors of the mean. With the t ratio, we will always be comparing *two* means, in this case a known sample mean with an assumed population mean.

Since the sample mean value of 103 is one of the possible means in an entire distribution of means, we may now ask what the probability is that a sample so chosen could deviate from the assumed μ of 100, in either direction, by an amount equal to 3 or more. We say "in either direction" because if the distribution of means is indeed normal, we could just as easily have selected a random sample mean of 97, or a value of 3 less than μ. When a sample is chosen randomly, its mean has the same probability of being less than μ as it has of being greater than μ.

Parameter Estimates as Hypotheses. Whenever we assume what a parameter value might be, what we are doing, in effect, is guessing or hypothesizing its value. For our conformity problem we assumed a parameter mean of 100. The null hypothesis would state that on the basis of chance, our sample mean of 103 can be readily expected as one of the possible sample means in a distribution of means whose μ is equal to 100.

H_0, the **null** or **"chance" hypothesis,** is written symbolically as

$$H_0: \mu_1 = \mu_0$$

where μ_1 represents the population mean being estimated by the sample and μ_0 is the assumed mean under the null hypothesis of the population at large. Also, there is the alternative hypothesis, H_a, which states that the sample estimate for μ_1 is some value other than that being assumed by the researcher under the null hypothesis. The alternative hypothesis is stating in this case that we cannot expect a sample mean of 103 if the true parameter mean is really equal to 100.

$$H_a: \mu_1 \neq \mu_0$$

Now, it is obvious that these two statements, H_0 and H_a, cannot both be simultaneously true. Either μ is equal to 100 or it isn't. We, thus, have to make a decision, called the statistical decision, to either reject H_0, in which case H_a is assumed to be true, or accept H_0, in which case H_a is assumed to be false. In either case, however, the statistical decision to reject or accept H_0 is a probability statement, and its accuracy can never be totally assured. But making the decision using probability values of .05 or .01 certainly tips the scales in our favor; we're going to make the correct decision far more often than the incorrect one. Our statistical decision, however, never has absolute certainty. Although not infallible, scientists believe that proof without certainty is far better than certainty without proof.

Recall that our t had a value of 1.49.

$$t = \frac{\bar{X} - \mu}{s_{\bar{x}}} = \frac{103 - 100}{2.01} = 1.49$$

Since we had 29 df, $30 - 1$, Table C tells us that the critical value of t under the .05 column is equal to 2.045, or rounded, 2.05. We then compare our calculated value of t with the table value for the appropriate degrees of freedom. If the calculated t is equal to or greater than the table value of t, the null hypothesis is rejected. In this case, since 1.49 is neither greater than nor equal to the table value, our statistical decision is to accept the null hypothesis or, in other words, to state that μ_1 does equal μ_0. We conclude that there is no real difference between the population mean being estimated by our sample and the population mean that was assumed by the researcher.

The Sign of the t Ratio

When making the t comparisons on Table C, the sign of the t ratio is disregarded. The importance of the t values on this table lies in their absolute values, since at this point we are testing whether a sample mean could have deviated by a certain amount from the population mean, not whether the sample mean was greater or less than μ, just whether it's different. A discussion of the other case, the directional or one-tail t, will be presented in the next chapter.

Writing the t Comparison

When the table value of t is taken from the .05 column, it is written as t sub .05, then (29) for the df:

$$t_{.05(29)} = \pm 2.05$$

This is read as "t at the .05 level with 29 degrees of freedom." Just below this value, the calculated value of t is written, with a statement regarding the statistical decision. Thus,

$$t_{.05(29)} = \pm 2.05$$
$$t = 1.49 \quad \text{Accept } H_0; \text{ difference is not significant.}$$

The fact that we accept H_0 indicates that our sample comes from the same population as the one whose parameter mean was assumed to be 100.

If, on the other hand, the calculated t value had been higher, say, 2.50, then we would have written

$$t_{.05(29)} = \pm 2.05$$

$t = 2.50$ Reject H_0; difference is significant.

The decision to reject, then, would lead to the conclusion that the difference between the sample mean and the assumed mean is so great that we must reject the null hypothesis. This sample mean could not, in all likelihood, have come from a distribution of means whose overall μ was indeed 100.

Sometimes the **single-sample t ratio** is used to compare sample results with a standard value which is known to be true. In testing for the Muller-Lyer illusion, for example, subjects are asked to adjust the length of one line until it appears to be equal in length to the standard line. The estimated lengths, sample values, produced by the subjects may then be compared, via the t test, with the physically true length of the standard line. If the t ratio is significant, the researcher may conclude that a perceptual illusion has indeed occurred.

Let's put things together with a fresh example. A researcher into flight safety assumes, on the basis of government-reported data, that the distribution of scores on the Federal Aviation Administration's private pilot's exam should be normally distributed around a population mean of 80. A random sample of 10 scores is obtained, and the null hypothesis that $\mu_0 = 80$ is tested. The distribution of sample scores is

X	X^2
98	9,604
85	7,225
82	6,724
81	6,561
81	6,561
80	6,400
79	6,241
79	6,241
78	6,084
64	4,096
$\Sigma X = 807$	$\Sigma X^2 = 65,737$

$$\bar{X} = \frac{\Sigma X}{N} = \frac{807}{10} = 80.70$$

Using estimated standard deviation of the population

$$s = \sqrt{\frac{\Sigma X^2 - [(\Sigma X)^2/N]}{N-1}}$$

Using true standard deviation of the sample

$$SD = \sqrt{\frac{\Sigma X^2}{N} - \bar{X}^2}$$

$$s = \sqrt{\frac{65,737 - [(80.70)^2/10]}{9}}$$

$$SD = \sqrt{\frac{65,737}{10} - 80.70^2}$$

$$s = 8.25$$

$$SD = 7.82$$

$$s_{\bar{x}} = \frac{s}{\sqrt{N}} = \frac{8.25}{\sqrt{10}} = 2.61$$

$$s = \frac{SD}{\sqrt{N-1}} = \frac{7.82}{\sqrt{9}} = 2.61$$

$$t = \frac{\bar{X} - \mu}{s_{\bar{x}}} = \frac{80.70 - 80}{2.61} = \frac{.70}{2.61} = .27$$

$$t_{.05(9)} = \pm 2.26$$

$t = .27$ Accept H_0; difference is not significant.

Significance

An extremely important statistical concept is that of **significance.** The research literature has many statements like "the means were found to be significantly different," or "the correlation was determined to be significant," or "the frequencies observed differed significantly from those expected." *Do not read too much* into the term "significance." Because of the connotations of the English language, the research reader sometimes assumes that anything that has been found to be significant must necessarily be profound, heavy, or fraught with deep meaning. The concept of significance is really based on whether or not an event could reasonably be expected to occur strictly as a result of chance. If we decide that we can exclude chance as the explanation, we say that the event is significant. If we decide that the event is the result only of chance, it is considered not significant. There are, therefore, many significant conclusions in the research literature that are not especially profound; some may even be trivial. It is necessary to read the word significant as probably not due to chance.

THE ALPHA ERROR

Researchers are aware that the decision to reject the null hypothesis, or chance, involves a degree of risk. After all, in the area of inferential statistics, chance never precludes anything; it positively insists that somewhere, sometime, someplace some very freaky and far-out things will indeed occur.

Currently, some research writers refer to certain rare events as having "defied chance." This is an erroneous way to phrase the slight probability concept, for in inferential statistics nothing can be said to defy chance. One researcher in the area of extrasensory perception wrote that a certain subject guessed 10 cards correctly out of a deck of 25 cards of 5 different suits and that chance could not possibly explain this phenomenal string of successes. Chance, in this instance, predicts that only 5 cards should be correctly identified. While it is

"Strange events permit themselves the luxury of occurring"—Charlie Chan.
(Springer/Bettmann Film Archive.)

true that this particular subject did beat the chance expectation, this is certainly no reason to suggest that chance has been defied. The probability of winning the Irish Sweepstakes is far lower than the probability of guessing 10 out of 25 cards, but someone does win it, probably without the aid of any occult forces. The odds against any four people being dealt perfect bridge hands (13 spades, 13 hearts, 13 diamonds, and 13 clubs) are, indeed, astronomical, being something on the order of 2,235,197,406,895,366,368,301,599,999 to 1, and yet such bridge hands do occur.*

In inferential statistics, chance never totally precludes anything. Aristotle once said that it was probable that the improbable must sometimes occur. Or as the famous Chinese detective, philosopher, and critic of the human condition Charlie Chan put it, "Strange events permit themselves the luxury of occurring." In short, any time we, as researchers, reject the possibility of chance (the null hypothesis), we must face the fact that we may be wrong—that the difference really was due to chance after all. Because of this, the researcher *always* prefaces the decision to reject chance with an important disclaimer. This is called the **alpha error** (or the Type 1 error), and it is defined as the probability of being wrong whenever the null hypothesis is rejected. That is, since we can never be absolutely

---

*These odds assume a random shuffle and dealing of the cards, a condition that is not usually met by most card players. Statisticians tell us that it takes at least seven of the standard riffle shuffles to get a good mix, whereas the average card player shuffles only three or four times.

certain that the rejection of chance as the controlling factor is a correct decision, we must add a probability statement specifying the degree of risk involved.

The Levels of Alpha Error

Two levels of alpha error are typically used, .05 and .01. These levels have been established by convention, and the choice of which one to use is a function of both the researcher's personality and the area of research. Some researchers are so cautious that they never reject the null hypothesis unless the probability of their being wrong is only .01 or less. Also, in some areas of medical research, in testing drugs in which a patient's life might be at stake, it is important to limit the alpha error to as small a value as possible. In most social science studies, however, a level of .05 for the alpha error is considered to be stringent enough. One should probably not set the acceptable level of alpha error higher than .05. A line has to be drawn somewhere, and .05 is usually that line. To make this clear, suppose that researchers were free to disregard that line. Someone could then do a study and decide to reject the null hypothesis with an alpha error set at .50. That is, the decision then would be to reject chance with the probability of being wrong a rousing 50 chances out of 100. The researcher in this obviously extreme example could have saved the time and trouble of doing the experiment by simply flipping a coin.

"It is probable that the improbable must sometimes occur"—Aristotle. (Brown Brothers.)

Remember, the alpha error by definition applies only when the null hypothesis has been *rejected.* If the null hypothesis is accepted, the alpha error is irrelevant. In a later chapter, the concept of beta error (or Type 2 error), will be introduced. Beta error is the probability of being wrong when the null hypothesis is accepted and is an important concept in determining the power of a test.

In each of the previous examples, the statistical decision was to accept the null hypothesis. Because of this, we have not yet had to be concerned with the alpha error, since, by definition, the alpha error only becomes an issue when H_0 is rejected. In short, we have not yet had to deal with a significant result.

The Case of a Significant Difference

Assume that the blood cholesterol levels for the population of U.S. adult males are normally distributed around a mean of 200. A random sample of 30 vegetarian adult males was selected from throughout the country, and their blood cholesterol levels were measured. The sample produced a mean of 180, with an estimated population standard deviation of 35 (or a true sample standard deviation of 34.42). We now address the question of whether this sample of vegetarians is representative of the population of adult males in general.

First, calculate the estimated standard error of the mean.

From the estimated population standard deviation: *or*   From the true standard deviation of the sample:

$$s_{\bar{x}} = \frac{s}{\sqrt{N}} = \frac{35}{\sqrt{30}} = \frac{35}{5.48} = 6.39 \qquad s_{\bar{x}} = \frac{SD}{\sqrt{N-1}} = \frac{34.42}{\sqrt{29}} = \frac{34.42}{5.39} = 6.39$$

Second, calculate the one-sample t ratio.

$$t = \frac{\bar{X} - \mu}{s_{\bar{x}}} = \frac{180 - 200}{6.39} = \frac{-20}{6.39} = -3.13$$

The null hypothesis states that there is no difference between the population mean being represented by the sample and the assumed mean of the population at large.

$$H_0: \mu_1 = \mu_0$$

where μ_1 represents the estimated population mean of the sample of vegetarians and μ_0 is the assumed population mean at large.

The alternative hypothesis states that there is a difference:

$$H_a: \mu_1 \neq \mu_0$$

Looking up the critical values of t on Table C, we find that for 29 degrees of freedom (df = $N - 1$), the rounded t value is 2.05 for an alpha error of .05 and 2.76

for an alpha error of .01. Since our obtained value of t is greater than either of these (remember to compare the absolute values), we will use the t value for the .01 level and reject the null hypothesis.

$$t_{.01(29)} = \pm 2.76$$
$$t = -3.13 \quad \text{Reject } H_0; \text{ significant at } P < .01.$$

The latter expression, $P < .01$, states that the alpha error has a probability value of less than .01 or that there is less than one chance in a hundred of our being wrong in rejecting the null hypothesis. We conclude, therefore, that, regarding blood cholesterol levels, this sample of vegetarian men is not representative of the population of men at large.

INTERVAL ESTIMATES

As stated earlier, population means may be inferred by setting up a range of values within which there can be some degree of confidence that the true parameter mean is likely to fall. That is, once the sample mean has been used to predict that μ is equal to, say, 100 (the point estimate), the interval estimate then suggests how often μ will fall somewhere between two values, say 95 and 105. In this case, the two values, 95 and 105, would define the limits of the interval, and are, therefore, called confidence limits. Predictions of **confidence intervals** are probability predictions. A confidence interval having a probability of .95, for example, brackets an area of the sampling distribution of means which can reasonably be assumed to contain the population mean 95 times out of every 100 predictions. Similarly, a confidence interval set at a probability level of .99 (a somewhat wider interval) can be assumed to contain the parameter mean 99 times out of every 100 predictions.*

Since the parameter may only be inferred in this case, not known for certain, we can never determine whether any particular interval contains it, as tempting as that prospect might be.

Confidence Intervals and Precision

Of course, the larger the predicted range of mean values, the more certain we can be that the true mean will be included. For example, if we are trying to predict the average temperature for the next Fourth of July, we can be very sure of being correct by stating limits of 120°F on the high side and 0°F on the low side. While we may feel very confident about this prediction, we are not being precise at all. This is not the kind of forecast to plan a picnic around.

*Although technically the probability of a parameter falling within a given interval is .50 (it either does or it doesn't), the concept of "credible intervals" allows for the definition of a probability value for a given parameter being contained within a calculated interval. For a further discussion of this technical issue, see Hays and Winkler.[3]

The same holds true for predicting the parameter mean—the more the certainty of being right, the less precise the prediction. Typically, statisticians like to be right at least 95% of the time, which gives a prediction a 95% confidence interval. Some statisticians demand a confidence interval of 99%, but as we shall see, this creates a broader, less precise range of predicted values. Confidence and precision must be constantly traded off, the more we get of one, the less we get of the other.

Also, the larger the sample size, the narrower becomes the confidence interval for a given set of values and, thus, the more precise the eventual prediction. This is due to the fact that as the sampling distribution approaches normality, the area out under the tails of the curve becomes smaller.

Calculating the Confidence Interval

The procedure involved in calculating a confidence interval is actually very similar to that used for the z score problems in Chapter 6, under the heading "inclusion area." You have already learned how to handle the following type of problem.

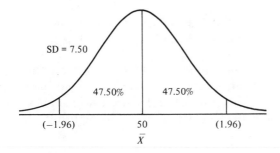

The mean on a normal distribution is 50 with an *SD* of 7.50. Which two raw scores include the middlemost 95% of all scores?

$$X = zSD + \bar{X}$$
$$X_1 = (1.96)(7.50) + 50 = 14.70 + 50 = 64.70$$
$$X_2 = (-1.96)(7.50) + 50 = -14.70 + 50 = 35.30$$

Now, using the same general procedures, consider the following example.

Example. With a sample of 121 cases, a sample mean of 7.00, and a predicted standard error of the mean of .56, between which two mean values can the true mean be expected to fall 95% of the time? Since these are sample data, we must first find the *t* value that corresponds to the *z* of ±1.96, that is, the *t* values that include 95% of all the cases. Table C shows that for 120 df ($N - 1$), the *t* ratio, under .05 on the table, which excludes the extreme 5%, and therefore includes the middlemost 95%, is equal to ± 1.98. Because the sample is so large, this particular t value is extremely close to the 1.96 value found under the normal curve.

The predicted population mean, μ, falls within an interval bounded by the product of the table value of t times the estimated standard error of the mean plus the mean of the sample. The reason we use the mean of the sample is that it is the point estimate of μ.

$$\text{confidence interval} = \pm t s_{\bar{x}} + \bar{X}$$

In this equation we don't have the population parameters, so we can't use z or the population standard error of the mean, $\sigma_{\bar{x}}$, but we do have the next best thing to these parameters. We have their estimates, t and $s_{\bar{x}}$. Using the previous data, we have

$$\text{at .95 and 120 df, } \mu = (\pm 1.98)(.56) + 7.00$$

$$\text{upper-limit } \mu = (+1.98)(.56) + 7.00 = 1.11 + 7.00 = 8.11$$

$$\text{lower-limit } \mu = (-1.98)(.56) + 7.00 = -1.11 + 7.00 = 5.89$$

Thus, we are estimating that 95% of the time, the true mean value will fall within an interval bounded by 8.11 on the high side and 5.89 on the low side.

A frequency of 95% generates a probability value of .95. Therefore, at a probability level of .95, the confidence interval is between 5.89 and 8.11. If we want to predict at the 99% confidence interval, we simply extend the range by going farther into the tails of the distribution. For the next two examples, remember that two mean values must be determined in order to establish the limits for each interval.

Example.   Find the .95 confidence interval for the following sample measures. A random sample of 25 persons yields a mean of 200 and an $s_{\bar{x}}$ of 2.86.

$$\text{at .95 and 24 df, } \mu = (\pm 2.06)(2.86) + 200$$

$$\text{upper-limit } \mu = (+2.06)(2.86) + 200 = \quad 5.89 + 200 = 205.89$$

$$\text{lower-limit } \mu = (-2.06)(2.86) + 200 = -5.89 + 200 = 194.11$$

The limits for the .95 confidence interval are, therefore, 194.11 and 205.89.

Example.   Find the .99 confidence interval for the following sample measures. A random sample of 41 persons yields a mean of 72 and an $s_{\bar{x}}$ of 1.39.

$$\text{at .99 and 40 df, } \mu = (\pm 2.70)(1.39) + 72$$

$$\text{upper-limit } \mu = (+2.70)(1.39) + 72 = \quad 3.75 + 72 = 75.75$$

$$\text{lower-limit } \mu = (-2.70)(1.39) + 72 = -3.75 + 72 = 68.25$$

The limits for the .99 confidence interval are, therefore, 68.25 and 75.75.

We are able to use our sample mean in the confidence interval equation because, since it is a sample mean, we know that it is one of the possible values in the distribution of sample means. Further, since the central limit theorem states that the distribution of sample means is normal, we know that most of the values in the distribution fall close to the true mean. It is, of course, possible that our particular sample mean is one of those rarities that falls way out in one extreme tail of the distribution. It is possible but not very probable. That, after all, is the main point of inferential statistics—estimates are made on the basis of probability, not on the basis of proclaiming eternal truth. Thus, although we are at a .95 or .99 confidence level about our prediction of the true mean, there is always the possibility of being wrong. When we are at the .95 confidence level, we will be wrong at a probability level of .05, and when we are at the .99 confidence level, obviously we will only be wrong at a probability level of .01. There's no sure thing in statistics, but with chances like these, it sure beats flipping a coin.

Confidence and "the Long Run"

As we talk about being right or wrong regarding the various confidence intervals, it must be remembered that sample means and estimated standard errors are statistics—sample measures—and as such are subject to random variation. If we were to draw a second or third or fiftieth sample, we would expect, due to sampling error, to obtain slightly different values. Each of these different sample values could then produce a somewhat different confidence interval. However, if we were to use a confidence level of .99, then 99% of these predicted intervals should contain μ, while 1% will not. Thus, in the words of the noted American statistician Allen Edwards, "in the long run we may expect .99 of our estimates about μ, based upon a 99% confidence interval, to be correct."[4]

The Single Sample *t* and the Confidence Interval

Choosing to spotlight the single sample *t* and the confidence interval in the same chapter was, as you may have suspected, no accident. The two procedures bear a marked similarity, and are, in fact, two sides of the same coin. With the *t* ratio, the attempt was made to determine whether a given sample mean could be representative of a population whose mean was known. With the confidence interval, a given sample mean was used as a point estimate, around which we hoped to bracket the true population mean.

Let's look back at the example of the single-sample *t* on page 155. In that problem we rejected the null hypothesis and stated that the vegetarian's sample mean of 180 could probably not come from the general population, whose mean was known to be 200. Using the data from that problem, a .95 confidence interval will be set up.

at .99 and 29 df, $\mu = (\pm 2.76)(6.39) + 180$

upper limit $\mu = +17.64 + 180 = 197.64$

lower limit $\mu = -17.64 + 180 = 162.36$

As you now can see, the known population mean of 200 *does not fall* within the .99 limits produced by the sample mean of 180. Thus, the two means, 180 and 200, probably represent different populations. Notice also that had the known population mean been within the confidence interval, that is, anywhere between 197.64 and 162.36, the single-sample *t* would have led us to accept the null hypothesis, and the two means would have been evaluated as representing the same population.

SUMMARY

Although the *SD* of the sample is indeed a true measure of the standard deviation for that particular set of scores, it does (especially with very small samples) tend to underestimate the standard deviation of the population from which the sample was drawn. Researchers correct for this by using *s*, the unbiased estimate of the population standard deviation. The true standard error of the mean (the standard deviation of the entire sampling distribution of means) may also be estimated on the basis of the information contained in a single sample, its size, and its variability. Since the central limit theorem states that under certain conditions the distribution of sample means approaches normality, then the estimated standard error of the mean may be used to make probability statements regarding where specific sample means might fall.

Methods for predicting where the true mean of the population might fall were presented and were based on the *t* distributions, a family of distributions that deviates from normality as a function of the sample size. The *t* ratio may be used for assessing the likelihood of a given sample being representative of a particular population. The null, or "chance," hypothesis states that the sample is representative of the population, whereas the alternative hypothesis states that it is not. Since, in inferential statistics, chance never precludes, the decision to reject the null hypothesis could be erroneous—in which case the alpha error has been committed. Point estimates are hypothesized parameters based on sample measures. The point estimate is a sample value which is supposed to provide for the "best" estimate of a single parameter value. Due to sampling error, however, even this "best" estimate can never fully assure us of the true parameter value. Confidence intervals attempt to predict the parameter mean by bracketing, between high and low estimates, a range of mean values within which the population mean is presumed to fall.

KEY TERMS

alpha error
confidence interval
degrees of freedom
estimated standard deviation
estimated standard error of the mean

null or "chance" hypothesis
point estimate
significance
single-sample t test
unbiased estimator

REFERENCES

1. HEFFERNAN, P. M. (1988). New measures of spread and a simpler formula for the normal distribution. *The American Statistician, 42,* pp. 100–102.
2. COWLES, M., & DAVIS, C. (1982). On the origin of the .05 level of statistical significance. *American Psychologist, 37.* 553–558.
3. HAYS, W., & WINKLER, R. (1975). *Statistics: Probability, inference and decision.* (2nd ed.). New York: Holt, Rinehart & Winston. p. 490.
4. EDWARDS, A. L. (1967). *Statistical methods.* (2nd ed.), New York: Holt, Rinehart & Winston, p. 209.

PROBLEMS

1. A random sample of 25 elementary school children was tested and yielded a sample mean mental age of 6.25 years, with an s (estimated population standard deviation) of .90 years (or a true sample SD of .88). On the basis of this information, estimate whether the hypothesized mental age mean of the population could indeed be 6.0 years. Test H_0 at the .05 level.

2. A researcher in sports medicine hypothesizes a mean resting pulse rate of 60 beats per minute for the population of long-distance college runners. A random sample of 20 long-distance runners was selected from the college track team. The mean resting pulse rate for the sample was 65 with an s of 3.10 (or a true sample SD of 3.01). On the basis of the sample information, estimate whether the population mean could be equal to 60. Test H_0 at the .05 level.

3. On a nationally standardized reading comprehension test, the norm for the population mean is designed to yield a scaled score value of 50 for the population of fifth-grade students. A random sample of 10 elementary school children was selected and given the test. Their scores were 48, 42, 55, 35, 50, 47, 45, 45, 39, and 42. Could this sample be representative of the population for whom the test was designed? Test the hypothesis at both the .05 and .01 levels.

4. Assume that a researcher discovers that the national mean on the Verbal SAT is 451. A sample of 121 seniors from a high school in an affluent suburb of Chicago yields a mean of 470 and an estimated standard error of the mean of 8.86. Could this sample be representative of the national population?

5. Based on the data from problem 4, use the sample measures and
    a. Produce a point estimate of the population mean.
    b. Construct a .95 confidence interval for this estimate.
    c. Would the national mean of 451 be included within the interval?
    d. What conclusion would you have drawn had the national mean been equal to 452.50?

6. Given a sample mean of 35, an N of 41, and an estimated standard error of the mean of .80,
   a. Calculate the .95 confidence interval.
   b. Calculate the .99 confidence interval.
7. A researcher selects a random sample from a population of truck drivers and gives them a driver's aptitude test. Their scores are 22, 5, 14, 8, 11, 5, 18, 13, 12, 12.
   a. Calculate the estimated standard error of the mean.
   b. At the .95 probability level, calculate the confidence interval.
8. A random sample of 11 female college students is selected and given the "Women's Attitude Scale" (high scores are pro-lib, low scores are anti-lib). The scores are 9, 8, 8, 4, 3, 7, 5, 6, 6, 3, 6.
   a. Calculate the estimated standard error of the mean.
   b. What is the confidence interval at a probability level of .95?
   c. What is the confidence interval at a probability level of .99?
9. A random sample of nine recreational vehicles is selected from the manufacturer's production line. Each is given a road test, and its fuel efficiency is computed. The following are the miles-per-gallon rates achieved by the vehicles: 15, 14, 7, 5, 5, 10, 12, 9, 9.
   a. Calculate the estimated standard error of the mean.
   b. Find the .95 confidence interval.
   c. Find the .99 confidence interval.

Fill in the blanks in problems 10 through 22.

10. For the single-sample t ratio, degrees of freedom are equal to _____.
11. As the degrees of freedom increase (as sample sizes get larger), the t distributions approach what other important distribution? _____
12. When the calculated value of t is equal to or greater than the table value of t for a given number of degrees of freedom, the researcher should _____ the null hypothesis.
13. The statement that the sample is representing the same parameter as that assumed for the population at large is called the _____ hypothesis.
14. To establish a "significant difference," what decision must first be made regarding the null hypothesis? _____
15. When the null hypothesis is rejected, even though it should have been accepted, the _____ error has been committed.
16. With a confidence interval of .95, the probability of not including the parameter mean is equal to _____ or less.
17. Assuming a sample of 12 subjects, what t value would be needed to reject H_0 at an alpha level of .01? _____
18. With infinite degrees of freedom, what t ratio is needed to reject H_0 at the .05 alpha level? _____
19. The range of mean values within which the true, parameter mean is predicted to fall is called the _____.
20. For a given sample size, setting the alpha error at .01 rather than .05 makes it (more or less) likely to reject the null hypothesis. _____

21. A point estimate is a hypothesized parameter value which is based on a _____ value.
22. Changing the confidence interval from .95 to .99 _____ (increases or decreases) the range of mean values within which μ is assumed to fall.

True or False—Indicate either T or F for problems 23 through 27.

23. The larger the standard deviation of the sample, for a given N, the larger is the estimated standard error of the mean.
24. For a given value for the sample's standard deviation, increasing the sample size decreases the estimated standard error of the mean.
25. Because of sampling error, a point estimate can never be guaranteed to produce the true population parameter.
26. A point estimate is a sample value which is said to provide the "best" estimate of a single parameter value.
27. In the long run we may expect .99 of our estimates, based on a 99% confidence interval, to be correct.

chapter 9

The Hypothesis of Difference

Up to this point, we have dealt exclusively with samples selected one at a time. Just as we previously learned to make probability statements about where individual raw scores might fall, we next learned to make similar probability statements regarding where the mean of a specific sample might fall. Note the emphasis of part of the definition of the estimated standard error of the mean: "an estimate made on the basis of information contained *in a single sample.*" In many research situations, however, we must select more than one sample. Any time the researcher wants to *compare* sample groups to find out, for example, if one sample is quicker, or taller, or wealthier than another sample, it is obvious that at least two sample groups must be selected.

Compared to What?

When the social philosopher and one-line comic Henny Youngman was asked how his wife was, he always replied, "Compared to what?" The researcher must constantly ask and then answer the same question. In Chapter 1, we pointed to the fallacious example used to argue that capital punishment does not deter crime. The example cited was that when pickpockets were publicly hanged, other pickpockets were on hand to steal from the watching crowd. Since pockets were picked at the hanging, so goes the argument, obviously hanging does not deter pickpocketing. Perhaps this is true, but we can still ask, "Compared to what?" Compared to the number of pockets picked at less grisly public gatherings? For such an observation of behavior to have any meaning, then, a control or comparison group is needed. We cannot say that Sample A is different from Sample B if, in fact, there is no Sample B. In this chapter, then, the focus will be on making comparisons between *pairs* of sample means. The underlying aim of the entire chapter is to discover the logical concepts involved in selecting a pair of samples and then to determine whether or not these samples can be said to represent a single population.

THE SAMPLING DISTRIBUTION OF DIFFERENCES

Again, we need that large fishbowl containing the names of all the students at Omega University, a total population of 6000 students. We reach in, only instead of selecting a single random sample, this time we select a pair of random samples. Say we select 30 names with the left hand and another 30 names with the right. Again, we give each student selected an IQ test and then calculate the mean IQ for each sample group. These values, as we have seen, are statistics, that is, measures of samples. In Group 1 (the names selected with the left hand), the mean turns out to be 118, whereas in Group 2 (selected with the right hand), the mean is 115. We then calculate the *difference* between these two sample means.

$$\bar{X}_1 - \bar{X}_2 = \text{difference}$$
$$= 118 - 115 = +3$$

Back to the fishbowl, and another pair of samples is selected. This time the mean of Group 1 is 114, and the mean of Group 2 is 120. The difference between these sample means is calculated, $114 - 120 = -6$. We continue this process, selecting pair after pair of sample means, until the population is exhausted, that is, until there are no more names left in the fishbowl. Since we started with 6000 names and drew out pairs of samples of 30 names, we end up in this instance with a long list of 100 mean difference values ($30 \times 2 \times 100 = 6000$).

SELECTION	$\bar{X}_1 - \bar{X}_2 = DIFFERENCE$
1	$118 - 115 = +3$
2	$114 - 120 = -6$
⋮	⋮
100	$119 - 114 = +5$

Each of the values in the "difference" column represents the difference between a pair of sample means. Since all pairs of samples came from a single population (the names in the fishbowl were of *all* the students at Omega University, and *only* those at Omega University), we expect that the plus and minus differences will cancel each other out. That is, the chance of Group 1 having a higher mean than Group 2 is exactly the same as that of Group 1 having a lower mean than Group 2. For any specific mean difference, then, the probability of that difference having a plus sign is identical to the probability of its sign being minus, that is, $P = .50$. After all, when selecting pairs of random samples, pure chance determines which names are included in Sample 1 and which in Sample 2.

The Mean of the Distribution of Differences

If we add the column of differences and divide by the number of differences, we can, theoretically, calculate the mean of the distribution. Furthermore, since all of the pairs of samples came from a single population, the value of this mean should approximate zero. (Since there will be about the same number of plus differences as there are minus differences, the plus and minus differences will cancel each other out.) To get this mean value of zero, however, we had to add the differences between *all* of the pairs of samples in the entire population. This mean value of zero, then, is a parameter, since to get it we had to measure the whole population. *Whenever the distribution of differences is constructed by measuring all pairs of samples in a single population, the mean of this distribution should approximate zero.*

Also, although there will be a few large negative differences and a few large positive differences, most of the differences will be either small or nonexistent; thus, *the shape of this distribution will tend toward normality.* (See Fig. 9.1.)

Note that each point on the abscissa of this distribution of differences represents a value based on measuring a pair of samples. This, like the distribution of means, is a sampling distribution, and each point on the abscissa is a statistic. The only exception to the latter fact is that the very center of the distribution, the mean, is a parameter, since *all* pairs of differences in the entire population had to be used to obtain it. It must constantly be kept in mind that to create this normal distribution of differences with a mean value of zero, all pairs of samples had to come from the same population.

The Distribution of Differences with Two Populations

Assume that we now select our samples from two different populations; that is, we have two fishbowls, one containing the names of all the students at the Albert

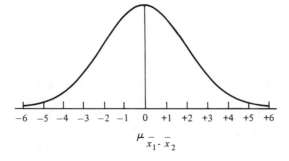

$$\mu_{\bar{x}_1 - \bar{x}_2}$$

Figure 9.1
A distribution of differences. Where $\mu_{\bar{x}_1 - \bar{x}_2}$ equals the population mean of the entire distribution of the differences between pairs of randomly selected sample means.

Einstein School of Medicine and the other the names of all the students enrolled at the Acme School for Elevator Operators. For each sample pair, Group 1 is selected from the population of medical students and Group 2 from the elevator operator trainees. It is a virtual certainty that in this case the mean IQs in Group 1 will be consistently higher than the mean IQs in Group 2. Thus,

SELECTION	$\bar{X}_1 - \bar{X}_2 = DIFFERENCE$
1	$125 - 100 = +25$
2	$120 - 105 = +15$
3	$122 - 103 = +19$
$\vdots$	
100	$130 - 100 = +30$

Obviously, the mean difference for such a sampling distribution of differences does not approximate zero, but is instead a fairly high positive value (or negative value if we had arbitrarily designated the medical students as group 2). Thus, *whenever the mean of the sampling distribution of differences does not equal zero, we can be fairly certain that the sample pairs reflect different populations.*

The Standard Deviation of the Distribution of Differences

As we have seen over and over again, whenever we have a normal distribution, the only values needed to make a probability statement are the mean and the standard deviation. To get the standard deviation of the distribution of differences, we square all the differences, add them, divide by N, the number of differences, subtract the mean difference squared, and take the square root ($\bar{x}_1 - \bar{x}_2$ designates the sampling distribution of differences).

$$\sigma_{\bar{x}_1 - \bar{x}_2} = \sqrt{\frac{\Sigma(\bar{X}_1 - \bar{X}_2)^2}{N} - \mu^2_{\bar{x}_1 - \bar{x}_2}}$$

That is, we can theoretically get the actual standard deviation of the distribution of differences in the same way that we calculated the standard deviation of a distribution of raw scores. We do this by treating all the differences as though they are raw scores and cranking them through the standard deviation equation. However, since the differences between all pairs of sample means in the entire population must be used to calculate this standard deviation, the resulting value is a parameter. Once it is known, there is nothing left to predict. This parameter is called the **standard error of difference.**

THE ESTIMATED STANDARD ERROR OF DIFFERENCE

Since it is exceedingly rare for anyone actually to measure all sample pairs in a given population, a technique has been devised for inferring the population parameters with the use of statistics. This, as has been pointed out, is really what inferential statistics is all about—measuring samples and inferring population parameters. The standard deviation of the distribution of differences can be predicted on the basis of the information contained in only two samples. The resulting value, called the **estimated standard error of difference,** can be used to predict the true parameter standard error of the entire distribution. The estimated standard error of difference is based on the information contained in just two samples, and it is used to predict the value of the standard deviation of the entire distribution of differences between the means of successively drawn pairs of random samples. Since we only measure one pair of samples to generate this value, the estimated standard error of difference is a statistic and not a parameter. The symbol for the estimated standard error of difference is $s_{\bar{x}_1 - \bar{x}_2}$. We use the statistic $s_{\bar{x}_1 - \bar{x}_2}$ to estimate the true parameter, $\sigma_{\bar{x}_1 - \bar{x}_2}$ (standard deviation of the distribution of differences). The equation for the estimated standard error of difference, based on independently selected pairs of samples, is

$$s_{\bar{x}_1 - \bar{x}_2} = \sqrt{s_{\bar{x}_1}^2 + s_{\bar{x}_2}^2}$$

The estimated standard error of difference, therefore, is based on the estimated standard errors that have been obtained from each of the two samples. As stated, then, this standard error of difference estimates the variability of the entire distribution of differences. To get it, we combine, or pool together, the two variability estimates of the sampling distribution of means.

If this sounds like a lot of estimating, it's because it is. We rarely, if ever, know the true population parameters, and the main job of the statistician is to provide educated guesses as to what those parameters *probably* are. It has even been said, tongue-in-cheek, that if you ask a statistician what time it is, the reply would be an estimate.

Example. Calculate the estimated standard error of difference for the following two independently selected samples.

Group 1: 17, 16, 16, 15, 14.
Group 2: 13, 12, 11, 10, 9.

Using the estimated population standard deviation, the procedure is as follows:

Group 1		Group 2	
X_1	X_1^2	X_2	X_2^2
17	289	13	169
16	256	12	144
16	256	11	121
15	225	10	100
14	196	9	81
$\Sigma X_1 = 78$	$\Sigma X_1^2 = 1222$	$\Sigma X_2 = 55$	$\Sigma X_2^2 = 615$

1. Find the mean for each group.

$$\bar{X}_1 = \frac{\Sigma X_1}{N_1} = \frac{78}{5} = 15.60 \qquad \bar{X}_2 = \frac{\Sigma X_2}{N_2} = \frac{55}{5} = 11.00$$

2. Find the estimated population standard deviation for each group.

$$s_1 = \sqrt{\frac{\Sigma X_1^2 - (\Sigma X)^2/N_1}{N_1 - 1}}$$

$$s_1 = \sqrt{\frac{1222 - 78^2/5}{5 - 1}} = \sqrt{\frac{122 - 6084/5}{4}}$$

$$s_1 = \sqrt{\frac{1222 - 1216.80}{4}} = \sqrt{\frac{5.20}{4}}$$

$$s_1 = \sqrt{1.30} = 1.14$$

$$s_2 = \sqrt{\frac{\Sigma X_2^2 - (\Sigma X)^2/N_2}{N_2 - 1}}$$

$$s_2 = \sqrt{\frac{615 - 55^2/5}{5 - 1}} = \sqrt{\frac{615 - 3025/5}{4}}$$

$$s_2 = \sqrt{\frac{615 - 605}{4}} = \sqrt{\frac{10}{4}}$$

$$s_2 = \sqrt{2.50} = 1.58$$

3. Find the estimated standard error of the mean for each group.

$$s_{\bar{x}_1} = \frac{s_1}{\sqrt{N}} = \frac{1.14}{\sqrt{5}} \qquad s_{\bar{x}_2} = \frac{s_2}{\sqrt{N}} = \frac{1.58}{\sqrt{5}}$$

$$s_{\bar{x}_1} = \frac{1.14}{2.24} = .51 \qquad s_{\bar{x}_2} = \frac{1.58}{2.24} = .71$$

Stop at this point and do an "is-it-reasonable?" check. Make sure that each esti-mated standard error of the mean has a smaller value than the standard deviation (found in step 2) it came from. Remember, the sampling distribution of means has less variability than does the distribution of raw scores.

4. Find the estimated standard error of difference for both groups combined.

$$s_{\bar{x}_1 - \bar{x}_2} = \sqrt{s_{\bar{x}_1}^2 + s_{\bar{x}_2}^2}$$

$$s_{\bar{x}_1 - \bar{x}_2} = \sqrt{(.51)^2 + (.71)^2} = \sqrt{.26 + .50} = \sqrt{.76} = .87$$

If the use of the sample's true standard deviation is preferred, the procedure is as follows:

Group 1		Group 2	
X_1	X_1^2	X_2	X_2^2
17	289	13	169
16	256	12	144
16	256	11	121
15	225	10	100
14	196	9	81
$\Sigma X_1 = 78$	$\Sigma X_1^2 = 1222$	$\Sigma X_2 = 55$	$\Sigma X_2^2 = 615$

1. Find the mean *for each* group.

$$\bar{X}_1 = \frac{\Sigma X_1}{N_1} = \frac{78}{5} = 15.60 \qquad \bar{X}_2 = \frac{\Sigma X_2}{N_2} = \frac{55}{5} = 11.00$$

2. Find the actual sample standard deviation for each group.

$$SD_1 = \sqrt{\frac{\Sigma X_1^2}{N_1} - \bar{X}_1^2} \qquad\qquad SD_2 = \sqrt{\frac{\Sigma X_2^2}{N_2} - \bar{X}_2^2}$$

$$SD_1 = \sqrt{\frac{1222}{5} - 15.60^2} \qquad\qquad SD_2 = \sqrt{\frac{615}{5} - 11.00^2}$$

$$SD_1 = \sqrt{244.40 - 243.36} \qquad\qquad SD_2 = \sqrt{123 - 121}$$

$$SD_1 = \sqrt{1.04} = 1.02 \qquad\qquad SD_2 = \sqrt{2} = 1.41$$

3. Find the estimated standard error of the mean for each group.

$$s_{\bar{x}_1} = \frac{SD_1}{\sqrt{N_1 - 1}} \qquad\qquad s_{\bar{x}_2} = \frac{SD_2}{\sqrt{N_2 - 1}}$$

$$s_{\bar{x}_1} = \frac{1.02}{\sqrt{5 - 1}} = \frac{1.02}{\sqrt{4}} \qquad\qquad s_{\bar{x}_2} = \frac{1.41}{\sqrt{5 - 1}} = \frac{1.41}{\sqrt{4}}$$

$$s_{\bar{x}_1} = \frac{1.02}{2} = .51 \qquad\qquad s_{\bar{x}_2} = \frac{1.41}{2} = .71$$

4. Find the estimated standard error of difference for both groups combined.

$$s_{\bar{x}_1 - \bar{x}_2} = \sqrt{s_{\bar{x}_1}^2 + s_{\bar{x}_2}^2}$$

$$s_{\bar{x}_1 - \bar{x}_2} = \sqrt{(.51)^2 + (.71)^2} = \sqrt{.26 + .50}$$

$$s_{\bar{x}_1 - \bar{x}_2} = \sqrt{.76} = .87$$

THE *t* TEST FOR INDEPENDENT SAMPLES

Everything in this chapter has been building toward the calculation of the ***t* ratio** for independent samples. This *t* test allows us to make a probability statement regarding whether or not two independently selected samples represent a single population. By independently selected samples, we mean that the choice of one sample does not *depend in any way* on how the other sample is chosen. For instance if the subjects in the two samples are matched on some relevant variable, or if the same sample is measured twice, we cannot use the independent *t* test for establishing any possible differences. (There are statistical tests available for such situations, as we shall see in Chapter 15.)

Calculating the *t* Ratio

The *t* ratio is calculated with the following equation:

$$t = \frac{\bar{X}_1 - \bar{X}_2 - \mu_{\bar{x}_1 - \bar{x}_2}}{s_{\bar{x}_1 - \bar{x}_2}}$$

That is, *t* is a ratio of the difference between the two sample means and the population mean of the entire sampling distribution of differences to the estimated standard error of that distribution.

Notice that the actual numerator of the *t* ratio is $\bar{X}_1 - \bar{X}_2 - \mu_{\bar{x}_1 - \bar{x}_2}$. Since we expect that the value of $\mu_{\bar{x}_1 - \bar{x}_2}$ (the mean of the entire distribution of differences) is really equal to zero when both samples have been selected from a single population, then our assumption is that $\mu_{\bar{x}_1 - \bar{x}_2} = 0$. Obviously, subtracting zero from the value $\bar{X}_1 - \bar{X}_2$ does not change the numerator's value at all. Nevertheless, it is still a good idea to think about that zero, at least at first, because it teaches an important lesson about the real meaning of the *t* ratio. It also allows us to view the *t* ratio in the context of something with which, by now, we do have some familiarity—the z score.

The Relationship Between the *t* Ratio and the *z* Score

We are familiar with the z score equation, which states that

$$z = \frac{X - \bar{X}}{SD.}$$

The z value tells us how far the raw score is from the mean in units of a measure of variability, or the standard deviation. Our total focus with z scores, then, is on the *raw score* and its relationship to the other values.

The equation for the *t* ratio is as follows:

$$t = \frac{\bar{X}_1 - \bar{X}_2 - \mu_{\bar{x}_1 - \bar{x}_2}}{s_{\bar{x}_1 - \bar{x}_2}} = \frac{\bar{X}_1 - \bar{X}_2 - 0}{s_{\bar{x}_1 - \bar{x}_2}}$$

Here our total focus is on the difference between the sample means and how this difference relates to the mean of the distribution of differences, also in units of variability.

What is the mean of the distribution of differences? We have already learned that it is equal to zero if both samples come from a single population. That, then, is the reason for including the zero in the numerator of the *t* ratio; zero is the expected mean of the distribution. In the z score equation, we translate our numerator difference into units of standard deviation. For the *t* ratio, since it is based on measuring samples, we do not have the true parameter standard deviation. However, we do have the next best thing—the estimated standard error of difference, which is estimating the true standard deviation of the distribution of differences. Therefore, the *t* ratio is in fact a kind of z score used for inferring parameters.

$$z = \frac{X - \bar{X}}{SD} \qquad t = \frac{(\bar{X}_1 - \bar{X}_2) - 0}{s_{\bar{x}_1 - \bar{x}_2}}$$

The *t* ratio, then, tells how far, in units of the estimated standard error of the difference, the specific difference between our two sample means deviates from the mean of the distribution of differences, or zero. We can think of it as just a plain old z score wrapped up in a new package.

The Independent *t* Ratio with Samples of Equal Size

Example. Calculate the independent *t* ratio for the following two sets of sample scores.

Group 1: 13, 12, 12, 9, 8, 8.
Group 2: 8, 8, 5, 3, 3, 2.

Group 1		Group 2	
X_1	X_1^2	X_2	X_2^2
13	169	8	64
12	144	8	64
12	144	5	25
9	81	3	9
8	64	3	9
8	64	2	4
$\Sigma X_1 = 62$	$\Sigma X_1^2 = 666$	$\Sigma X_2 = 29$	$\Sigma X_2^2 = 175$

If the use of the estimated standard deviation is preferred, the procedure is as follows:

1. Find the mean for each group.

$$\bar{X}_1 = \frac{\Sigma X_1}{N_1} \qquad\qquad\qquad \bar{X}_2 = \frac{\Sigma X_2}{N_2}$$

$$\bar{X}_1 = \frac{62}{6} = 10.33 \qquad\qquad\qquad \bar{X}_2 = \frac{29}{6} = 4.83$$

2. Find the estimated standard deviation for each group.

$$s_1 = \sqrt{\frac{\Sigma X_1^2 - (\Sigma X_1)^2/N_1}{N_1 - 1}}$$

$$s_1 = \sqrt{\frac{666 - 62^2/6}{6 - 1}} = \sqrt{\frac{666 - 3844/6}{5}}$$

$$s_1 = \sqrt{\frac{666 - 640.67}{5}} = \sqrt{\frac{25.33}{5}} = \sqrt{5.07} = 2.25$$

$$s_2 = \sqrt{\frac{\Sigma X_2^2 - (\Sigma X_2)^2/N_2}{N_2 - 1}}$$

$$s_2 = \sqrt{\frac{175 - 29^2/6}{6 - 1}} = \sqrt{\frac{175 - 841/6}{5}} = \sqrt{\frac{175 - 140.17}{5}}$$

$$s_2 = \sqrt{\frac{34.83}{5}} = \sqrt{6.97} = 2.64$$

3. Find the estimated standard error of the mean for each group.

$$s_{\bar{x}_1} = \frac{s_1}{\sqrt{N}} = \frac{2.25}{\sqrt{6}} = \frac{2.25}{2.45} = .92 \qquad s_{\bar{x}_2} = \frac{s_2}{\sqrt{N}} = \frac{2.64}{\sqrt{6}} = \frac{2.64}{2.45} = 1.08$$

4. Find the estimated standard error of difference for both groups combined.

$$s_{\bar{x}_1 - \bar{x}_2} = \sqrt{s_{\bar{x}_1}^2 + s_{\bar{x}_2}^2} = \sqrt{.92^2 + 1.08^2} = \sqrt{.85 + 1.17} = \sqrt{2.02}$$

$$s_{\bar{x}_1 - \bar{x}_2} = 1.42$$

5. Find the t ratio.

$$t = \frac{\bar{X}_1 - \bar{X}_2}{s_{\bar{x}_1 - \bar{x}_2}} = \frac{10.33 - 4.83}{1.42} = \frac{5.50}{1.42}$$

$$t = 3.87$$

Or, using the actual standard deviations of the sample, *SD*, then

1. Find the mean for each group.

$$\bar{X}_1 = \frac{\Sigma X_1}{N_1} \qquad\qquad\qquad \bar{X}_2 = \frac{\Sigma X_2}{N_2}$$

$$\bar{X}_1 = \frac{62}{6} = 10.33 \qquad\qquad\qquad \bar{X}_2 = \frac{29}{6} = 4.83$$

2. Find the standard deviation for each group.

$$SD_1 = \sqrt{\frac{\Sigma X_1^2}{N_1} - \bar{X}_1^2}$$

$$SD_1 = \sqrt{\frac{666}{6} - 10.33^2} = \sqrt{111 - 106.71}$$

$$SD_1 = \sqrt{4.29} = 2.07$$

$$SD_2 = \sqrt{\frac{\Sigma X_2^2}{N_2} - \bar{X}_2^2}$$

$$SD_2 = \sqrt{\frac{175}{6} - 4.83^2} = \sqrt{29.17 - 23.33}$$

$$SD_2 = \sqrt{5.84} = 2.42$$

3. Find the estimated standard error of the mean for each group.

$$s_{\bar{x}_1} = \frac{SD_1}{\sqrt{N_1 - 1}} \qquad\qquad\qquad s_{\bar{x}_2} = \frac{SD_2}{\sqrt{N_2 - 1}}$$

$$s_{\bar{x}_1} = \frac{2.07}{\sqrt{6 - 1}} = \frac{2.07}{\sqrt{5}} \qquad\qquad s_{\bar{x}_2} = \frac{2.42}{\sqrt{6 - 1}} = \frac{2.42}{\sqrt{5}}$$

$$s_{\bar{x}_1} = \frac{2.07}{2.24} = .92 \qquad\qquad\qquad s_{\bar{x}_2} = \frac{2.42}{2.24} = 1.08$$

4. Find the estimated standard error of difference for both groups combined.

$$s_{\bar{x}_1 - \bar{x}_2} = \sqrt{s_{\bar{x}_1}^2 + s_{\bar{x}_2}^2}$$

$$s_{\bar{x}_1 - \bar{x}_2} = \sqrt{.92^2 + 1.08^2} = \sqrt{.85 + 1.17} = \sqrt{2.02}$$

$$s_{\bar{x}_1 - \bar{x}_2} = 1.42$$

5. Find the *t* ratio.

$$t = \frac{10.33 - 4.83}{1.42} = \frac{5.50}{1.42} = 3.87$$

The Independent *t* Ratio with Samples of Unequal Size

The technique used in the preceding example for finding the *t* ratio assumes that the two samples are of equal size. When conducting research, it is best to come as close to this ideal as possible. A researcher obviously should not place 100 persons in the first group and only 1 person in the second group. However, there are times when it is simply not possible to achieve identical sample sizes. When this does occur, the following variation must be used for calculating the standard error of difference.*

Using the estimated population standard deviations, *s*, then

$$s_{\bar{x}_1 - \bar{x}_2} = \sqrt{\left[\frac{(N_1 - 1)s_1^2 + (N_2 - 1)s_2^2}{N_1 + N_2 - 2} \right] \left(\frac{1}{N_1} + \frac{1}{N_2} \right)}$$

Or, if the use of the actual standard deviation of the sample is preferred, the procedure is

$$s_{\bar{x}_1 - \bar{x}_2} = \sqrt{\left[\frac{N_1 SD_1^2 + N_2 SD_2^2}{N_1 + N_2 - 2} \right] \left(\frac{1}{N_1} + \frac{1}{N_2} \right)}$$

SIGNIFICANCE

As was the case with the single sample *t* ratio, the independent *t* for two samples will also be evaluated for **significance**. As we learned in the previous chapter, a significant difference is one for which the chance explanation has been rejected. Again don't read the word "significant" as being synonymous with "profound." Significant differences are not always especially meaningful, even though the probability is small that, for example, they could have occurred by chance.

For example, a study of personality differences among married couples indicated that "nonpossessive wives" are significantly less likely to feel their relationship has much of a future, when compared to a control group of wives judged as "possessive."[1] Could it be, as implied, that the nonpossessive wife doesn't care if her husband is unfaithful, since she feels the marriage isn't going to last anyway? Let's look at the numbers.

Among the possessive wives, 87% say that their marriages will last, whereas of the nonpossessive wives, only 73% make the same forecast. Now, this difference may be significant, but it may not tell us much. After all, almost three out of four of these so-called nonpossessive wives look forward to a successful marriage.

*Although this technique may also be used with equal sample sizes, it does tend to obscure the logic of the relationship between the standard error of difference and the standard errors of the means.

The Null Hypothesis: No Real Difference

As we have seen, statisticians use the term **null hypothesis,** symbolized as H_0, to refer to the idea that the events in question are due only to chance. A significant difference is, therefore, one for which we have excluded chance as the explanation, or one for which we have *rejected the null hypothesis*. Again, be careful here of the connotations of the English language. The word reject has a negative connotation; it sounds as though it might be associated with failure. In statistics, however, the word reject refers only to the researcher's *decision* regarding chance, or the null hypothesis. If the null hypothesis is rejected, the event is significant. If the null hypothesis is accepted, the event is due to chance and is not significant.

Suppose that we conduct a research project in which we are endeavoring to prove that a certain carefully phrased communication will change political attitudes toward a certain candidate. We test this by selecting two random samples from a population of individuals who are opposed to the candidate. Group 1 receives the special message, while Group 2 does not. Both groups are then given a questionnaire designed to measure attitudes toward the candidate. We compare the mean attitude score for each group. On the basis of the results of the *t* test, we decide to reject the null hypothesis. This is anything but failure! By rejecting chance, we have established a significant difference between the groups and, furthermore, have established that our special message does indeed work. The decision to reject, then, should not necessarily be viewed as a discouraging development. It is often the very decision the researcher is hoping to be able to make. After all, not many Nobel Prizes have been awarded for accepting the null hypothesis, that is, concluding that all the findings are simply due to chance.

The Null Hypothesis for the t *Test.*   For the independent *t* test, the null hypothesis is written as follows:

$$H_0: \mu_1 = \mu_2$$

This is read as "mu one is equal to mu two." It means that Sample 1 is representative of a population whose mean is identical with the mean of the population being represented by Sample 2. In other words, H_0 states that the two samples represent populations with the same parameter mean, μ. Even though the sample means may differ ($\bar{X}_1 \neq \bar{X}_2$), they do not differ by a large enough amount to reject the possibility that the samples are taken from a single population.

The null hypothesis is *not* stating that the sample means are the same, rather that they simply are not different enough to reject the conclusion that they represent a common population mean. After all, as we saw in the last chapter during the discussion of the sampling distribution of means, we must expect sample means selected from a single population to differ somewhat from each other. The null hypothesis is always made with reference to the population parameters, and does not refer to the sample statistics. Also, the statistical decision as to whether to accept or reject is always based on the null hypothesis.

The Alternative Hypothesis

The opposite of the null hypothesis is the **alternative hypothesis** (sometimes called the research hypothesis), symbolized as H_a. Whereas the null hypothesis states that the parameter means are equal, the alternative hypothesis states that they are different.

$$H_a: \mu_1 \neq \mu_2$$

H_a is saying that Sample 1 is representative of a population whose mean is not equal to the mean of the population being represented by Sample 2. In short, H_a states that the two samples represent *different* populations.

Thus, whenever we reject H_0, we are betting that H_a is true, and whenever we accept H_0, we are betting that H_a is false. The researcher, though always basing the statistical decision to accept or reject on the null hypothesis, in doing so, makes inferences about the alternative hypothesis. The null and alternative hypotheses can never both be true or false. Either μ_1 is equal to μ_2 or it is not. There is no in-between choice. This is not a "shades-of-gray" issue. H_0 and H_a are qualitatively different positions.

Since the alternative hypothesis for the t test states that the parameter means are different, it is also called the *hypothesis of difference*. The t test is, therefore, said to be designed to test the hypothesis of difference, even though the statistical decision uses only the null hypothesis. This is because the rejection of the null hypothesis indicates that the samples represent different populations, whereas the acceptance of the null hypothesis means that the samples represent a single population.

H_0, H_a, and the Distribution of Differences

Earlier in this chapter, we discovered that the mean of the distribution of differences is equal to zero whenever all pairs of random samples are selected from a single population. The statistical decision regarding the null hypothesis, then, goes right to the heart of this earlier discussion. Accepting H_0 tells us that the two samples in question are both taken from a single population and that the difference between the sample means is part of a normal distribution of differences with a mean value of zero. Conversely, rejecting H_0 indicates that the two samples represent different populations and that the difference between the obtained sample means is part of a larger distribution of differences whose mean is not equal to zero. The size of the t ratio is the determining factor in deciding whether the samples represent the same population or different populations. The larger the t ratio, the more likely it is that the mean of the distribution of differences is not equal to zero and that the null hypothesis can be rejected.

Degrees of Freedom for the Independent *t* Test

In Chapter 8 we learned that when finding the mean from a set of sample scores, there are $N - 1$ degrees of freedom. Since the independent *t* always compares the means from two sets of sample scores, the **degrees of freedom** in this case must equal the size of the first sample minus one, plus the size of the second sample minus one. Thus, for the *t* ratio,

$$df = N_1 - 1 + N_2 - 1$$

or, perhaps, more conveniently,

$$df = N_1 + N_2 - 2$$

This rule holds true for both equations for the *t* ratio—for equal sample sizes and for unequal sample sizes. Again, we see that the larger the sample sizes, the more degrees of freedom allowed, and the smaller the sample sizes, the fewer the degrees of freedom.

For other statistical tests, the calculation of the degrees of freedom will be different. Thus, the equation is only valid for the independent *t* test.

In general, degrees of freedom are based on the number of values in any set of scores which are free to vary once certain restrictions are set in place.

THE TWO-TAIL *t* TABLE

As we saw in Chapter 8, the two-tail *t* table, Appendix Table C, presents the critical values of *t* for alpha errors of either .05 or .01 and for the various degrees of freedom.

To use this table, we must first calculate the *t* ratio and then compare our obtained value with the table value of *t* listed beside the appropriate degrees of freedom. In an example earlier in this chapter, the calculated value of *t* was 3.87. Since the sample sizes were 6 for each group in that example, the degrees of freedom equal 10.

$$N_1 + N_2 - 2 = 6 + 6 - 2 = 10$$

The rule for the statistical decision is that if our calculated value of *t* is equal to or greater than the table value of *t*, we reject the null hypothesis.

We look down the df (degrees of freedom) column until we find 10. Then we compare the obtained value of *t*, 3.87, with the table values of *t* in that row.

The Sign of the *t* Ratio

When making the comparison on this two-tail table, we have already learned (in Chapter 8) to disregard the sign of the *t* ratio. The importance of both the calculated *t* and the table value of *t* lies in their absolute values, since it is an arbitrary decision as to whether the larger of the two mean values is placed first in the numerator. The table values for our example are 2.228, which rounds to 2.23, in the .05 column and 3.169, which rounds to 3.17, in the .01 column. Our calculated value is certainly higher than the value of 2.23, but it is also higher than the value of 3.17. We, thus, reject the null hypothesis, and state that our calculated value of *t*, 3.87, is *significant* at the .01 level. If the calculated *t* ratio were -3.87, we would also reject the null hypothesis. Remember, the table values of *t* in Table C can be compared with either plus or minus values.

Writing the *t* Comparison

When the table value of *t* is taken from the .01 column, it is written as

$$t_{.01(10)} = \pm 3.17$$

This is read as "*t* at the .01 level with 10 degrees of freedom." Just below this value, we then write our calculated value of *t*, with a statement regarding our statistical decision. Remember, if we reject the null hypothesis, the *t* ratio is significant. If we accept the null hypothesis, it is not significant. Thus,

$$t_{.01(10)} = \pm 3.17$$

$t = 3.87$ Reject H_0; significant at $P < .01$.

If our calculated value of *t* is 2.25, instead of 3.87, we have to shift to the .05 column and write

$$t_{.05(10)} = \pm 2.23$$

$t = 2.25$ Reject H_0; significant at $P < .05$.

Suppose, however, that the calculated value of *t* is equal to 1.28. Looking at the table values of *t* for 10 df, we see that our value is less than either one. In this case we write

$$t_{.05(10)} = \pm 2.23$$

$t = 1.28$ Accept H_0; not significant.

ALPHA ERROR AND CONFIDENCE LEVELS

Since **alpha error** is the probability of being wrong whenever the null hypothesis is rejected, then the alpha error must be inversely related to the level of confidence. That is, if the alpha error is .05, we have a confidence level of .95. In other words, if there are only 5 chances in 100 of being wrong, there are 95 chances in 100 of being right. Similarly, if the alpha error is at .01, the confidence level jumps to .99.

Statistical inference does not produce eternal truth, but the statistician's ability to limit the probability of being wrong and thereby to maximize the probability of being right is far more accurate than random guessing. If, however, we conduct 100 experiments and in each one reject the null hypothesis with an alpha error of .05, the cold realization will eventually dawn that in 5 of those experiments we probably are making the wrong decision. What is even more frustrating is that we don't know which 5!

The Normality of the *t* Distribution

As we saw in Chapter 8, when sample sizes are sufficiently large, *the t distribution approaches normality.* This is illustrated in Fig. 9.2. Since the *t* ratios of ±1.96 *exclude* the extreme 5% of all possible *t* ratios, any *t* ratio as extreme or more ex-

"I'm sorry, but you've been rejected at the .05 level!"

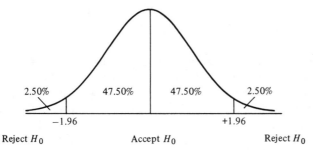

<table>
<tr><td>2.50%</td><td>47.50%</td><td>47.50%</td><td>2.50%</td></tr>
</table>

−1.96   +1.96

Reject H_0 Accept H_0 Reject H_0

Figure 9.2
The t distribution with infinite degrees of freedom; t ratios of ±1.96 exclude the extreme 5%.

treme than ±1.96 can occur by chance at a probability level of .05 or less. This is why we say that rejecting the null hypothesis at the .05 level means that the decision to reject will be wrong 5% of the time. Why? Because t values of this magnitude can occur by chance, infrequently perhaps, but they can occur.

Similarly, with t ratios of ±2.58, the extreme 1% of the curve is now being excluded, and the middlemost 99% is being included. Figure 9.3 illustrates this situation.

Just as the alpha error concept is based on z score exclusion problems, confidence is based on inclusion problems. Remember, if the t value is greater than ±2.58, the probability level for the alpha error is only .01 or less, whereas the confidence we have in our result has a probability value of .99.

The Effect of the Degrees of Freedom

Recall from Chapter 8 that as the degrees of freedom are reduced, the t ratios needed to exclude the extreme percentages of the curve get higher and higher. For example at 10 degrees of freedom, we need a t value of at least ±2.23 to exclude the extreme 5% and ±3.17 to exclude the extreme 1%. The most inflated

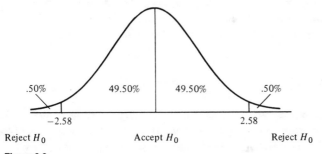

.50%   49.50%   49.50%   .50%

−2.58   2.58

Reject H_0 Accept H_0 Reject H_0

Figure 9.3
The t distribution with infinite degrees of freedom; t ratios of ±2.58 exclude the extreme 1%.

example is found for *t* values at 1 degree of freedom. At this point, a *t* value of at least ±12.71 is needed to exclude the extreme 5% and ±63.66 is needed to exclude the extreme 1%.

Every degree of freedom value is specifying a different t distribution. As the degrees of freedom are reduced, each of these *t* distributions is deviating more and more from normality. As sample sizes decrease, relatively fewer cases fall around the middle of the distribution, and relatively more cases fall out in the tails.

The more degrees of freedom there are, the more the *t* distribution approaches the *z* distribution and the easier it becomes to reject the null hypothesis. Rejection of null depends, remember, on the calculated *t* being equal to or greater than the table value of *t*. A *t* ratio of 1.96 (which rejects the null hypothesis at .05 with an infinite number of degrees of freedom) is easier to obtain for a given difference between the sample means than a *t* ratio of 12.71 (needed to reject at .05 with only one degree of freedom). Other things being equal, then, the larger the sample sizes, the more likely is the prospect of obtaining a *t* value that will turn out to be significant. This issue will be addressed again later during the discussion of "power."

THE ONE-TAIL *t* TEST

The *t* test completed earlier in this chapter was based on the assumption that the alpha error was occurring in both tails of the *t* distribution; hence, it is called a **two-tail *t* test.** The alternative hypothesis states that $\mu_1 \neq \mu_2$, that is, that the population means are *different*—not that one population mean is greater than the other, just that they are different. This is why we disregard the sign of the *t* ratio when using the two-tail *t* table, since it is irrelevant if the first mean is greater than the second mean or less.

Sometimes, however, a researcher not only assumes that a mean difference between samples will occur, but also predicts *the direction of the difference*. When this happens, the statistical decision is not based on both tails of the distribution, but only on one. In this instance, the sign of the *t* ratio is crucial. When conducted in this way, the *t* test is called **a one-tail *t* test.** *The calculation of a one-tail t test is identical to that of the two-tail t.* The only changes occur in the way the alternative hypothesis is written and in the method used for looking up the table value of *t*.

The One-Tail *t* Test and the Alternative Hypothesis

Since to use the one-tail *t* test the researcher must predict *beforehand* the direction of the difference between sample means, the alternative hypothesis, which simply states that a difference exists ($\mu_1 \neq \mu_2$), is no longer the full story. The researcher is now defining how the samples differ, for example, that Sample 2 is greater than Sample 1. The alternative hypothesis is still stated in terms of parameter means, but now the direction of the difference must be shown. The researcher will state that $\mu_1 > \mu_2$, meaning that the first sample reflects a population whose mean is greater than the mean of the population reflected by the second sample. The statement still implies that the samples represent different populations, but now the way the populations differ is also being predicted. Of course, depending on the logic behind the research, the alternative hypothesis can instead be written as $\mu_1 < \mu_2$.

Using the One-Tail *t* Table

When the direction of the mean difference is predicted, we must turn to the one-tail *t* table, Appendix Table D, to make the statistical decision as to whether or not to reject H_0. On the bottom of this table, where degrees of freedom are infinite, note the value of 1.65. We know that with infinite degrees of freedom the *t* distribution approaches the *z* distribution. A *z* score of 1.65 is exceeded by only 5% of the cases. This is pictured in Fig. 9.4.

The probability, therefore, of receiving a *z* score of 1.65 or higher is .05 or less. This is also true for the *t* ratio when the distribution is normal. But as the *t* distributions depart from normality, that is, as sample sizes decrease, we must compensate for this by utilizing the degrees of freedom found in the one-tail *t* table. Again, as degrees of freedom decrease, the value of *t* necessary to reject H_0 increases.

In the one-tail *t* table, instead of splitting the extreme 5% of the distribution into the lowest 2½% and the highest 2½% (as was the case with the two-tail *t*), we are concerned here with only the most extreme 5% *on one tail of the curve*. That is, we are now dealing with only half of the distribution.

Example. Assume that we are researching the theory that training on a balance beam increases reading achievement among learning disabled children. We randomly select 20 learning disabled children, placing 10 of them in the training group and the other 10 in a group receiving no training. Since our hypothesis stipulates an *increase* in reading achievement, the one-tail *t* test can be used in this

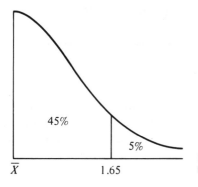

45%

5%

$\overline{X}$ 1.65

Figure 9.4
Percentage of cases falling above a *z* score of 1.65.

instance. Assume that the mean for Sample 1 (the group receiving the training) is 55 and that the mean for Sample 2 (the group that is not trained) is 50. Assume further that the t ratio is $+1.87$. (It has to be plus since $\bar{X}_1$ is greater than $\bar{X}_2$.) The degrees of freedom are found from $N_1 + N_2 - 2$, or $10 + 10 - 2 = 18$. Looking up 18 degrees of freedom on the one-tail t table, we find that the table value of t is 1.73 at the .05 alpha error level. By comparing the obtained t of 1.87 with the table value of 1.73, the null hypothesis is rejected. We can conclude that the two samples originally selected from a single population now represent two separate populations. There is indeed a significant difference. Therefore, training has been shown to increase achievement.

Negative **t** *Ratios and the One-Tail* **t** *Table.*   Since using the one-tail t table requires predicting the direction of the mean difference, we must accept the null hypothesis for all t ratios less than 1.73, including *all negative t ratios.* Even a t ratio of -3.00, which with the 18 degrees of freedom in the preceding example is more than enough to reject H_0 with the two-tail t table, produces an accept of H_0 with the one-tail t table.

If, however, we predict a negative difference between the sample means, then the one-tail t table is read as though all values are negative. Returning to the previous example, suppose that we had labeled the group not receiving the training as Sample 1 and the group receiving the training as Sample 2. We then predict a lower mean for Sample 1, or $\bar{X}_1 < \bar{X}_2$. The numerator of the t ratio would now be written as

$$\bar{X}_1(50) - \bar{X}_2(55) = -5$$

This negative numerator, of course, produces a negative t ratio. (The denominator of the t ratio can never be negative, since the estimated standard error of difference is a variability estimate, and there can never be less than zero variability.) We again assume a resulting t ratio of 1.87, but now it must be written as -1.87. With 18 degrees of freedom, the table value for the one-tail t is 1.73 at the .05 level. We therefore reject the null hypothesis. Because we are now concerned with only the negative tail of the curve, all positive values of t, no matter how high, must necessarily lack significance (cannot yield a reject of H_0).

Advantage and Disadvantage of Using the One-Tail **t** *Table: Good News and Bad News.*   Using the one-tail t table has one decided advantage, but also an important disadvantage. The advantage is that we do not have to obtain as high a t value to reject the null hypothesis as we do when using the two-tail t table. However, the other side of the coin is that t values that would have been significant on the two-tail t table, where the direction of difference is not important, are not significant on the one-tail t table—simply because they do not vary in the predicted direction.

The bottom line on the one-tail t test is as follows: when predicting a positive mean difference, only plus t ratios can possibly be significant, and when predicting a negative mean difference, only minus t ratios can possibly be significant.

Deciding Which t Table to Use.   When deciding which t table to use, one-tail or two-tail, we must be consistent. If we can logically predict the direction of the mean difference and decide to use a one-tail t, we must then stick with the one-tail table, no matter how the t ratio comes out.

For example, if we state the alternative hypothesis as $\mu_1 > \mu_2$, and then obtain a highly negative t ratio, we cannot restate the alternative hypothesis as $\mu_1 \neq \mu_2$ just to get a reject of H_0. Similarly, if we start out doing a two-tail t test with H_a as $\mu_1 \neq \mu_2$, and then obtain a t ratio that does not reach significance on the two-tail t table, we cannot suddenly switch to the one-tail t table just to get the rejection of the null hypothesis.

Because of the possibility of statistical sleight-of-hand, some statisticians believe that the one-tail t test should never be used. They insist that the only legitimate alternative hypothesis for the t test is the tried-and-true $\mu_1 \neq \mu_2$.

After all, if a t ratio is significant, but not in the predicted direction, it is still considered to be a nonchance event and should be interpreted as such. It might very well be important for other researchers to know about. These statisticians also argue that the one-tail t opens the door to the ethical problem of exactly when the researcher actually decided to make the prediction—before or after discovering that the two-tail t was not significant.

If your instructor happens to be one of those holding this opinion, then forget about the one-tail t test, and do all t problems as though they are two-tail, nondirectional t ratios. Remember, in terms of technique, the only difference between a one-tail and a two-tail t is in how we look up the significance level.

THE IMPORTANCE OF HAVING TWO SAMPLES

The full significance of Henny Youngman's question "Compared to what?" can now be more fully appreciated. When doing research on questions of difference, there must be at least two groups of subjects. If the claim is made that persons using Brand X toothpaste have 27% fewer cavities, the logical question is "compared to what?" Compared to persons that do not brush at all? Compared to persons that brush less frequently? Compared to the same group's cavity record before switching to Brand X?

"t for Two."   In Chapter 11, we will discuss in some detail the techniques researchers use to create equivalent groups of subjects. But for now, the main message when testing the hypothesis of difference is that there must be at least two sets of sample scores. Many years ago, there was a popular song-and-dance hit titled "Tea for Two." Keep that title in mind. Whenever we want to establish whether there are differences between the sample means of two independent sets of interval scores, the t test is for two. As we saw in the previous chapter, even the one-sample t ratio was comparing two means, the sample mean with a population mean.

Some Limitations on the Use of the t Test

To calculate the t ratio, both samples must have been measured *on the same trait.* We can use the t test to find out if Group 1 is taller than Group 2 or smarter than Group 2. We cannot use the t test for comparing qualitatively different measures—for comparing apples and oranges. As a blatant example (haven't they all been?), it makes no sense to compare the mean height of one group with the mean weight of another group. A statement such as Americans are heavier than they are tall is absurd. Also, we cannot use the t test to compare different measures taken on the same group. We cannot say that a sample of Republicans has more dollars than IQ points.

Requirements for Using the t Test

To use the t test, the following requirements must be met:

1. The samples have been randomly selected.
2. The traits being measured do not depart significantly from normality within the population.
3. The standard deviations of the two samples must be fairly similar.
4. The two samples are independent of each other.
5. Comparisons are made only between measures of the same trait.
6. The sample scores provide at least interval data. (Any test that utilizes interval data can also be used with ratio data.)

POWER

Up to this point our interest in the statistical decision has been riveted on the alpha error, or the probability of being wrong when rejecting the null hypothesis. This is indeed as it should be, because whenever null is rejected you are, in effect, going out on a statistical limb. By rejecting chance, you're saying that you have discovered a significant result, a public fact that may be of interest to a rather wide audience. When null is accepted, on the other hand, it may be more of a private decision. The public isn't always as interested in studies whose results can be explained on the basis of chance. But when null is accepted, another possible source of difficulty arises.

Whenever a statistical decision is made, be it an accept or a reject, there is a possibility of being wrong. Since all statistical decisions are based on a probability model, there must always be both the probability of being right or of being wrong when making the decision. If the null hypothesis is erroneously rejected, we are committing the alpha error; but if the null hypothesis is erroneously accepted, we are committing the beta (or Type 2) error. That is, whenever the null hypothesis is accepted when it really should be rejected, the beta error is committed. The **beta error** is truly important, because it means that perfectly valid research hypotheses may have been needlessly thrown away whenever it is committed. And this is too bad, since good research hypotheses are not always that easy to come by. In fact, it has recently been stated that there is a definite "bias among editors and reviewers for publishing almost exclusively studies that reject the null hypothesis."[2] Couple this with the fact that most research studies in which null has been accepted are never sent to journals in the first place and we suspect that only a tiny fraction of the possible beta errors ever see the light of day—most of them simply languishing in someone's file drawers.

The **power** of a statistical test is based on the beta-error concept. Power is equal to 1 minus the beta error $(1 - \beta)$. Thus, anything that helps to reduce the beta error also helps to increase the power of the test. For example, if the beta error is equal to .10, then the power of the test is equal to $(1 - .10)$ or .90. However, if the beta error is reduced to .05, then the power increases to $(1 - .05)$ or .95.

A powerful test is one that rarely inadvertently commits the beta error. That is, when there is significance, a powerful test detects that significance. The more powerful a test is, therefore, the higher is the likelihood of rejecting null when null really should be rejected. What kinds of techniques can be used to increase a test's power?

1. Increasing Sample Sizes.   A quick look at the tables of significance confirms the fact that the likelihood of rejecting null increases as sample sizes increase. For example, with a t ratio, when the df are equal to 5, a t of 2.57 is needed to reject null at the .05 level. However with 120 df, a t of only 1.98 is needed. The message—always use the largest sample available to you, of course, under the constraints of practicality.

2. Increasing the Alpha Error.   A test's power is also increased by increasing the alpha error from .01 to .05. The significance tables again confirm the fact that with a given number of degrees of freedom, the statistical value needed to reject H_0 at .05 is smaller than the value needed at .01. The researcher should rarely slavishly insist on .01 significance, for by so doing, the beta error has to be increased. Typically, however, statisticians do not increase the alpha error above .05. A line has to be drawn somewhere, and, by convention, that's usually the bottom line.

3. Using All the Information the Data Provide.   When you have the opportunity to use interval-ratio data, use those data, and, also, use the statistical tests which have been designed to take advantage of that level of data. Interval measures provide more information than do ordinal or nominal measures, and the more information contained in the measurement the more sensitive can be the statistical analysis. In the next chapter, we will find that both the Pearson r and the rank order r_s can be used to assess the possibility of a linear correlation between two sets of measures. Now, although the r_s is a direct, ordinal derivation of the Pearson r, it is still a less powerful test.

4. Fitting the Statistical Test to the Research Design.   Your chances of rejecting null, and thereby increasing test power, can be greatly increased by being extremely careful in fitting the statistical analysis to the research design. Much more will be said on this important topic after we've covered the material on research designs in Chapter 11. In fact, this will be a key issue in Chapter 14 when we focus on the relationship between the statistical analysis and the appropriate research design.

In summary, then, the power of a statistical test is a measure of its sensitivity. In some tests, such as t, power may even be increased by shifting from the two-tail to the one-tail analysis. As the value of a test's power increases, the possibility of accepting the null hypothesis and thereby possibly committing the beta error decreases. Since a powerful test leads, more often, to a reject of null, it allows smaller differences (or, as we shall see in the next chapter, smaller correlations) to show through as being significant.*

SUMMARY

When successive pairs of random samples are selected from a single population and the differences between the paired means are calculated, the mean of the resulting distribution of differences is assumed to be zero. Furthermore, if the mean differs significantly (a nonchance difference) from zero, it is then assumed that the sample pairs were not drawn from a single population. Probability statements can be made as to how large this difference must be in order to assume significance, if the variability of the distribution of differences is known. This variability can be inferred through the use of a statistic called the estimated standard error of difference, which is an estimate of the standard deviation of the entire distribution of differences. This estimate is made on the basis of the information contained in just two samples. All that needs to be known are the estimated standard errors of the mean for each sample.

Using the estimated standard error of difference, an independent t ratio can be calculated. This is the *ratio of the difference between sample means minus the mean of the distribution of differences to the estimated standard error of difference.* The value we

---

*The direct calculation of power as a specified value can be found in more advanced texts.

get tells us by how much the difference between the means deviates from zero (the assumed mean of the distribution of differences) in estimated standard error of difference units.

The size of the two samples determines the degrees of freedom, and, with the appropriate degrees of freedom, the calculated value of t can be checked for significance by comparing it with the table values of t. When the obtained value of t is equal to or greater than the table value, the null hypothesis (H_0) is rejected.

Whenever the null hypothesis is rejected, however, there is still some possibility that the wrong statistical decision has been made. The alpha error expresses the probability of being wrong whenever the null hypothesis is rejected. The alpha error should be kept to a value of .05 or lower.

Whenever the researcher is testing the straight hypothesis of difference (that one sample mean is simply different from the other), the obtained t must be compared to the critical table values of t for the two-tail distribution. If, on the other hand, the researcher predicts *beforehand* the direction of the difference between the sample means, a one-tail check of significance may be made. Power refers to a statistical test's ability to find significant differences among samples when there really are true differences in the population. Power is based on the beta-error concept, or the probability of being wrong when accepting null. Thus, power equals 1 minus the beta error (the smaller the beta error, the higher the power). Several methods are available for increasing power. Among them are: (1) increasing sample sizes, (2) increasing the alpha error (not usually above .05, however, (3) using all the information the data provide, (4) fitting the statistical test to the research design, and (5) when appropriate, shifting from a two-tail to a one-tail analysis.

KEY TERMS

alpha error	one-tail (directional) t test
alternative hypothesis	power
beta error	significance
degrees of freedom	standard error of difference
estimated standard error of difference	t ratio
null hypothesis	two-tail (nondirectional) t test

REFERENCES

1. BLUMSTEIN, P., & SCHWARTZ, P. (1983). *American couples.* New York: William Morrow.
2. KUPFERSMID, J. (1988). Improving what is published. *American Psychologist 43,* pp. 635–642.

PROBLEMS

1. Calculate a two-tail t ratio between the scores of the following randomly selected sample groups.

GROUP A	GROUP B
12	8
10	10
12	10
14	12
12	11
10	6
8	7

a. Do you accept or reject H_0?
b. If you reject H_0, state the probability level of the alpha error.
c. Had this been a one-tail t test, would you have accepted or rejected H_0?
d. For a one-tail t test, state the probability level of the alpha error.

2. Hypothesis: Democrats and Republicans differ with respect to the personality trait of "need dependence." Random samples of 10 Republicans and 10 Democrats are selected and each subject is given the Edwards Personal Preference Schedule (EPPS). The scores on the scale for "need dependence" are then compared.

DEMOCRATS	REPUBLICANS
14	13
8	10
10	11
12	14
3	8
10	12
12	14
10	12
9	10
8	10

a. Do you accept or reject H_0?
b. Do the two groups represent a common population with respect to the trait of "need dependence"?
c. Do you think party affiliation causes people to develop different personality characteristics, or could different personality types be attracted to different political parties?

3. Hypothesis: Increasing illumination increases speed of productivity among piece-work employees. Two groups are randomly selected from among the piecework employees at the Hawthorne Electric Company. Each subject is given 100 transistors to solder to the connecting relays. The time taken (in hours) is recorded for each subject to complete the task. Group A works under

normal plant lighting conditions, while for Group B, the illumination level is increased by 50%. Their scores are as follows:

GROUP A	GROUP B
6.5	5.0
5.0	4.0
3.9	3.0
4.2	3.5
4.5	3.7
6.2	4.2
5.3	3.7

   a. Do the groups represent a common population of productivity speeds?
   b. If H_0 is rejected, what is the probability of alpha error?
   c. Is this a one- or a two-tail test?

4. A researcher is interested in whether a certain hour-long film which portrays the insidious effects of racial prejudice will affect attitudes toward a minority group. Two groups of 31 subjects each were randomly selected and randomly assigned to one of two conditions. Group A watched the movie, whereas Group B spent the hour playing cards. Both groups were then given a racial-attitude test, high scores representing a higher level of prejudice. The data are as follows:

Group A	Group B
$N = 31$	$N = 31$
$\bar{X} = 42.60$	$\bar{X} = 39.62$

The estimated standard error of the difference between the means was 1.36. Was there a significant difference between the two sample means?

5. A random sample of 122 delinquent boys was selected and randomly divided into two groups. The researcher was interested in discovering whether a six-week, nondirective, individual therapy program would affect levels of measured anxiety. The boys in Group 1 all received the therapy, whereas those in Group 2 did not. Both groups were then given an anxiety-level test (high scores indicating more anxiety). The data were as follows:

Group 1	Group 2
$N = 61$	$N = 61$
$\bar{X} = 98.06$	$\bar{X} = 102.35$
$s_{\bar{x}} = 1.98$	$s_{\bar{x}} = 2.02$

Was there a significant difference between the two sample groups?

6. If you are doing a t test with one degree of freedom, how many subjects are needed among the two sample groups?

7. With extremely small samples, the opportunity for rejecting H_0 becomes (*more* or *less*) likely.

8. When all pairs of samples come from a single population, the mean of the distribution of differences assumes what numerical value?

9. To calculate a t ratio, the data must come from which scale of measurement?

10. Setting the alpha error at .01, rather than at .05, makes it (*more* or *less*) likely that H_0 will be rejected.

11. The t ratio tells us by how much the difference between the sample means deviates from a mean difference of zero in units of what?

12. The standard error of difference is literally the standard deviation of which important sampling distribution?

Indicate what term is being defined by each of problems 13 through 17.

13. An estimate of the standard deviation of the entire distribution of differences.

14. The probability of being wrong when the null hypothesis is rejected.

15. $\mu_1 = \mu_2$.

16. The two samples represent a single population.

17. $\mu_1 \neq \mu_2$.

True or False—Indicate either T or F for problems 18 through 24.

18. When the two sample means differ, no matter by how much, we always reject H_0.

19. When specifying the direction of the difference between two sample means, we use the two-tail t.

20. Alpha error becomes zero only when both samples reflect different populations.

21. The mean of the distribution of differences becomes zero only when all pairs of samples represent a single population.

22. The alternative hypothesis for the t test always specifies the direction of the differences between the means.

23. Reducing the alpha error from .05 to .01 increases the power of a statistical test.

24. If null is accepted when it should have been rejected, the beta error has been committed.

chapter 10

The Hypothesis of Association: Correlation

We now enter the treacherous and murky waters of correlation. Perhaps no other area of research demands more caution or is fraught with more danger. Too many of us assume that because the word "correlation" is easy to say, that it's likewise easy to understand. In some respects, it is. Mathematically, the correlation coefficient is rather straightforward and easily calculated. What can legitimately be inferred from its numerical value, however, is quite another story.

The meaning of the word correlation comes literally from its parts: "co" means with, together, or jointly, and "relation" means association. Thus, when two events regularly occur together, then they are said to be correlated, as with blond hair and blue eyes. Also, when changes in one set of events are regularly accompanied by changes in another set of events, correlation is said to exist; for example, as children get taller, they also tend to get heavier. As we shall soon see, the correlation coefficient provides us with a numerical value that states the extent to which two events, or two sets of measurements, tend to occur or to change together—the extent to which they covary.

CAUSE AND EFFECT

So far, correlation seems straightforward, easy, and obvious. Why then the warning in the first paragraph about "treacherous and murky waters"? Consider the following findings. A great number of studies have shown that there is a large, positive correlation between reading speed and IQ. People who read quickly tend to have higher IQs than do people who read slowly. Therefore, enrolling in a speed-reading course will lead to an increase in IQ. Right? Wrong! Because the same factors that make a person a fast reader (verbal ability, alertness, quick reactions) probably also enable that person to perform well on IQ tests. Another study reported a positive correlation between coffee consumption and heart attacks among men. Men who had heart attacks were asked how much coffee they drank. This rate was compared to the rate of coffee consumption among a comparable group of men (matched on such characteristics as age, weight, etc.) who had never had heart attacks. The results tallied. Therefore, drinking coffee causes heart attacks. Right? Wrong! It may only be that elements of the pre-heart attack condition, such as fatigue or stress, cause the person to seek relief by turning to the stimulating effects produced by coffee. It's like saying that just because your alarm clock went off at 7:30, it must have been the sound of the alarm that caused it to be 7:30.

In the foregoing examples, legitimate correlational statements were made, but illegitimate inferences were drawn. It's not that the existence of cause and effect has been ruled out in these cases; however, the given correlations, in and of themselves, do not prove it. It may be true that drinking coffee causes heart attacks or that increased reading speed causes a better IQ test performance. The point is that correlation does not directly address the cause-and-effect issue.

The Problem of Isolating the Cause. Suppose that a researcher is interested in finding out whether or not parental rejection causes juvenile delinquency among teen-age boys. Furthermore, the researcher has developed an accurate method for assessing perceived parental rejection; that is, the researcher has a tool for determining a given adolescent's perception of his parents' rejection-acceptance attitudes toward him. After gathering that data, the researcher checks police records and finds that a relationship does indeed exist. The more a boy perceives himself the victim of parental rejection, the more frequent have been his confrontations with those in blue uniforms. But has a cause-and-effect relationship been established? We can use A to symbolize parental rejection, B to indicate juvenile delinquency, and an arrow to indicate the causal direction. (Using letter symbols when analyzing correlational studies is usually a good idea, since such symbols do not have any of the literary overtones that are often inherent in word descriptions of the variables.) Three hypotheses are possible to account for the findings:

1. $A \rightarrow B$. It is possible, though not proven by this study, that A (parental rejection) does cause B (juvenile delinquency).
2. $B \rightarrow A$. It is possible that B (juvenile delinquency) causes A (parental rejection). A parent may not project warm feelings of affection toward a son who is brought home night after night in a patrol car.
3. $X \rightarrow A + B$. It is further possible that X (some unknown variable) is the real cause of both A and B. For instance, X could be an atmosphere of frustration and despair that permeates a given neighborhood, leading both parents and children to generate feelings of hostility.

All three of these explanations are possible; the point is that the correlation alone is not enough to identify which is the real explanation.

The problem with using correlation as a tool for establishing causation is due to the difficulty in predicting *the direction* of the relationship. For example, research on married couples has found that persons who find their partners attractive have happier sex lives. There is no guarantee in this finding, however, that personal attractiveness increases sexual pleasure, for it is just as likely that persons with happier sex lives tend to view their partners as being more attractive. Beauty may indeed reside in the eye of the beholder. Or perhaps a third variable, say, an optimistic approach to life in general, causes people both to find their partners more attractive and view their sex lives as more complete. In short, perhaps both variables are the results of a person's rosy, self-fulfilling prophecy.

In the field of psychology, the literature on the condition known as agoraphobia (the fear of having a panic attack away from a safe place) shows high and significant correlations between agoraphobic symptoms and passive-dependent personality types. Perhaps conventional wisdom might interpret this relationship to mean that the passive-dependent personality tends to produce agoraphobic symptoms—even to the point of being rendered house-bound. On further analysis, however, it should be pointed out that the data may just as likely be saying that a person with agoraphobic symptoms might then become a passive-dependent—that is, that the symptoms might be producing the personality type. Or it might be that some unknown third factor is causing the individual both to have panic attacks and to be a passive-dependent.

Using Correlation.   If correlational research is so limited as to cause and effect, then why do we bother? The reason is that correlational research can yield better than chance predictions. If two events are correlated, then a knowledge of one of those events allows a researcher to predict the occurrence of the other. This is helpful, because if we can predict an event, we may later be able to control it.

THE PEARSON *r*

During the latter half of the nineteenth century, an English scholar and mathematician, **Karl Pearson,** became impressed with the fact that individuals vary so widely in such characteristics as height, weight, and reaction time. At the time Pearson was working with **Sir Francis Galton,** considered to be the father of the very important concept of individual differences. Pearson reasoned that since there was such wide variety to human characteristics, it would be extremely useful if the measurement of these characteristics could be expressed in relational terms rather than in absolute units of measurement. After all, it is just as useful to know that a woman's height places her in the *relative* position of exceeding 50% of the adult female population as it is to know that, in *absolute* terms, she is 65 inches tall.

Karl Pearson (1857–1936) produced the correlation coefficient, now called the Pearson *r*. (Brown Brothers.)

The idea of relative position also makes it possible to create a common ground of comparison between qualitatively different measurements. Even though measurements of height and weight cannot be compared directly, because the two are conceptually different and are expressed in different units, such measurements can be compared in terms of how much they each vary *from their own average.* In effect, this means that the proverbial apples and oranges can be compared on the basis of whether a given orange and a given apple are both larger, smaller, juicier, or riper than the average orange and apple. Thus, the relationship between two qualitatively different objects can be expressed in quantitative terms. Pearson called this expression the **correlation coefficient.** The correlation coefficient, then, is a numerical statement of the relationship between two or more variables. A single value can be calculated to express this relationship.

Although correlation does not necessarily imply causation, it is still a useful tool for making predictions. The amount of correlation tells us the extent to which two or more variables are associated, or the extent to which they covary or occur together. Although the previous sentence mentions two *or more* variables, the focus in this chapter is on two-variable correlations, or bivariate analyses. Later in the book, we will present a technique for handling multivariable correlations.

Positive, Negative, and Zero Correlations

Correlations take three general forms: positive, negative, and zero. Positive correlations are produced when persons or things having high scores on one variable also have high scores on a second variable and those with low scores on the first variable are also low on the second. For example, a positive correlation between height and weight means that those individuals who are above average in height are also above average in weight and that those who are below average in height are correspondingly below average in weight.

Negative correlations are produced when high scores on one variable are associated with low scores on a second variable and low scores on the first correspond with high scores on the second. A negative correlation between college grades and absenteeism means that those who are above average in college grades tend to have fewer than the average number of absences, while those who are below average in grades tend to have more than the average number of absences.

Finally, zero correlations are produced when high scores on one variable are as likely to correspond with high scores on a second variable as they are with low scores on the second or when low scores on the first are just as apt to be associated with either high or low scores on the second.

Interpreting Correlation Values

To express the degree to which variables are associated, or correlated, a single number is used. This number may vary from $+1.00$ through zero to -1.00. A value of $+1.00$ indicates the maximum positive correlation. A maximum positive correlation occurs when two measures associate perfectly. For example, if there were a perfect positive correlation between height and weight, a person taller than another person would always be heavier too, with no exceptions. Needless to say, perfect positive correlations are extremely rare in the social sciences.

In math and physics, there are numerous examples of perfect, 1.00, correlations. The radius of a circle bears an exact relationship to the circle's circumference. A greater radius will always coincide with a greater circumference. Perfect negative correlations are also easily found. The greater the distance an object is from its light source, the less illumination will fall on the object, hence, a correlation of -1.00. In psychology and education, where the variables are almost always empirically found, things just aren't that simple.

A value of zero indicates no relationship at all, a zero correlation. A value of -1.00 indicates a maximum negative correlation, meaning every single individual in a group who is higher than another on one measure is lower than that other on a second measure. The closer the correlation value is to either ± 1.00, the more accurate is the resulting prediction, and prediction, after all, is the name of the game in correlational research. The predictive efficiency of a correlation, then, increases as the correlation value differs from zero, *regardless of sign*. A correlation of $-.95$ predicts more accurately than does a correlation of $+.85$. Remember, for correlation research, that a negative value does not mean a lack of correlation; it simply refers to an inverse relationship.

Graphs and Correlation

For a visual representation of how much two distributions of scores correlate, statisticians use a graphic format known as a **scatter plot.** The scatter plot allows the visual representation of two separate distributions on a single diagram.

Table 10.1 Data on hours per week spent studying and GPA for seven students.

Student	Hours/week studying, X	GPA, Y
A	40	3.75
B	30	3.00
C	35	3.25
D	5	1.75
E	10	2.00
F	15	2.25
G	25	3.00

A Positive Correlation. In Table 10.1 are some data (some of which might be propaganda) from seven students who have been measured for both hours per week spent studying and grade-point average (GPA).

In Fig. 10.1 the data from the table are plotted. Note that each point represents two scores. For Student A, the point is directly above 40 on the X axis and directly across from 3.75 on the Y axis. Also, note that the points drift upward, from lower left to upper right. This configuration represents a positive correlation between the variables.

A Negative Correlation. Scatter plots can also portray negative correlations. For example, suppose that we select a sample of seven students and measure them on both hours per week of Frisbee playing and grade-point average. Our data are summarized in Table 10.2

We again plot the data, using a single point to represent the pair of scores for each student. (The point directly above 2 on the X axis and across from 3.50 on the Y axis represents the pair of scores for Student A.) Our plot is shown in Fig. 10.2. Note that high scores on one measure correspond with low scores on the other. This situation constitutes, as stated earlier, a negative correlation. The scat-

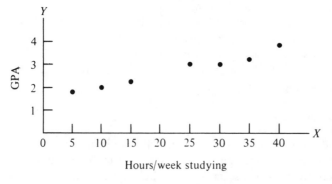

Figure 10.1
Scatter plot of study hours versus GPA showing positive correlation.

Table 10.2 Data on hours per week spent playing Frisbee and GPA for seven students.

Student	Frisbee hours/week, X	GPA, Y
A	2	3.50
B	8	2.50
C	10	2.25
D	12	1.75
E	16	.50
F	6	3.00
G	4	3.25

ter plot allows us to see this relationship. When the correlation is negative, the general slope of the points in the scatter plot falls from upper left to lower right.

A Reminder About Cause and Effect.   No direct statement regarding cause and effect can or should be made about either of the preceding. For the data on the relationship between study time and grades, although it is certainly a viable hypothesis that hours spent studying affect grades ($A \rightarrow B$), other hypotheses are at least possible. Perhaps the student who receives high grades is encouraged by them to spend more time studying ($B \rightarrow A$). Finally, it may be a third variable, perhaps old-fashioned ambition, that drives a student to want to study and also to want high grades ($X \rightarrow A + B$). The negative correlation between hours spent playing Frisbee and grade-point average does not mean that playing Frisbee causes a lower GPA (although that is one possible explanation; $A \rightarrow B$). Perhaps, instead, students who are doing poorly in school use Frisbee tossing as an outlet for pent-up frustrations ($B \rightarrow A$). Or, perhaps, a personality factor of a gregarious nature exists in some students, who need constant social stimulation and so seek the companionship of Frisbee partners *and* avoid the lonely hours involved in academic preparation ($X \rightarrow A + B$).

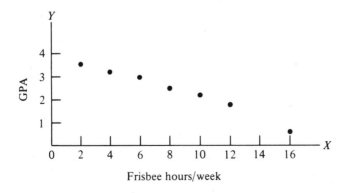

Figure 10.2
Scatter plot of Frisbee-playing hours versus GPA showing negative correlation.

Scatter Plot Configurations.   Figure 10.3 shows how the three general types of correlations are revealed by the scatter plots formed when large numbers of subjects are measured on two variables and the pairs of scores graphed. The scatter plot on the left portrays a positive correlation; the array of points tilts from lower left to upper right, telling us that as one variable increases, so does the other. The scatter plot in the middle portrays a negative correlation; the array of points falls from upper left to lower right, telling us that as one variable increases, the other decreases. Finally, the scatter plot on the right portrays a zero correlation; as one variable changes, there is no related change in the other.

It is important to note that the array of points on a scatter plot generally tends to form an oval shape. This is due to central tendency. Since most scores in any distribution tend to cluster around the middle of that distribution, the configuration of points in a scatter plot, which reflects two distributions, typically shows a bulge of scores in the middle and a tapering off at the extremes.

The *z* Score Method for Calculating the Pearson *r*

Karl Pearson called his equation the product-moment correlation coefficient, but in his honor it is more familiarly known as the **Pearson *r*.** Pearson based his equation on the concept of *z* scores, defining *r* as the mean of the *z* score products for the *X* and *Y* variables, as follows:

$$r = \frac{\Sigma \, z_X z_Y}{N}$$

$$z_X = \frac{X - \bar{X}}{SD_X} \quad \text{and} \quad z_Y = \frac{Y - \bar{Y}}{SD_Y}$$

To calculate *r*, each raw score for both the distributions must be converted into a *z* score. The *z* score pairs are then multiplied. These products are added, and, to get the mean, this product sum is divided by the number of *pairs* of *z* scores. This is admittedly a rather complicated process; however, it will be shown

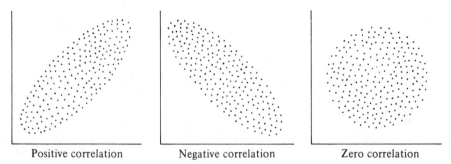

Positive correlation   Negative correlation   Zero correlation

Figure 10.3   The three types of correlation.

because it illustrates beautifully the tie-in between the logic of correlation and its calculation. To make this second foray into z scores as painless as possible, both the standard deviations and the z scores will be computed for you and presented as "givens."

> ***Example.*** We can use the data on hours spent studying and grade-point averages presented earlier (in Table 10.1).

Student	Hours of study, X	z_X	GPA, Y	z_Y	$z_X z_Y$
A	40	1.40	3.75	1.51	2.11
B	30	.59	3.00	.42	.25
C	35	1.00	3.25	.78	.78
D	5	−1.46	1.75	−1.39	2.03
E	10	−1.05	2.00	−1.03	1.08
F	15	−.64	2.25	−.67	.43
G	25	.18	3.00	.42	.08
	$\Sigma X = 160$		$\Sigma Y = 19.00$		$\Sigma z_X z_Y = 6.76$

$$\bar{X} = 22.86 \qquad \bar{Y} = 2.71$$

$$SD_X = 12.20 \qquad SD_Y = .69$$

$$r = \frac{\Sigma z_X z_Y}{N} = \frac{6.76}{7} = .97$$

The Pearson r, therefore, is calculated to be .97, an extremely high correlation. Note that the correlation is positive, meaning that the association is direct—the higher the amount of study time, the higher the grades.

The Sign of the Correlation. In any distribution, raw scores above the mean must yield plus z scores. Therefore, when both scores in a pair yield positive z scores, the resulting product, $z_X z_Y$, must be positive (plus times plus equals plus). Furthermore, since all scores below the mean yield negative z scores, pairs of low scores also create positive z score products (minus times minus equals plus).

When most of the z score products are positive, the sum of these products will be positive and the resulting correlation will have to be positive. As defined earlier, a positive correlation occurs when high scores on one variable associate with high scores on the other *and* when low scores on one variable associate with low scores on the other. In other words, when plus z scores occur with plus z scores and minus z scores occur with minus z scores, the resulting products are positive and *so is the correlation.*

Had the relationship been inverse, high scores pairing off with low scores, the z score products would have been negative (plus times minus equals minus). With negative products, the correlation would then be negative.

The Computational Method for Calculating the Pearson *r*

Although doing the correlation via the *z* score method is instructive, it is also extremely time consuming. Getting all those *z* scores is laborious indeed, especially
when there are long lists of paired scores. Fortunately, a computational equation
has been derived that makes the calculation of *r* far easier. If you can calculate a
mean and standard deviation, you can easily calculate the Pearson *r*. (In fact, there
are a number of Pearson *r* derivations, all of which, of course, produce the same
answer. However, unlike the others, the one chosen here does allow for more internal checks, since the means and standard deviations become available for direct viewing. For example, if you were to press the wrong key on your calculator,
or, worse, the batteries began to die, you might wind up with an impossible answer, an *r*, say, that was greater than ±1.00. With this equation you can go back
over your work, and, keeping in mind what the "reasonable" values should be for
the mean and standard deviation, you can probably locate the source of your
trouble.)

$$r = \frac{\dfrac{\Sigma\, XY}{N} - (\bar{X})(\bar{Y})}{SD_X\, SD_Y}$$

Example. Calculate the Pearson *r* value for the data from Table 10.1.

Student	Hours of study, X	X^2	GPA, Y	Y^2	XY
A	40	1600	3.75	14.06	150.00
B	30	900	3.00	9.00	90.00
C	35	1225	3.25	10.56	113.75
D	5	25	1.75	3.06	8.75
E	10	100	2.00	4.00	20.00
F	15	225	2.25	5.06	33.75
G	25	625	3.00	9.00	75.00
	$\Sigma\, X = 160$	$\Sigma\, X^2 = 4700$	$\Sigma\, Y = 19.00$	$\Sigma\, Y^2 = 54.74$	$\Sigma\, XY = 491.25$

$$\bar{X} = \frac{\Sigma\, X}{N} = \frac{160}{7} = 22.86$$

$$SD_X = \sqrt{\frac{\Sigma\, X^2}{N} - \bar{X}^2} = \sqrt{\frac{4700}{7} - 22.86^2}$$

$$SD_X = \sqrt{671.43 - 522.58} = \sqrt{148.85} = 12.20$$

$$\bar{Y} = \frac{\Sigma\, Y}{N} = \frac{19.00}{7} = 2.71$$

$$SD_Y = \sqrt{\frac{\Sigma\, Y^2}{N} - \bar{Y}^2} = \sqrt{\frac{54.74}{7} - 2.71^2}$$

$$SD_Y = \sqrt{7.82 - 7.34} = \sqrt{.48}$$

$$SD_Y = .69$$

$$r = \frac{\Sigma\,XY/N - (\bar{X})(\bar{Y})}{SD_X\,SD_Y} = \frac{491.25/7 - (22.86)(2.71)}{(12.20)(.69)} = \frac{70.18 - 61.95}{8.42}$$

$$r = \frac{8.23}{8.42} = .977 = .98$$

1. *The means.* Add all the X's, $\Sigma\,X$, and Y's, $\Sigma\,Y$, and divide each by N. Stop now, and examine the means. Make sure that they're reasonable, that is, that both are well within their respective ranges.
2. *The standard deviations.* Square all X's and Y's and add each column of squared values. Divide the sum of the squared scores by N, subtract the squared mean, and extract the square root for each distribution. Stop, and examine the SD's. Assure yourself of reasonable values—that neither is greater than half its range.
3. *The XY term.* Multiply each X score by the corresponding Y score. Then, sum the XY column.
4. *The equation.* Plug all values into the Pearson r equation. Divide the $\Sigma\,XY$ value by N and then subtract the product of the means. Be careful of the sign in this step; if the $(\Sigma\,XY)/N$ term is smaller than the product of the means, the numerator will be negative and a negative correlation will result. Divide by the product of the standard deviations to get the Pearson r.

If you prefer using the estimated population standard deviation, the equation becomes

$$r = \frac{\dfrac{\Sigma\,XY - (\Sigma\,X)(\Sigma\,Y)/N}{N - 1}}{s_x s_y}$$

$$s_X = \sqrt{\frac{\Sigma\,X^2 - (\Sigma\,X)^2/N}{N - 1}} = \sqrt{\frac{4700 - 160^2/7}{6}} = \sqrt{\frac{4700 - 3657.14}{6}}$$

$$= \sqrt{173.81} = 13.18$$

$$s_Y = \sqrt{\frac{\Sigma\,Y^2 - (\Sigma\,Y)^2/N}{N - 1}} = \sqrt{\frac{54.74 - 19^2/7}{6}} = \sqrt{\frac{54.74 - 51.57}{6}}$$

$$= \sqrt{.53} = .73$$

$$r = \frac{\dfrac{491.25 - [(160)(19)]/7}{6}}{(13.18)(.73)} = \frac{(491.25 - 434.29)/6}{9.62} = \frac{9.49}{9.62} = .99$$

Discrepancy Among Calculated Values

There is a slight discrepancy among the r values obtained from the methods just shown. Using the z score equation, the resulting r rounded to .97, whereas with the first computational method, the r rounded to .98. The other computational method produced a rounded r of .99. This does *not* mean the equations logically differ from one another. Algebraically, they are all the same. The discrepancy is due to the fact that by rounding to two places (and we did a lot of that in the first method when calculating all those z scores), we lost a tiny degree of numerical accuracy. If we were to carry out all values to six or eight places, all the methods would produce identical answers. In fact, within rounding errors, they did! (The *PH-STAT* computer program prints out an r of .99.)

Problem 10a: Using the data in Table 10.2, calculate the Pearson r correlation coefficient between Frisbee hours per week and grade-point average. Remember, in this case the relationship between the paired scores is inverse.
Answer: $-.97$ (or $-.98$ using s instead of SD).

Testing the Pearson r for Significance

Although the focus in this chapter will be on correlation as an inferential procedure, the correlation coefficient can also be viewed simply as another descriptive statistic. As a descriptive measure, the correlation is used to portray the relationship between two sets of scores, and the findings are then concerned only with the observed data and the variables being represented by these data. In most research situations, especially in the behavioral sciences, the importance of the correlation lies in its generalizability, not in its straight description. Thus, our concern throughout this chapter will be aimed at discovering whether the observed correlation, as shown by the sample, can be extrapolated to the population being represented by that sample.

Therefore, as we did with the t test, we must determine whether or not the correlation, which was obtained from sample data, can be legitimately generalized to the entire population. When applying the t test, differences between sample means may or may not reflect population differences. To find out, we have to test for the significance of our obtained value. Similarly, with the Pearson r, an association between sets of sample scores may or may not reflect an association that truly exists in the population. We must also, therefore, test r for significance. If the results of any statistical test are going to be used to infer population characteristics, those results must be subjected to a significance test.

The Null Hypothesis. We have learned that the null hypothesis is the hypothesis of chance. No matter what results have been obtained on a sample, the null hypothesis insists that these results are strictly due to the vagaries of chance and, therefore, cannot be generalized to the population. For the Pearson r, the null hypothesis states that $\rho = 0$; this means that there is no correlation in the population, regardless of the value that has been obtained for the sample. The alternative hypothesis, on the other hand, is that there is indeed a correlation in the population, or that $\rho \neq 0$. The Greek letter ρ (rho) is used to refer to the correlation that may or may not exist in a population, whereas r always refers to a sample.

Using the Pearson r Table

To make the statistical decision, we use Table E, the Pearson r table. This table is set up like the t table, with degrees of freedom in the far left column and the critical values of r in two columns, one headed .05 and the other headed .01. As in the t table, .05 and .01 represent the probability of alpha error—the probability of rejecting the null hypothesis when we should have accepted it. Degrees of freedom are assigned on the basis of $N - 2$, the number of *pairs* of scores minus 2. We compare the obtained value of r with the table value of r for the appropriate degrees of freedom. If the absolute value of the obtained r is equal to or greater than the critical table value, we reject the null hypothesis. If the absolute value calculated is less than the table value, we accept the null hypothesis.

Example.   In the last example (found on page 203), the obtained value of r was .98. Is that a significant correlation?
Since we have 7 pairs of scores, the degrees of freedom ($N - 2$) are $7 - 2$, which equals 5. At 5 degrees of freedom, the table values of r are .75 for .05 and .87 for .01. Our obtained r is higher than the .05 value of .75, but it is also higher than the .01 value of .87. We, thus, reject the null hypothesis and state that our calculated value of r is significant at the .01 level.
Beside the value of r from the table, we write r with a subscript of .01 for the level of alpha error and, in parentheses, 5 for the degrees of freedom. Below that, we write the obtained value of r and the statistical decision.

$$r_{.01(5)} = .87$$
$$r = .98 \quad \text{Reject } H_0; \text{ significant at } P < .01.$$

The statistical reasoning behind this procedure is that since the obtained correlation coefficient equals .98, its probability is less than .01 of occurring by chance when the assumed population correlation is zero. Since the population distribution of correlations under the null hypothesis is assumed to center around zero, the obtained correlation of .98 has a probability of occurring by chance less than one time in a hundred.
In making the table comparison, it is the absolute value of r that determines significance. Although the sign of the correlation is crucial in terms of the interpretation of the results, it is totally irrelevant to the significance check.
Again, with five degrees of freedom, if the obtained r were equal to .77, we would have written

$$r_{.05(5)} = .75$$
$$r = .77 \quad \text{Reject } H_0; \text{ significant at } P < .05.$$

Finally, again with 5 df, if the obtained r were equal to .50, we would have written

$$r_{.05(5)} = .75$$
$$r = .50 \quad \text{Accept } H_0.$$

When determining significance, there are only three possible outcomes:

1. Reject H_0 at the .01 level.
2. Reject H_0 at the .05 level.
3. Accept H_0.

The Meaning of a Significant Correlation. Even though a correlation is significant, it does not necessarily mean that it is communicating an especially profound message. A significant correlation is one that is *not* likely to be a result of chance. Scanning the df column of the *r* table reveals that as sample size increases, extremely small correlations attain significance. With 400 degrees of freedom, for example, an *r* of only .10 may be significant. An *r* of .10 is indicative of only a very slight associational trend; when it is found to be significant, this slight trend does indeed exist in the population.

Studies show that there is a small, but significant, correlation between hair color and frequency of temper tantrums among children. Yes, it's true; redheads do have slightly more temper outbursts than do their blonde and brunette siblings. However, we cannot leap to the conclusion that this association is genetic; it could just as easily be environmental. Perhaps some parents permit only their redheaded children to act out in this fashion. Also, we cannot assume that red hair causes temper tantrums. Although this conclusion might increase hair color sales in other shades, it is simply not justified on the basis of the evidence.

For reasons of space, all possible df values for the significance of *r* could not be included in Table E. For a rough, and as we shall see extremely conservative, estimate, however, you may evaluate the *r* by using the next smaller df value. For example, if the number of pairs of scores were 122, which produces 120 degrees of freedom, you might safely use the table value for 100 degrees of freedom, or a significance value of .195. If an exact value is required, use the following equation:

$$t = \frac{r\sqrt{(n-2)}}{\sqrt{1-r^2}}$$

and then evaluate the *t* ratio using the two-tail *t* table, and, of course, the appropriate degrees of freedom for the Pearson *r*.

For example, assume that you obtained an *r* of .16, with 122 pairs of scores. Using the conservative approach outlined above, this value would not be found significant. However, by using the equation in order to obtain an exact evaluation, the following would result:

$$t = \frac{.16\sqrt{122-2}}{\sqrt{1-.16^2}} = \frac{.16\sqrt{120}}{\sqrt{1-.40}} = \frac{(.16)(10.9544)}{\sqrt{.60}}$$

$$t = \frac{1.7527}{.7746} = 2.26$$

$$t_{.05(120)} = 1.980 \quad \text{Reject } H_0 \text{; significant at } P < .05.$$

Restricted Range

The Pearson r has the potential for yielding its highest value when both sampling distributions, X and Y, represent the entire range of normally distributed values. If the range of either or both sampling distributions is in any way restricted, the Pearson r may seriously underestimate the true population correlation. As stated, the Pearson r was designed to show the strength of the relationship between high and low scores on one variable with the high and low scores on the other. If either of the variables fails to contain the high or low end of its distribution, the resulting correlation will tend to be closer to zero than it otherwise would be. For example, suppose a researcher were interested in correlating IQ and SAT scores among students at a special private school for gifted adolescents. In this case, both the IQ and SAT distributions might show severely restricted ranges, especially if the school's admission policy were to select students largely on the basis of IQ. The correlation, because this group was so homogeneous, would tend to be considerably lower than it would for unspecified cases taken from the secondary school population at large.

   As another example of how a restricted range can affect the correlation, it has been reported that the correlation between IQ scores and grades in school drops dramatically as the size of the group being tested becomes smaller and more homogeneous. The correlations average in the .60s for elementary-school children (the largest and most heterogeneous of the groups tested), to the .50s for high school students, to the .40s for college students, to only the .30s for graduate students (the smallest and most homogeneous group).[1] The groups tested became smaller and obviously more homogeneous since the lower IQs are consistently and systematically weeded out as the students progress toward more intellectually demanding experiences.

Interpreting the Pearson r

As we have seen, the value of the Pearson r varies between -1.00 and $+1.00$. Since r gives a precise numerical value within this range, it can express an enormous variety of associational meanings. Extremely subtle degrees of associational strength are communicated with the Pearson r. Despite this, certain broad categories should be kept in mind for attempting to explain any given correlation. One such set of verbal tags has been supplied by the distinguished American statistician J. P. Guilford (see Table 10.3).

   Guilford points out that these interpretations may only be used when the correlation coefficient is significant. Also, the same interpretations "apply alike to negative and positive r's of the same numerical size."[2] An r of $-.90$ indicates the same degree of relationship as an r of $+.90$ does.

   The Coefficient of Determination.   Although Table 10.3 gives a few correlational bench marks with which to interpret the value of r, far more precise methods have been devised. In Chapter 14, for example, we will utilize the Pearson r as

Table 10.3 Guilford's suggested interpretations for values of *r*.

r value	Interpretation
Less than .20	Slight; almost negligible relationship
.20–.40	Low correlation; definite but small relationship
.40–.70	Moderate correlation; substantial relationship
.70–.90	High correlation; marked relationship
.90–1.00	Very high correlation; very dependable relationship

a means for making specific predictions and will also master techniques for assessing the accuracy of those predictions. For now, however, an easily calculated statistic, called the **coefficient of determination,** can help bridge the gap. The coefficient of determination is simply the square of the Pearson *r*.

$$\text{coefficient of determination} = r^2$$

The coefficient of determination is used to establish the proportion of the variability among the *Y* scores that can be accounted for by the variability among the *X* scores. That is, anytime there is a correlation between *X* and *Y*, then *X* contains some information about *Y*. When the correlation is a maximum of 1.00, then the coefficient of determination is also 1.00 (since one squared equals one). It tells us that *X* is carrying all there is to know about *Y*. In fact, when *r* = 1.00, then *X* and *Y* may be measuring a single underlying trait, and knowing the *X* score automatically gives the *Y* score. By multiplying the coefficient of determination by 100, we get the *percentage* of information about *Y* that is contained in *X*. For example, a correlation value of .70 yields a coefficient of determination of .49, meaning that 49% of the information about *Y* is contained in *X*. Therefore, although a correlation value of .70 does *not* mean 70% accuracy, the square of the correlation, or .49, can be used to give us a percentage estimate of accuracy.

Requirements for Using the Pearson *r*

To use the Pearson *r*, the following requirements must be met:

1. The sample has been randomly selected from the population.
2. The traits being measured do not depart significantly from normality. This is not as severe a limitation as it might appear. The Pearson *r* is a "robust" test, meaning that rather large departures from normality still allow for its use. The distribution forms, however, must be unimodal and fairly symmetrical.
3. Measurements on both distributions are in the form of, at least, interval data. Of course, any test that can handle interval data can also be used with ratio data.
4. The variation in scores in both the *X* and *Y* distributions must be similar. This property is known as **homoscedasticity,** and it may be assumed unless either of the distributions is markedly skewed.

5. The association between X and Y is linear. That is, the Pearson r cannot be used unless the relationship forms a straight line. Curvilinear relationships (as when an increase in X is accompanied by an increase in Y up to a point and is then accompanied by a decrease in Y) cannot be assessed by the Pearson r.

***Limitations on the Use of the Pearson* r.**   The Pearson r, as we have seen, assesses the strength of a linear relationship between two sets of sample scores in interval form and representing normal distributions. Sometimes the data do not meet these requirements.

In Chapter 1, it was pointed out that interval data demand that the distances between successive scale points be equal. Therefore, with interval data, we know not only that one observation is greater than another, but also how much greater. Ordinal data, on the other hand, which come in the form of ranks, present no information regarding the distances between scale points. With ordinal data, we know that if one observation has a higher rank than another, then the first is greater than the second, but we do not know how much greater.

Therefore, whenever the distribution of interval scores is definitely not normal, or whenever the data come to us in ordinal form, then we should not use the Pearson r.* For example, the Pearson r can almost never be used on income data, since the income distribution in the population is usually skewed. Similarly, the Pearson r should not be used on a distribution of rank-ordered scores, for example, the order of finish of a group of race horses.

THE SPEARMAN r_s

When the Pearson r is inapplicable, the correlation value is obtained using an equation derived by **Charles Spearman.** The resulting value is called the **Spearman r,** or, more commonly, the r_S. (In some texts, this statistical test is called the Spearman rho.) Since the Spearman r_S does not make any assumptions regarding either the parameter mean or the parameter standard deviation, it is called a nonparametric test. The Pearson r, on the other hand, is a parametric test. Much more will be said later, especially in Chapters 16 and 18, concerning the difference between parametric and nonparametric statistical tests, especially with regard to power.

Three examples of calculating the r_S are to follow: one in which both sets of scores are originally given in ordinal form (rank order); one in which the two sets of scores are given in different forms, one interval and the other ordinal; and one in which both of the distributions of interval scores are known not to be normal.

*Although in the case of no tied ranks, the Pearson r will yield the same numerical value as the r_S, it still should not be used on ordinal data. As shown previously, interval data tests such as the Pearson r are more powerful than ordinal tests such as the r_S.

Calculating r_S with Data Originally Given in Ordinal Form

In this first example, both sets of raw scores are in the form of ordinal data.

   Example. Suppose that a researcher wishes to establish whether or not there is a significant correlation between aggressiveness and leadership ability among female military recruits. A random sample of 10 recruits is selected. The soldiers are watched closely by a trained observer who rank orders the women on the basis of aggressiveness exhibited. Independently, another trained observer watches the women and rank orders them on the basis of leadership abilities. (Independent observers are required in this study to prevent a type of research error known as the halo effect; see Chapter 11.) The data, with R_1 being the rank on aggressiveness and R_2 the rank on leadership, are as follows:

SUBJECT	R_1	R_2	$d = |R_1 - R_2|$	d^2
1	2	3	1	1
2	3	1	2	4
3	7	5	2	4
4	6	9	3	9
5	1	2	1	1
6	5	6	1	1
7	10	8	2	4
8	8	10	2	4
9	9	7	2	4
10	4	4	0	0
				$\Sigma d^2 = 32$

We use the following steps for calculating the value of r_S:

1. Pair off the two ranks for each subject. In this case, Subject 1 ranks second in aggressiveness and third in leadership.
2. Obtain the absolute difference, d, between each subject's pair of ranks. In the case of Subject 1, d is equal to 1, the difference between 2 and 3.
3. Square each difference.
4. Calculate Σd^2 by adding the squared differences.
5. Substitute this value, Σd^2, into the r_S equation, along with N, the number of paired ranks. These are the only variables in the equation, the numerical terms are constants.

$$r_S = 1 - \frac{6\Sigma d^2}{N(N^2 - 1)}$$

$$r_S = 1 - \frac{6(32)}{10(100 - 1)} = 1 - \frac{192}{10(99)} = 1 - \frac{192}{990}$$

$$r_S = 1 - .19 = .81$$

The calculated correlation is .81. It must next be evaluated for significance. The null hypothesis is $\rho_s = 0$ (that there is no correlation in the population from which the sample was drawn), and the alternative hypothesis is $\rho_s \neq 0$ (that there is a correlation in the population).

With the r_S equation, there is no need to establish the degrees of freedom. To check for significance, all we need to know is the value of N, the number of paired ranks. We turn to Appendix Table F. For an N of 10, we find correlations of .65 for an alpha error level of .05 and .79 for an alpha error level of .01. We compare our obtained value with the table values. If the absolute value of r_S is equal to or greater than the table value, we reject the null hypothesis.

$r_{S.01(10)} = .79$

$r_S = .81$ Reject H_0; significant at $P < .01$.

This correlation tells us that, at least within the military, women who are aggressive are more apt to be leaders. We must be careful in interpreting this finding, however. Despite the use of different observers, the same behaviors may have been seen both as aggressive and as signs of leadership.

Calculating r_S with Data Originally Given in Interval (or Ratio) and Ordinal Form

For our second example, we take the quite common situation in which the data are given in different measurement scales, one interval and one ordinal. The solution to this apparent dilemma is really quite simple—we convert the interval data to ordinal data and then calculate the r_S. Since interval data gives information as to greater than or less than status, it can therefore be rank ordered. Since ordinal data contains no information regarding how much greater or less, it cannot be converted into interval data. This is, of course, also true of ratio data, and in the example to follow, the "years served" distribution is actually composed of ratio values.

Example.   Assume that a researcher wishes to discover if there is a significant correlation between a congressional member's physical attractiveness and years in office. A random sample of eight members from throughout the country is selected and rank ordered on the basis of physical attractiveness. These rankings are paired with the number of years each has served in the House of Representatives.

These data are measures on two different scales. The rank order for physical attractiveness is clearly an example of ordinal scaling, whereas the number of years served represents at least an interval scale. Before calculating the r_S, these measures must be rank ordered. We assign a rank of 1 for the most years served, 24, and a rank of 8 for the fewest years, 6.

MEMBER	RANK FOR PHYSICAL ATTRACTIVENESS	YEARS SERVED
A	1	6
B	2	8
C	3	12
D	4	14
E	5	12
F	6	20
G	7	20
H	8	24

The Case of Ties.   Note that there are two ties, one at 12 years served and the other at 20. When ranking these data, we handle ties by adding the ranks at the given positions and dividing by the number of tied scores. The tied scores are then assigned the same average rank. For example, Member H, with 24 years, receives the rank of 1, first place, but Members F and G, each with 20 years, are tied for second and third place. Add the ranks 2 + 3; divide by the number of tied scores, 2; and assign both F and G the rank of 2.5. Also, Members C and E are tied for fifth and sixth place, each are given the average rank of (5 + 6)/2 or 5.5. The ranks, then, are as follows:

MEMBER	YEARS SERVED	RANK FOR YEARS SERVED
A	6	8.0
B	10	7.0
C	12	5.5
D	14	4.0
E	12	5.5
F	20	2.5
G	20	2.5
H	24	1.0

Note that Member D is assigned the rank of 4. This is because ranks 1, 2, and 3 have already been filled. Similarly, Member B is assigned the rank of 7.

Now that both measures are clearly ordinal (R_1 is the rank for attractiveness and R_2 is the rank for length of service), the r_S can be calculated as before.

As with the Pearson r, the null hypothesis is rejected whenever the *absolute* value of r_S is equal to or greater than the table value. This r_S is therefore significant, and we reject the null hypothesis.

Thus, it seems that more attractive congressional members are less likely to be reelected. We must be very careful of this interpretation, however. In this study, another extremely important variable, age, may be influencing or being influenced by the other two variables. Probably the members of Congress with the

MEMBER	R_1	R_2	$d = |R_1 - R_2|$	d^2
A	1	8.0	7.0	49.00
B	2	7.0	5.0	25.00
C	3	5.5	2.5	6.25
D	4	4.0	0.0	0.00
E	5	5.5	0.5	0.25
F	6	2.5	3.5	12.25
G	7	2.5	4.5	20.25
H	8	1.0	7.0	49.00

$$\Sigma d^2 = 162.00$$

$$r_S = 1 - \frac{6\Sigma d^2}{N(N^2 - 1)}$$

$$r_S = 1 - \frac{(6)(162)}{8(64 - 1)} = \frac{972}{(8)(63)} = \frac{972}{504}$$

$$r_S = 1 - 1.93 = -.93$$

$$r_{S.01(8)} = .88$$

$$r_S = -.93 \quad \text{Reject } H_0; \text{ significant at } P < .01.$$

longest service are also the oldest and, therefore, alas, perhaps considered the least attractive.

Problem 10b: In 1985, a list of the 10 highest-paid major league baseball players (excluding pitchers) was announced, along with their batting averages for the previous season. Although batting average is just one measure of a player's team value (not accounting for home-run production, defensive skills, etc.), determine the correlation, r_s, between salary rank and batting average. (This problem requires converting the batting averages into ordinal ranks.)

Answer: − .02

PLAYER	SALARY RANK	1984 BATTING AVERAGE
Mike Schmidt	1	.277
Jim Rice	2	.280
George Foster	3	.269
Dave Winfield	4	.340
Gary Carter	5	.294
Dale Murphy	6	.290
Bob Horner	7	.274
Rickey Henderson	8	.293
Eddie Murray	9	.306
Ozzie Smith	10	.257

Calculating r_S for Nonnormal Distributions of Interval Data

In the next example, the last for this chapter, we present a situation which, on the surface, may appear to be appropriate to the Pearson r rather than the r_s. That is, scores on two distributions, both of which are made up of at least interval data, are being assessed for the possibility of correlation. In this case however, the interval distributions deviate markedly from normality, thus violating one of the Pearson r's important assumptions. The rule is that if one or both sets of interval scores lack normality, then all the interval scores *must be converted into ordinal ranks,* and the r_s, not the Pearson r, must be chosen as the correlation equation.

Example. A university researcher wishes to assess whether or not a significant correlation exists between the amount of money contributed to the school by an alumnus and the age of the alumnus. It is obvious to the researcher that the distribution of alumni donations is skewed. The mean donation last year was more than \$100, but the mode was less than \$20 and the median less than \$40. Thus, a few large donors jacked up the mean and skewed the entire distribution toward the high end. The age distribution is also skewed, with most of the alumni less than 40 years old but a few in their 80s and even 90s. Both sets of interval scores, then, have to be translated into ordinal data before an analysis can take place. A random sample of 12 alumni is selected and both the size of their donations and their ages are noted.

ALUMNUS	AMOUNT OF DONATION	AGE (YEARS)
A	\$ 0	25
B	10	30
C	400	80
D	200	90
E	20	25
F	10	35
G	5	30
H	35	25
I	40	55
J	10	30
K	35	40
L	10	25

The rankings of the two sets of scores (R_1 is the rank for the size of donation and R_2 is the rank for the age of the donor) are as follows:

ALUMNUS	R_1	R_2	$d = \lvert R_1 - R_2 \rvert$	d^2
A	12.0	10.5	1.5	2.25
B	8.5	7.0	1.5	2.25
C	1.0	2.0	1.0	1.00
D	2.0	1.0	1.0	1.00
E	6.0	10.5	4.5	20.25
F	8.5	5.0	3.5	12.25
G	11.0	7.0	4.0	16.00
H	4.5	10.5	6.0	36.00
I	3.0	3.0	.0	.00
J	8.5	7.0	1.5	2.25
K	4.5	4.0	.5	.25
L	8.5	10.5	2.0	4.00

$$\Sigma d^2 = 97.50$$

$$r_S = 1 - \frac{6\Sigma d^2}{N(N^2 - 1)}$$

$$r_S = 1 - \frac{6(97.50)}{12(144 - 1)}$$

$$r_S = 1 - \frac{585}{12(143)} = 1 - \frac{585}{1716} = 1 - .34$$

$$r_S = .66$$

$$r_{S.05(12)} = .59$$

$r_S = .66$ Reject H_0; significant at $P < .05$

Thus, there is a significant positive correlation between alumni age and amount donated. A correlation of .66 is far from perfect, but predicting from it is better than chance.

Requirements for Using the r_S

To use the r_S, the following requirements must be met:

1. The sample has been randomly selected from the population.
2. Both distributions of scores must be in ordinal form.
3. The relationship between the two measures must be linear.

Deciding Which r to Use

The r_S was derived directly from the Pearson r. In fact, in situations where there are no tied scores, the r_S and the Pearson r yield exactly the same value. However, whenever the requirements for the Pearson r are met, it should definitely be used.

Converting interval scores into ordinal ranks because the r_S is easier to calculate is a mistake. Interval data contain more information than do ordinal data, and one should not throw away perfectly good information.

The Pearson r is a more sensitive test than the r_S. For a given sample size, an r of a certain value is more apt to be significant than is an r_S of the same value. That is, if there is in fact a correlation existing in the population, the Pearson r is more likely than the r_S to detect that correlation.

A quick look at the tables for r and r_S (Appendix Tables E and F) shows that for a given sample size, the Pearson r value can be consistently lower than the r_S value *and still attain significance.* With a sample of, say, seven pairs of scores (five degrees of freedom for the Pearson r), a Pearson r value of .75 reaches significance at the .05 level. For that same size sample, the r_S value needed for significance at .05 is a larger .786. Thus, the Pearson r value reaches significance, whereas the same r_S does not.

In Chapter 9 we discussed the concept of power. The Pearson r, a parametric test of correlation, is a more powerful test than is the nonparametric r_S, since with the Pearson r the likelihood of committing the beta error is reduced.

AN IMPORTANT DIFFERENCE BETWEEN THE CORRELATION COEFFICIENT AND THE *t* TEST

During the discussion of the t test in the preceding chapter, it was emphasized that both of the distributions being compared have to be composed of measures of the same trait. We cannot do a t test if the data are weight versus height scores, since the t test is only concerned with comparisons on a single measured variable. With correlation, however, where the focus is on the association rather than the difference between measures, qualitatively different measures can be compared. A correlation coefficient can be used to compare the same types of measurements, as in correlating heights of fathers with heights of sons. It can also be used to compare different measures, as in correlating heights and weights. Correlation is used to assess whether measures change together in some systematic fashion. Correlation does not ferret out a causal factor, but it can be utilized to make better than chance predictions.

SUMMARY

Correlation is defined as the degree to which two or more variables are associated. This chapter focuses on two-variable, or bivariate, relationships. When the association is direct, that is, high scores on the first measure pair off with high scores on the second *and* low scores on the first measure pair with low scores on the second, the correlation is said to be positive. When the association is inverse, that is, high scores on the first measure pair off with low scores on the second or low scores on the first pair off with high scores on the second, the correlation is said to be negative. When there is no consistent pairing of the measured scores, a zero correlation is said to exist. Through the use of correlation, even qualitatively different measures can be compared and stated as a precise quantitative value.

The major hazard when interpreting correlational studies lies in assuming that merely because two events are associated, one of the events must necessarily be the cause of the other. Although correlation does not imply causation, it is still an extremely useful tool for establishing better than chance predictions.

Correlations can be detected on a scatter plot, in which two distributions are graphed simultaneously. Each point on a scatter plot represents a pair of scores. When the array of points slopes from lower left to upper right, a positive correlation is portrayed. When the array slopes from upper left to lower right, a negative correlation is shown.

The procedure for calculating the correlation coefficient for interval data was introduced by Karl Pearson, who called it the product-moment correlation coefficient. It is now generally known as the Pearson r. The Pearson r ranges in value from $+1.00$ (a perfect positive correlation) through zero (no correlation) to -1.00 (a perfect negative correlation). The more the value of r deviates from zero, the greater is its predictive accuracy. The value of the r calculated on the basis of sample scores must be assessed for significance before being extrapolated to the population.

The coefficient of determination, r^2, provides specific information about a given correlation's predictive accuracy. By multiplying r^2 by 100, the approximate percentage of information about one variable that is supplied by the other variable can be determined.

If the data are in ordinal form or the interval scores are skewed, a different correlation coefficient, the r_S, is used. It was developed by Charles Spearman. The r_S, however, is a nonparametric test and is not as powerful as the Pearson r. Therefore, the r_S should not be used unless the requirements for the use of the Pearson r are not met.

KEY TERMS AND NAMES

coefficient of determination	homoscedasticity	scatter plot
correlation coefficient	Pearson, Karl	Spearman, Charles
Galton, Sir Francis	Pearson r	Spearman r_S

REFERENCES

1. FANCHER, R. E. (1985). *The intelligence men.* New York: W. W. Norton, p. 145.
2. GUILFORD, J. P. (1956). *Fundamental statistics in psychology and education,* (3rd. ed.). New York: McGraw Hill, p. 145.

PROBLEMS

1. Hypothesis: Among Republicans, there is a dependable relationship between party loyalty and number of hours per week spent working for their candidate. A 20-item "Party Loyalty Test" is devised and given to a random sample of 10 Republicans (high scores indicate greater loyalty). These scores were then compared with the average number of hours per week being volunteered to the candidate.

SUBJECT	PARTY LOYALTY SCORE	VOLUNTEER HOURS/WEEK
1	18	30
2	3	5
3	7	6
4	10	11
5	9	9
6	10	7
7	12	10
8	8	6
9	16	25
10	5	6

Test the hypothesis and state conclusions.

2. Hypothesis: Among elementary school children, there is a significant association between height and running speed. A random sample of eight children is selected. Each child is measured for height (in inches) and is timed in the 40-yard dash (in seconds).

SUBJECT	HEIGHT (INCHES)	RUNNING SPEED (SECONDS)
A	60	8
B	55	11
C	56	10
D	52	12
E	48	14
F	44	16
G	47	13
H	52	12

Test the hypothesis and state conclusions. Does being taller cause one to run faster? If not, what other factor might account for the relationship?

3. An investigator studying the relationship between anxiety and school achievement selects a random sample of 15 fifth-grade students, all aged 10 years. Each student is given an anxiety test (high scores signifying high anxiety) and

then these measures are paired with the student's score on an academic achievement test. The data are as follows:

SUBJECT	ANXIETY SCORE	ACHIEVEMENT SCORE
1	10	2
2	2	10
3	8	7
4	6	5
5	9	1
6	5	6
7	2	10
8	6	5
9	4	6
10	9	3
11	5	6
12	6	5
13	8	2
14	5	4
15	1	9

   a. Calculate the Pearson r correlation.
   b. Is it significant?
   c. Interpret these results. Does anxiety prevent some children from performing academically? Does a poor academic performance tend to produce anxiety in some children?

4. A group of 12 third-grade students is randomly selected and given a standardized arithmetic test. A trained observer watches the children for a period of 1 week and then rank orders them in terms of the amount of extraversion each child displays. The data are as follows:

STUDENT	ARITHMETIC SCORE	EXTRAVERSION RANK
1	100	12
2	50	8
3	90	7
4	65	6
5	87	5
6	75	9
7	80	10
8	95	11
9	80	3
10	75	4
11	60	1
12	75	2

   a. Find the rank-order correlation between arithmetic ability and extraversion.
   b. Is it significant?

5. A random sample of seven junior high students is selected, and each student is given both a math and a spelling test. Their scores are as follows:

MATH	SPELLING
15	13
5	6
16	14
10	13
11	11
3	5
12	10

Is math ability related to spelling ability?

6. Hypothesis: There is a significant correlation in the population between income and health. A random sample of 10 men, all 30 years old, is selected and given complete physical examinations. The men are then rank ordered by the physician in terms of their overall health. Each man's yearly income is also ascertained. The data are as follows:

SUBJECT	RANK FOR HEALTH	INCOME
A	2	$70,000
B	7	18,000
C	9	20,000
D	8	20,000
E	4	31,000
F	10	20,000
G	6	25,000
H	3	29,000
I	5	27,000
J	1	40,000

Test the hypothesis and state conclusions. Do men tend to earn more income because they are healthy, or are they healthy because they can afford better medical care, or might there be another explanation?

7. Some physiologists believe that language skills are controlled by one brain hemisphere and musical skills by the other. To test this, a researcher is studying whether musical ability correlates inversely with reading ability among 6-year-old children. A random sample of 6-year-olds is selected and given a reading comprehension test. Each child is also evaluated by a musicologist (for pitch, rhythm, etc.) and then rank ordered on musical ability. The data are as follows:

SUBJECT	READING SCORES	RANK FOR MUSICAL ABILITY
A	120	10
B	100	12
C	100	11
D	100	13
E	95	9
F	90	6
G	80	8
H	80	7
I	78	3
J	76	5
K	70	4
L	62	1
M	41	2

Test the hypothesis and state the conclusions.

8. Incomes (in millions of dollars) were announced for the highest paid entertainers for the past two years (1987–1988). A random sample of college students was selected and asked to rank order these performers in terms of their perceived "entertainment value." The incomes and median ranks for each performer were as follows:

NAME	INCOME (IN MILLIONS)	ENTERTAINMENT RANK
1. Michael Jackson	97	4
2. Bill Cosby	92	5
3. Sylvester Stallone	63	12
4. Eddie Murphy	62	1
5. Bruce Springsteen	61	3
6. Madonna	46	8
7. Johnny Carson	40	6
8. Kenny Rogers	26	2
9. Tina Turner	25	9
10. Prince	24	13
11. Jane Fonda	23	7
12. Steve Martin	22	15
13. Wayne Newton	21	14
14. Bruce Willis	17	10
15. Frank Sinatra	16	11

    a. Determine the correlation between salary rank and entertainment value.
    b. Test the correlation for significance.
9. The correlation between how much each state spends on education and the IQs of its school children is .70. Data are taken from J. N. Spuhler and G. Lind-

zey, "Racial Differences in Behavior," in J. Hirsch, ed., *Behavior Genetic Analysis* (New York: McGraw-Hill, 1967).

    a. Does this mean that increased state spending on education leads to an increase in the IQs of the children in the state?

    b. What other factors might produce this correlation?

10. Why does the array of points in a scatter plot usually form an oval shape?

11. When the data are in interval form and the distributions are normal, which correlation coefficient should be used?

12. When the data are in ordinal form, which correlation coefficient should be used?

13. When the data are in interval form, but the distribution is known to be heavily skewed, which correlation coefficient should be used?

14. When possible (that is, all of its requirements are met), why should a researcher choose the Pearson r over the r_S?

True or False—Indicate either T or F for problems 15 through 25.

15. If the correlation between X and Y is high and the correlation between X and Z is high, then the correlation between Y and Z must be high.

16. Significant correlations always predict better than chance.

17. Negative correlations, even when significant, never predict better than chance.

18. The higher the correlation between X and Y, the more information about Y is contained in X.

19. Whenever a correlation is significant, the possibility of a cause-and-effect relationship is totally ruled out.

20. The higher the Pearson r, the higher is the coefficient of determination.

21. A Pearson r of .90 means that the percentage of information about Y contained in X is roughly 81%.

22. If sample sizes are equal, the Pearson r and the r_S predict with exactly the same degree of accuracy.

23. If the range of either sets of sample scores is in any way restricted, the Pearson r will overestimate the degree of correlation.

24. In order to use the Pearson r, both sets of paired scores must be composed of at least interval data.

25. The Pearson r assumes that the association between X and Y is always curvilinear.

chapter 11

The Fundamentals
of Research Methodology

Without some knowledge of research methodology, the bare calculation of statistical tests is an empty exercise in mathematical procedures. Without some background as to how and when they are used and how their results are interpreted, *t* tests and correlation coefficients are just so many sterile and meaningless calculations. This chapter, therefore, provides an overview of the fundamental concepts of research methodology. The importance of these fundamentals cannot be overemphasized. Basic knowledge of the logic of research procedures puts you in a position to understand a goodly portion of the statistical research in your field. Without this knowledge, research literature may be somewhat like a credit card debt—easy to get into, but almost impossible to get out of. Besides, the fundamental research methods are not difficult to comprehend.

It's the ability to analyze and understand the data that breathes warm life into cold numbers. Merely reading research conclusions, without attempting to find out if they are justified, puts you at the mercy of the researcher's often subtle and perhaps even unconscious biases. This chapter, then, is a mixture of one part technical procedures and one part honest skepticism.

We often hear the derogatory comment, "Statistics can be used to prove anything." Sometimes we are addressed with the equally cynical, and somewhat plaintive, reproach, "But that's only a statistical proof." If it were true that one can prove anything with statistics, then, of course, there would be a real question as to the value of statistical analysis. The fact is, however, that the only time one can prove *anything* with statistics is when the audience is totally ignorant of statistical procedures. To the uninitiated, liars can indeed figure, and their figures may be seductively plausible. By the time you finish this chapter, they will not easily be able to lie to you.

RESEARCH STRATEGIES

Although there are a variety of general research techniques, such as naturalistic observation, descriptive, historical, case study, surveys, and so forth, the emphasis in this chapter will be on two of the major research strategies that demand inferential techniques of hypothesis testing: experimental and post-facto. **Experimental research** offers the best opportunity for drawing cause-and-effect conclusions, whereas **post-facto research** really doesn't. In fact, post-facto research doesn't truly address the issue of cause and effect; it doesn't prove or disprove causation. Thus, the problem boils down to this: if the researchers have used experimental methodology, then they are allowed by the rules of science to discuss the possibility of having isolated a causal factor, but if the research is post-facto, they should not. And as a person new to the field, you must keep in mind when reading research that it's not enough just to read the report's conclusions. You must examine the entire methodology to be sure that the author's conclusions are indeed justified. Unfortunately you just can't take the author's word for it when it comes to the "conclusions" section of the report. You must bore in yourself, and carefully read the "methods and procedures" sections with a critical eye. Above all, don't assume that something was done simply because it seems obvious to you that it should have been done. You'll soon learn that what ought to be isn't always what is!

Variables and Constants

A variable is anything that can be measured, and, although it might seem to go without saying, that *varies*. A person's height, weight, shoe size, intelligence, and attitudes toward war may all be measured, and, in comparison with measures on other people, all have the potential to vary. Indeed, the fact that people differ on a whole host of personal measures gave rise in the late 1800s to the study of what Sir Francis Galton called "individual differences." No two people are exactly alike on all human measures (even identical twins reared together are not exact personality clones), and, thus, the study of individual differences became the major theme song in psychology, education, and the social sciences. But, just because people differ on a wide range of personal measures, it does not mean that each and every study of potential differences has indeed allowed these measures to vary. Or, to state it differently, potential variables are not always treated by the researcher as variables. This is, after all, the way it should be. Each single study should not be obliged to allow all possible measures to vary simultaneously. Some measures should be left as *constants* or measures that are not allowed to vary. Here's an example.

It is known to all researchers in the area of growth and development that among elementary school boys there is a dependable relationship between height and strength—that is, taller boys tend to be stronger than shorter boys. Now, although this is a perfectly valid piece of research evidence, and it certainly allows the researcher to make better than chance strength predictions using height as the predictor, it can lead to some fuzzy interpretations if other variables are not controlled. For example, on the basis of the height-strength correlation alone, one might speculate that longer muscles produce more strength than do shorter muscles. The problem with this interpretation so far is that all other measures have been free to vary, and it may be that some of these other measures correlate even higher with strength than does the height measure. One that comes quickly to mind is age. Certainly 12-year-old boys tend to be both taller and stronger than 6-year-old boys. What would happen, then, if we were to hold age as a constant?

Suppose we were to select only boys of age 10 years 6 months and then attempt to assess the height–strength relationship. When this is done (and it has been), the original correlation between height and strength becomes minimal. Speculation now might be concerned with the maturity of the muscle, not its length, as the important component in strength. The point of all this is that when, as in this case, only subjects of the same age are used, age, although certainly measurable, is *no longer a variable*. In reading research articles, then, be especially careful to review how the subjects have been selected.

If the selection process has been designed to ensure that the sample is identical on some measured trait, then for that particular study the common trait is not a variable but a constant. This is also true of studies in which environmental conditions are under examination. For example, a researcher may suspect that warm-color room decors are more conducive to relaxation than are cool-color decors. Two groups of college students are selected, and one group is taken to a room with a red–yellow decor and the other to a blue–green room. The subjects are then connected to some biofeedback equipment which is supposed to assess their levels of relaxation. Here again, certain potential environmental variables should be held constant if indeed the researcher hopes to get at the effects of room color. The groups should be tested at the same time of day and under the same conditions of illumination, temperature, noise level, and so on. That is, the researcher should make every effort to control as many potential environmental variables as is possible—in short, to convert these other variables into constants to minimize their possible influence. Of course, a study of this type should also be designed to control individual difference variables as well. For example, one group should not be composed only of overweight, phlegmatic persons and the other, underweight, tense persons. Much more will be said on this topic later under the heading "confounding variables."

INDEPENDENT AND DEPENDENT VARIABLES

Scientists identify two major classes of research variables (variables, not constants) as independent and dependent. It is typically assumed that the **independent variable** precedes the **dependent variable,** and, thus, in any antecedent–consequent relationship, the independent variable (I.V.) is the antecedent and the dependent variable (D.V.) the consequent. Your ability to identify these variables is absolutely critical to your understanding of research methodology. In some studies, which as we will see later are called "experimental," the I.V. is truly antecedent in every sense of the word. In these studies, the I.V. is assumed to be the causal half of the cause-and-effect relationship, with the D.V. being the effect, or consequent, half. For example, in the study mentioned earlier on room color and relaxation, the room color (the presumed cause) would be the I.V. and the amount of resulting relaxation the D.V. In another type of study, which will be called *post-facto,* the I.V. is antecedent only to the extent that it has been chosen (sometimes arbitrarily) as the predictor variable, or the variable used from which to make the prediction. The D.V. in this type of research is the consequent variable to the extent that it is the *variable being predicted.* For example, in the TV violence studies, since the researchers were trying to use type of TV viewing to predict the extent of viewer aggression, the I.V. would be the type of TV that was viewed and the D.V. would be the amount of overt aggression being displayed. In general, then, the I.V.–D.V. relationship is similar to an input–output relationship, with the I.V. as the input and the D.V. as the output. Also, the D.V. is usually some measure of the subject's behavior, and in this case, behavior is being defined in its broadest possible context. The D.V. might be a behavioral measure as obvious as the subject's performance on an IQ or attitude test, or as subtle as the subject's incidence of dental cavities, or even the measured amount of acetylcholine at the neural synapse.

Researchers distinguish between two categories of independent variables, manipulated and subject. Moreover, this distinction is absolutely crucial to an understanding of research methodology. A decision regarding which type of I.V. is involved in turn determines which type of research is being conducted—and, hence, the kinds of conclusions which are permissible. A manipulated I.V. is one where the researcher has actively manipulated the environmental conditions to which the sample groups are being subjected. That is, with a manipulated I.V., the researcher makes the determination as to which groups of subjects are to be treated in which particular ways. Perhaps one group of subjects is being tested under low-illumination conditions and the other group under high illumination. Illumination would thus be the independent variable, and notice that it does and must vary, in this case from low to high. (If all subjects had been tested under identical conditions of illumination, then illumination could not be an independent or any other kind of variable.) Or, if the researcher in the field of learning were attempting to show that a fixed ratio reinforcement schedule of 3 to 1 (3 correct responses for each reinforcement) leads to more resistance to extinction than does a continuous reinforcement schedule (reinforcing each response), then the type of reinforcement schedule which has been set up by the researcher would be the manipulated I.V., while resistance to extinction would be the D.V. Thus, whenever the experimenter is fully in charge of the environmental conditions (the stimulus situation in which the subject works) and these conditions are *varied*, then the I.V. is considered to have been manipulated.

On the other hand, if the subjects are assigned to different groups, or are categorized in any way, on the basis of trait differences they already possess, the I.V. is considered to be a *subject variable*, or as it is sometimes called, an assigned variable. Thus, a subject I.V. occurs when the researcher assigns subjects to different categories on the basis of a measured characteristic and then attempts to discover whether these assigned variables either correlate with or differ from some measure of the subject's response, this response measure again being the D.V. For example, it might be hypothesized that women have higher IQs than do men. The subjects are selected and then categorized on the basis of gender, a subject I.V. The two groups then both take IQ tests, and the IQ scores (the D.V.) are compared. Or a researcher wishes to establish whether college graduates earn more money than do non-college graduates. The researcher selects a large sample of 35-year-old men, assigns them to categories on the basis of whether or not they graduated from college, and then obtains a measure of their incomes. Level of education would then be a subject I.V. and income level the D.V. Notice that in this example, both age and sex were not allowed to vary and are, thus, constants. Subject characteristics, such as age, sex, race, socioeconomic status, height, and amount of education may only be subject I.V.s. Such variables are simply not open to active manipulation. Some independent variables, however, can go either way, depending on how the researcher operates. For example, it might be hypothesized that fluoridated tooth paste reduces dental caries. The researcher might randomly select a large group of subjects and then *provide* half of them with fluoridated tooth paste, while also providing the other half with identical-appearing, nonfluoridated tooth paste. Then, perhaps a year later, the groups are compared regarding production of caries. In this case, then, the I.V., whether or not the tooth paste contained fluoride, would be manipulated. If, however, the researcher had simply asked the subjects what kind of tooth paste they usually used and categorized them according to whether the tooth paste did or did not contain fluoride, then the I.V. would have been a subject variable. The D.V., of course, would still be based on a comparison of the incidence of caries between the two groups.

THE CAUSE-AND-EFFECT TRAP

One of the most serious dangers lurking out there in the world of research is the so-called "cause-and-effect" trap. Too often the unwary reader of research is seduced into assuming that a cause-and-effect relationship has been demonstrated, when, in fact, the methodology simply doesn't support such a conclusion. For example, a study (which shall remain nameless, to protect both the innocent and the guilty) was conducted several years ago that purported to show that sleeping too long at night promoted heart attacks. Headlines throughout the country trumpeted this early-to-rise news, and America reset its alarm clocks. The message was stark—don't sleep too long or you'll die! The study, however, which led to all this sound and fury was done in the following way. A group of recent heart attack victims, all male, were questioned as to, among other things, their recent sleep habits. It was found that, as a group, they had been sleeping longer than had a comparison group of men of roughly the same age, weight, exercise patterns, and so on. That is, the independent variable, amount of sleep, was a subject variable and was *assigned after the fact*. Because of this, the study does not prove a causal link between sleep time and heart attacks. In fact, it's just as likely that the reverse is true, that is, that the same physiological conditions that led to the heart attack might also have led to feelings of fatigue and, therefore, a desire to stay in bed longer in the morning. The point is, that although both these explanations are possible, neither is proven by the study. When a cause-and-effect relationship is actually discovered, it must be unidirectional; that is, there must be a one-way relationship between the I.V. and the D.V. When a light switch is flipped on, the bulb lights up, and this is unidirectional since unscrewing the bulb doesn't move the light switch.

RESEARCH: THE TWO BASIC TYPES

The two basic types of research are experimental and post-facto, and only the experimental method truly allows for the possibility of isolating a causal factor. How do you tell the difference? Closely examine the independent variable. If the I.V. has been actively manipulated, the research is experimental. If not, it's post-facto. In the experimental method, then, the researcher always actively manipulates the independent (causal) variable to see if by doing so it produces a resulting change in the dependent (effect) variable.

In post-facto research, on the other hand, the researcher does not manipulate the independent variable. Rather, the independent variable is *assigned*. That is, the subjects are measured on some trait *they already possess* and then are *assigned to categories on the basis of that trait*. These trait differences (independent variable) are then compared with measures which the researcher takes on some other dimension (dependent variable).

Post-facto research precludes a direct cause-and-effect inference, because by its very nature it cannot identify the direction of a given relationship. For example, suppose that a researcher discovers that among students there is a significant relationship between whether or not algebra is taken in the ninth grade and whether or not the student later attends college. Since taking algebra in the ninth grade was the student's own decision, or perhaps that of the parents, this I.V. had to have been a subject variable, and the research, therefore, post-facto. Perhaps parents, or guidance counselors, who view a student as college bound encourage that student to elect algebra and also discourage the student whose professed goal is to become a garage mechanic. Or perhaps a highly intelligent student, knowing that he or she eventually wishes to attend college, is self-motivated to elect ninth-grade algebra. Since the direction of the relationship is often ambiguous in post-facto studies, isolating a causal factor becomes extremely difficult. Post-facto research does, however, provide the basis for better than chance predictions. (Correlational research is one form of post-facto research.)

In both psychology and sociology, experimental research is sometimes called S/R research, since an (S) stimulus (independent variable) is manipulated and a corresponding change in response (R, dependent variable) is sought. Similarly, post-facto research is sometimes called R/R research, since the responses (R), of a group of subjects are measured on one variable and then are compared with their measured responses (R) on another variable.

THE EXPERIMENTAL METHOD: THE CASE OF CAUSE AND EFFECT

In the experimental method, the relationship between the independent and dependent variables is unidirectional, since a change in the independent variable produces a change in the dependent variable. In its simplest form, the experimental method requires two groups, an experimental group which is exposed to one level of the independent variable and a **control group,** or comparison group, which is exposed to a different level of the independent variable. The two groups, experimental and control, must be as much alike as it's humanly possible to make them. The two groups are then compared with regard to the outcome, or dependent variable, and if a significant difference exists between the two groups, the independent variable can be said to have caused the difference. This is because all the other potential variables existing among the subjects in the two groups are presumed to have been held constant, or *controlled.* For example, suppose that you have been perusing the physiological literature and notice that a certain drug, magnesium pemoline, causes an increase in the production of one form of RNA in the cerebral cortex. Other studies then come to mind that seem to suggest that cortical RNA may be linked to human memory (through its role in protein synthesis). From reading all these studies and meditating on possible relationships, you begin to induce the hypothesis that perhaps the drug, magnesium pemoline, might lead to an increase in human memory. You decide to test the hypothesis by designing an experiment. First, you select two groups of students, and you deliberately attempt to make these two groups as much alike as possible. You try to hold constant all those variables that might possibly relate to memory, such as IQ, age, grade-point average, and so on. One of the groups, the experimental group, is then given the drug, whereas the other (control) group is not. It would also be important that both groups be situated in identical environmental conditions—

same type of room, same illumination, same temperature, and so on. That is, the two groups should be identical in every respect, *except* that one receives the drug and the other does not. Ideally, the subjects should not be aware of which group they are in, for it is possible that if subjects knew they were in the experimental group, that in itself might affect them, perhaps make them more motivated. For this reason, when the members of the experimental group are given a capsule containing the drug, the subjects in the control group are given a nonactive capsule, called a placebo. Actually, the person conducting the experiment shouldn't even know which group is which. This prevents any possible experimenter bias such as unconsciously encouraging one group more than the other. When neither subjects nor experimenter are aware of which group is which, the experiment is said to be a **double-blind.** (Obviously someone has to know which group received the drug. Otherwise, the results would be impossible to analyze.)

Finally, both groups would then be given a memory test of some sort, and if the scores of the subjects in the experimental group average out significantly higher than do those in the control group, a cause-and-effect relationship may legitimately be claimed. An inferential statistical test, such as the *t* test, could be used to determine if the difference between the means of the experimental and control groups is large enough to reject the null hypothesis.

In this example, whether or not the subjects received the drug would be the independent variable. This would be a *manipulated* independent variable, since it was the experimenter who determined which subjects were to receive how much of the drug. The subjects were not already taking the drug, nor were they given the opportunity to volunteer to take the drug. The dependent variable in this study would be the subjects' measured memory scores.

In experimental research, the independent variable

1. Is actively manipulated by the experimenter.
2. Is the causal half of the cause-and-effect relationship.
3. In the fields of psychology, sociology, and education, is always a stimulus—that is, some environmental change which impinges on the subject.

The dependent variable in experimental research

1. Is always the effect half of the cause-and-effect relationship.
2. In most social sciences, is usually a measure of the subject's response.

EXPERIMENTAL DESIGNS: CREATING EQUIVALENT GROUPS

In experimental research the subjects in experimental and control groups must be kept as nearly alike as possible. The reason is that if the groups of subjects were systematically different to begin with, then significant differences on the dependent variable would be impossible to interpret. One could not tell if these differences on the dependent variable resulted from the manipulation of the independent variable, or were merely due to the fact that the groups differed at the outset on some important dimension. If one wished to study the effects of training on reading speed, it would be obviously the height of folly to place all high-IQ subjects in one group and low-IQ subjects in the other. An experiment that is tightly controlled, that has no systematic differences between the subjects to begin with, is said to have **internal validity.**

Internal Validity

Researchers use the term "internal validity" to describe the extent to which the results of an experiment can unambiguously identify a cause-and-effect relationship. An experiment that is high in internal validity is considered to be one that is relatively free from the contaminating effects of confounding variables, or one in which the pure effects of the independent variable can clearly be isolated and interpreted. Perhaps the major factor in achieving internal validity is the implementation of an adequate control group.

External Validity

In experimental research, **external validity** describes the extent to which the experiment's results can be extrapolated to a real-life population. An experiment is considered to be high in external validity when it simulates real-life settings and its results, therefore, are generalizable to a naturally occurring (as opposed to an artificial) population. Examples of experiments low in external validity would be those conducted on rats whose results are then immediately extrapolated to the population of college sophomores. Or a study that is based on evaluating white, upper-class, prep school students would be low in external validity were its results then generalized to a lower-class, ghetto, student population. In other words, external validity addresses the question of whether or not the independent-dependent variable relationship found in the experiment would still hold in other settings to which the results are being generalized.

External validity demands, among other things, representative sampling.

Representative Sampling

The key to unlocking the sampling problem is to select samples that are *representative* of the defined population. A representative sample is one that truly reflects all the various characteristics of the population to which the results are to be generalized.

Random Samples and Equivalent Groups

Since a random sample is designed to reflect accurately the characteristics of the population from which it was selected, then selecting two random samples *from the same population* should give us confidence that these two samples are equivalent in all important respects. Since both samples share the same trait colorations of the population, it seems safe to assume, at least on a probability basis, that they must also contain the same gradations of traits as each other. Is this a 100% guarantee? No, since it is possible, even with random sampling, that the groups might differ from each other. Nevertheless, on balance, we can expect that the random sampling process has probably ironed out any original differences that might have existed between the sample groups.

Randomized Assignments

Another technique, closely allied to this, is called *randomized* assignment of subjects. In this technique the sample is not originally selected randomly from a single population (since this is often a lot easier to talk about than do) but is, instead, presented to you as a fact. For example, your professor says that you can use her 9 o'clock section of intro psych students or your fraternity volunteers as a group to participate in your study. Now, although this is not the ideal way to obtain a research sample, it is, alas, often the best that you're going to do. The solution is to, strictly by random assignment, divide the large group into two smaller samples. This is called randomized assignment, since, although the original sample was not randomly selected, the random assignment process is still used to divide the whole group into two, it is hoped, equivalent groups. The theory here is that even though your original group may not truly represent the population, at least the two smaller randomly assigned groups do mirror each other—and are, therefore, equivalent. In passing, note that in either the random samples or randomized assignment techniques, the sample groups are independent of each other. That is, the selection of one subject for one group in no way influences which subject is to be assigned to the other groups.

Dependent (or Correlated) Selection

Another method of obtaining equivalent groups is simply to select one group and then use that group under the different experimental conditions. The theory here is that no one on the planet is more like you than you, yourself, and so you are used in both the experimental and control(s) conditions. This method, although seemingly both simple and pure, actually opens up a veritable Pandora's box of pitfalls and, as we shall see later, should be used sparingly, if at all. When groups are formed in this way, the groups are obviously not independent of each other, since the same persons are used in each condition.

Finally, the last major method of creating equivalent groups is to select one group (ideally by random selection) and then create a second group by matching the subjects, person for person, on some characteristics that might relate to the dependent variable. For example, if one were to test the hypothesis that caffeine affects running speed, you would first attempt to list other variables that might influence the dependent variable or, in this case, running speed. Such variables as age, weight, height, sex, and physical conditioning come immediately to mind. Thus, if one of your subjects is a 16-year-old female who is 5 feet 5 inches tall, weighs 150 pounds, and typically runs 5 miles every day, you would attempt to find another person who closely resembles this subject on these four variables. Similarly, if one subject is a 30-year-old male who weighs 200 pounds, is 5 feet 8 inches tall, and hasn't done any running for several years, then you must again find a male subject (this should be easy) who is 30 years old, overweight, and out of shape. The members of a given matched pair, of course, should always be assigned to separate groups. Notice that these groups, then, will not be independent of each other, since the selection of one subject for one group totally determines who will be selected for the other group.

Experimental Designs

There are three major types of **experimental designs,** and each has as its primary goal the creation of equivalent groups of subjects. This does not mean that all the subjects in both groups will be absolute clones of each other. There will be individual differences *within* the groups, some subjects being taller or smarter, whatever, than other subjects within the same group. It does mean, however, that on balance, the two groups average out about equal on any characteristics that might influence the dependent variable.

The After-Only Experimental Design: Between Subjects

In this design the subjects are either randomly selected from a single population or else are placed in the separate sample groups by randomized assignment, or both. In any case, the presumption is that chance will assure the equivalence of the groups. Further, the subjects, in this design, are measured on the dependent variable *only* after the independent variable has been manipulated. This is deliberately done to avoid the possibility of having the D.V. measures possibly affect each other. For example, suppose we wish to find out whether the showing of a certain motion picture might influence the viewer's racial attitudes. In the after-only design, we only test the racial attitudes of the subjects *after* the I.V. has been manipulated, in this case, after the period of time when one group has seen the movie and the other group has not. The reasoning behind this procedure is that perhaps if the racial attitudes had been tested before the showing of the movie, the test itself may have influenced the way the viewers perceived the film, perhaps heightening viewer awareness of the racial content or sensitizing the viewers to the racial theme of the movie. Also, since the groups in the after-only design are set up on the basis of random selection and/or randomized assignments, the groups are known to be independent of each other, and since this is a completely randomized design, each subject is tested under only one treatment condition, so that each score represents a different subject. In this design there will always be the same number of scores as there are subjects. Hence, we say that the D.V. comparison is *between* subjects.

Finally, although we have been dealing with the after-only design as though it were always a two-group design, one experimental and one control group, this procedure can just as readily be used on multigroup designs, a topic we will cover in some depth in Chapters 12 and 16. The number of groups involved is a direct function of the number of levels of the independent variable being manipulated in any given study. If the I.V. is manipulated at only two levels, one group receiving zero magnitude of the I.V. (control group) and the other some specified magnitude of the I.V. (experimental group), then a two-group design is adequate. But, as you will soon see when reviewing the research literature, many experiments are produced in which the I.V. is manipulated at three, four, or even five levels.

For example, a researcher may wish to know not just whether or not seeing a film about racial prejudice will then reduce antiminority attitudes among the viewers, but whether increasing the number of racially intense scenes within the movie has even more impact on viewer prejudice. For each different version of the movie, then, another experimental group must be added. To be truly an after-only design, however, none of the groups, experimental or control, should be tested on the prejudice scale before seeing the movie.

The Before–After (Within-Subjects) Experimental Design: Repeated Measures on Same Subjects

On the surface, the design which is the most seemingly basic, the most obvious from a commonsense point of view, and yet also the one most fraught with the potential for dangerous ambiguities, is the before–after experimental design. Equivalent groups of subjects are formed by the simple expedient of using the same group twice. Rather than trying to establish differences between separate groups of subjects, the researcher focuses instead on the possible differences that might occur *within subjects* who are being measured under different treatment conditions. Could anything be more straightforward? But it sure can make the data interpretation a risky venture.

Advertisers constantly bombard us with examples of this basic design. You've seen the pictures in a before–after ad, the man looking fat, puffy, stupid, and sad in the "before" picture and, then, miraculously, looking slim, hard, lean, proud, and exhilarated in the "after" pose. You are told that you too can achieve this state of absolute bodily perfection by sending for the "Joe Hardguy" booklet and following its simple conditioning regimen, which, incidentally, only takes 10 minutes a week to perform. Now, look again at the pictures and count the number of variables that have changed, other than body shape. The poses are entirely different, first slouched, then ramrod straight. The facial expressions vary, from pathetically depressed to joyously self-confident. The clothes differ, the lighting differs, and on and on. In this example, trying to isolate the pure effects of the independent variable is like trying to find a needle in the proverbial haystack.

Let's look at how this design is presumed to work. In this design, a single group of subjects is selected, usually randomly, and then measured on the dependent variable. At this stage, the subjects are considered to be members of the control group. In this design, the subjects are obviously not independent of each other. Then, the independent variable is manipulated, and these same subjects are again measured on the D.V. At this point, the subjects are considered to be members of the experimental group. If the D.V. scores in the "after" condition differ significantly from those in the "before" condition, the independent variable is then assumed to have caused this difference. And this assumption, let us quickly add, is based on the allegation that nothing else (other than the I.V.) has changed in the lives of these subjects. It's this allegation that often contains this design's fatal flaw. Let's take an example. A researcher is interested in testing the hypothesis that teaching the "new math" increases math ability among sixth-grade, elementary school children. A random sample of sixth-graders is selected from the population of a large, metropolitan school district, and the chosen children are all given a standardized math ability test (measured on the D.V.). Then the I.V. is inserted, in the form of a 12-week "new math" program, replete with Venn diagrams and set theory. Then the math ability test is given again (D.V.), and the two sets of scores are compared. Let's assume that the second math measure was significantly higher than the first, thus, seeming to substantiate the efficacy of the teaching program. But was that the only variable in the child's life that changed during that 12-week time span? Of course not! First, the child had 12 weeks in which to grow, mature, and also practice his or her math skills, perhaps using the new math during school time and the old math after school (when making subtractions from his allowance to see whether he can afford to buy a new baseball or adding the total minutes of her piano practice time to prove to mother that she can go outside to play). Also, the mere fact of having been selected for this new program may have caused the children to feel somehow special, therefore increasing their feelings of motivation.

A final problem with the before–after design is that the very act of measuring the subjects on the D.V. the first time might influence how they respond to the D.V. the second time. For example, the subjects might become *sensitized* to the test, thus destroying the internal validity of the experiment. Thus, in the previous example, the act of taking the standardized math test the first time may have made some of the children more aware of what they didn't know or had forgotten, and therefore prodded them into reviewing the rules on adding fractions, or whatever. When subjects become sensitized to the first test, they become more responsive to the second test, even without the manipulation of the independent variable. Sometimes, on the other hand, subjects become fatigued by the first test. When this occurs, the subjects become less responsive to the second test. To take a blatant example, assume that a researcher wishes to study the effects of drinking coffee on running speed. A large sample of college students is selected and is told to report to the football field. They are then told to run eight laps and their running times are clocked. They are then given a cup of coffee to drink and are told to run another eight laps. Their running times are again clocked and are found to be significantly lower. Here we have the classic before–after design, running times being measured both before and after the introduction of the independent variable (coffee). However, the "before" measure so tired the subjects that the coffee probably had little to do with the lowered speed recorded in the "after" measure. This is the kind of study where the result could have been written before the data were even collected. Another hazard from the before–after design results from the passage of time. Since some amount of time must elapse between the pre- and postmeasures, perhaps the mere passage of time has changed the subjects in some important way. In short, literally dozens of other variables may have been impacting the lives of these children, and their higher math scores may have really resulted from these uncontrolled variables *and not* from the independent variable. (More will be said on this issue in our later discussion of the Hawthorne effect.) Suffice it to say here that whenever you come across a research study using a one-group, before–after design, be alert to the possibility of uncontrolled variables intruding on the results.

Before-After Designs with Separate Control Groups

An especially powerful variation of the before–after design occurs when completely separate control groups are used. In fact, the use of a separate control group greatly minimizes the dangers inherent in the traditional before–after design. For example, many of the problems we encountered with the "new math" research could have been diminished or even eliminated had a separate control group been added. That is, instead of selecting a single group of sixth-graders, the researcher could have (and should have) selected two equivalent groups. Both groups would then be measured on the math ability test, and then both groups would also enjoy the "heady" pride of being placed in an apparently special math program. At this point, the experimental group gets the 12-week "new math" course, and the other group, the control group, goes to a 12-week math course that is taught in the traditional manner. This should help in controlling the previously described uncontrolled variables—maturation, practice sensitization, fatigue, and motivation. Even here, however, there could still be a problem regarding the qualities of the teachers. If possible, to control for this variation, the same teacher should be used for both groups, a teacher, incidentally, who must consistently exhibit the same degree of enthusiasm for both instructional techniques.

In a study such as this, the analysis is based on discovering how much each subject changed from the pre- to the post-measure. If the members of the experimental group change significantly more than do the members of the control group, then the independent variable is assumed to have caused this difference. The analysis of the "change" scores is crucial when using this design, since in many of these studies the control group can also be expected to change. The question, then, should be: Which group changed the most?

When setting up the separate control group for this design, the researcher has two choices:

1. The control group may be independent of the experimental group, as when two separate random samples are selected from the targeted population or when a nonrandom sample is divided into two groups by randomized assignment.
2. The separate control group may depend on the experimental group, as when the subjects in the two groups are matched on the basis of some variable(s) deemed relevant to the dependent variable.

The before–after design should be used sparingly, and ideally should be restricted to those situations in which the researcher is investigating the effects of one or more treatment conditions on behavior under subsequent conditions. The subjects are measured and then are presented with Tasks A, B, and C before being measured again.

Sequencing Effects

When subjects in repeated-measure designs are given several tasks to perform, it becomes obligatory to ferret out the various effects that may be produced by the sequencing of the tasks. This becomes particularly important when those twin villains, practice and fatigue, are involved. The problem here is that when subjects are given several tasks to perform, that is, involved in several treatment conditions, *sequencing effects* can occur. The subjects may be just as affected by the sequence of experimental conditions as they are by the presumed independent variable. For example, suppose that subjects were offered a one-dollar reward for successfully completing Task A and a one-hundred-dollar reward for Task B. If the tasks are similar in difficulty, and if the subjects are not totally fatigued from completing Task A, we would suspect an increase in performance on the second task. However, if the one-dollar prize came second in the sequence, the subjects' performance might decrease as a result of the dollar being seen as far less rewarding than it would have been had the previous prize of one hundred dollars not been presented first.

Counterbalancing

To control for sequencing effects, researchers have created a technique called **counterbalancing.** When two experimental conditions are involved, each subject is faced with two tasks that are sequenced in the order ABBA. That is, the subjects are given Task A, then B, then B again, and finally A again. Although this technique does go a long way toward balancing out any possible sequencing effects, it is still suggested that a *separate control* group be used in which the order of presentation becomes B, A, A, B. In short, counterbalancing techniques, when used properly, tend to even out any possible sequencing effects, and thus help to prevent confounding the independent variable. Each experimental condition precedes and follows all other conditions the same number of times, and, ideally, each condition occurs the same number of times at each position in the sequence.

The Matched-Subjects Experimental Design

This design makes use of the aforementioned subject—matching technique for creating equivalent groups. The subjects in the control and experimental conditions are equated person for person on whatever variables the researcher assumes might possibly be related to the dependent variable. That is, each member of the experimental group has his or her counterpart in the control group. This design may be used as part of the after-only technique where the D.V. is only measured after the introduction of the I.V. (Again, this is done to prevent the first D.V. measure from, in any way, influencing a subject's performance on the second D.V. measure.) This method may also be used as part of a repeated-measure design, where a separate control group is set up on the basis of this person-for-person matching technique.

The matching process, especially when a large number of relevant variables is used, can become extremely difficult. It may even become virtually impossible to find a suitable matched pair, who resemble each other closely on all these important variables. As a general rule, the more variables used in the matching process, the more difficult becomes the task of selecting the subject's control counterpart. Another problem inherent in this design is caused by the fact that it is not always easy to know what the relevant matching variables should be.

Matched Groups

In an effort to simplify the matching process, some researchers have turned to the matched-group design, where instead of matching subjects on a one-to-one basis, the entire group, *as a group*, is matched with another, similar group. Typically, this takes the form of matching group averages, so that one group whose average IQ is, say, 103 is matched with another group whose average IQ is also 103 (or close to it). Since groups with similar averages might still show systematic differences in variability, it is important that the groups also have approximately equal standard deviations. Statisticians treat matched groups as though they were independent.

The Quasi-Experiment

Before leaving our discussion of experimental research, there is one variation on this theme that has been gaining in popularity. It is called the "quasi-experimental" method. With this technique, the independent variable is still under the full, active control of the experimenter, but the control and experimental groups are no longer clearly representative of a single population. That is, the researcher may use "intact" groups, or groups that have already been formed on the basis of a natural setting, and then, in some manner, subject them to different treatment conditions. Thus, the quasi-experiment is an attempt to simulate the true experiment, and is called by many researchers a compromise design.[1] For example, suppose a researcher wants to assess the possibility of establishing a difference in the

Television and Violence—A Research Example

A long-running controversy in the social sciences revolves around the issue of the possible impact of TV violence on viewer behavior. The suspicion, of course, has been that violence on TV seems to trigger aggressive behavior on the part of the viewer. This suspicion has been strengthened on the basis of the seemingly logical hypothesis that since TV viewing apparently affects a wide range of viewer responses (otherwise, advertisers wouldn't be spending huge sums in attempting to influence consumer behavior), why exclude aggressive behavior as a repertoire of responses so influenced? Further, a large number of studies which have been conducted over the past 30 years seem to lend empirical support to this notion. The typical study, summarized by Eron, has gone like this.*

Groups of teen-agers, almost exclusively males, have been questioned as to their TV viewing habits and have then been observed in terms of their overt or covert aggressiveness. The data clearly show that boys who indicate that they watch a lot of violence on TV are, themselves, more aggressive than are boys watching less violent TV shows. So there we have it; a relationship has been established between TV viewing and aggressive behavior. Case closed, right? Wrong! The studies, in fact, tell us very little regarding the possible link between TV viewing and behavior. Let's critically examine the procedure.

In each of these studies, the boys were *free* to choose what TV shows they watched, or, perhaps more important, they were free to choose which TV shows they wished to report having watched. (Nobody ever saw them sitting in front of their TVs.) It is true that their aggressive responses were observed, in the school yard, in the lunchroom, on the bus, and so on, but this in no way proves that these aggressive responses were caused by their choice of TV fare. It could just as easily be argued that a boy having had a fist fight on the way home from school is now more apt to find a violent TV show especially reinforcing on that particular evening. That is, maybe it's his overt aggression that causes a boy to select a violent TV show, maybe even as a rationalization for his actions. Or, perhaps a third hypothesis is possible. Perhaps boys whose lives are frustrating for one reason or another become more hostile and thus choose both to have fist fights *and also* to watch (or report having watched) more violence on TV. The question, then, becomes one of which came first, the chicken or the egg, or, even, were both the chicken and the egg produced together by some other cause? The danger in studies such as these is that the unwary reader (who is also sometimes a newspaper headline writer) assumes that a cause-and-effect relationship has been established when in fact the methodology of the study actually precludes such an interpretation. So, again, when reading research studies, be alert. Think out the procedures carefully, and, above all, don't just read the conclusions.

*L. D. Eron, "Relationships of TV Viewing Habits and Aggressive Behavior in Children," *Journal of Abnormal and Social Psychology,* vol. 67 (1963), pp. 193–196; and L. D. Eron, "Parent-Child Interaction, Television Violence, and Aggression of Children," *American Psychologist,* vol. 37, no. 2 (1982), pp. 197–211.

amount a student retains and the mode of presentation of the material to be learned. By a flip of a coin, it is decided that the students in Mrs. Smith's first-grade class will receive a visual mode of presentation, whereas the students in Mr. Shea's first-grade class (across the hall) will receive an auditory presentation of the same material. The measure of the D.V., scores on the retention test, are taken only after the groups have been differentially treated. This is done to prevent any pretest sensitization of the material being learned, and in this case, therefore would be an after-only design. Notice that the I.V. is being actively manipulated by the researcher; that is, the researcher has chosen which first-grade class is going to receive which treatment. However, the researcher did not create the groups on the basis of random assignment, but instead used groups that had already been formed. It is true that these two groups may have differed on a number of characteristics, some even that may have been related to the D.V., but for many real-world situations this is the best that one can do. School systems, clinics, even college deans, may not always allow groups to be formed artificially on the basis of random assignment. The quasi-experimental method may be set up according to after-only, before–after, or matched-subjects designs. The major problem with this type of research, however, focuses on the issue of how much equivalence the groups really have at the outset. The advantage of the quasi-experiment relates to its external validity. Since the groups have been naturally formed, extrapolation to real-world, natural settings becomes more convincing.

POST-FACTO RESEARCH

So far the discussion of the various experimental designs has focused on just that—experimental methodology, where the independent variable has been manipulated by the experimenter. In post-facto research, on the other hand, the independent variable is not some form of external stimulation being manipulated by the experimenter, but is, instead, based on a trait the subject already possesses, that is, a subject variable. Rather than attempting to place subjects in equivalent groups and then doing something to one of the groups in hopes of causing a change in that group, the post-facto method deliberately places subjects in nonequivalent groups—groups that are known to differ on some behavioral or trait measure. For example, the subjects may be assigned to different groups on the basis of their socioeconomic class, or their sex, or race, or IQ scores, whatever, and then the subjects are measured on some other variable. The researcher then attempts to ferret out either a correlation or a difference between these two variables. There's nothing wrong with this research. It's especially common in the social sciences. The problem lies not in the use, but in the misuse, of the inferred conclusions. You simply can't do a post-facto study and then leap to a direct cause-and-effect conclusion!

That type of conclusion commits what the statisticians call the *post-hoc fallacy*, which is written in Latin as "post hoc, ergo propter hoc" (translated as "because it came after this, therefore it was caused by this"). Many everyday-life examples of this fallacy come readily to mind. We have all heard someone complain that it always rains right after the car is washed, as though washing the car caused it to rain. Or the traffic is always especially heavy when you have an important meeting to go to, as though all the other drivers know when the meeting was planned and gang up in a contrived conspiracy to force you to be late. Or perhaps while watching a sporting event on TV you have to leave your TV set for a few minutes. When you return you discover that your favorite team has finally scored. You angrily hypothesize that the team deliberately held back until you weren't there to enjoy the moment. Some post-facto research conclusions have been very similar to these rather blatant examples. It all comes down to a matter of research control. In post-facto research, there are always a host of other, possibly influential, variables that are flying around in the background and are under nobody's control.

For example, several years ago Arthur Jensen reported that race makes a difference regarding measured intelligence. Jensen compared large groups of black and white children and found that whites performed significantly higher than did blacks on IQ tests.[2] This is obviously post-facto research, the I.V. (race) resulting from examining the children's skin color and then, on that basis, assigning them to the two groups. It is blatantly obvious, however, that these two groups also differ on myriad other important variables—all of which are left uncontrolled. Could anyone seriously argue that a black child, brought up, say, in the slums of Harlem will be provided with the same kinds of rich environmental and nutritional stimuli as a white child reared in Scarsdale? Such an argument would take us into the theater of the absurd, where the solution to every problem is to blame the victim. There are, in short, many systematic differences between white and black children other than skin color, any one of which could easily account for the IQ difference.

In another example of post-facto research, Jerome Bruner once did a study in which children were selected from differing socioeconomic backgrounds and were then compared with respect to their ability to estimate the sizes of certain coins.[3] It was found that poor children were more apt to overestimate the size of coins than were wealthier children. In this example, socioeconomic class was the independent variable and was clearly a subject variable, certainly not a variable the experimenter was able to manipulate. Bruner assigned the children to the economic categories on the basis of traits they already possessed. The dependent variable was the child's estimate of the coin size.

What can be concluded from this study? Various explanations have been put forth. Bruner felt that the poorer children valued the coins more highly and, thus, overestimated their size. Another suggestion was that richer children had more experience handling coins and were, thus, more accurate in estimating their size because of familiarity. The point is that no causal variable can legitimately be inferred from this study.

Of what use, then, is post-facto research? Prediction! Post-facto studies do allow the researcher to make better than chance predictions. That is, being provided with information about the independent variable puts the researcher in the position of making above-chance predictions as to performance on the dependent variable. If, in a two-newspaper city, there is a dependable relationship between which newspaper a person buys (liberal or conservative) and the political voting habits of that person, then one might predict the outcome of certain elections on the basis of the newspapers' circulation figures. We can make the prediction without ever getting into the issue of what caused what. That is, did the liberal stance of the newspaper cause the reader to cast a liberal vote, or did the liberal attitudes of the reader cause the selection of that newspaper? We don't know, nor do we even have to speculate on causation to make the prediction. In short, accurate predictions of behavior do not depend on the isolation of a causal factor. One need not settle the chicken–egg riddle to predict that a certain individual might choose an omelet over a dish of cacciatore.

Researching via Post-Facto

In the early stages of researching some new area, it is often necessary to use post-facto techniques and, then, as information accumulates follow up with the experimental method. Post-facto research is often quick and easy to do, and it may also lead to educated speculation concerning independent variables that might then be manipulated in experimental fashion. For example, in researching the problem of infertility among men, it has been noted that infertile men showed lower levels of vitamin C in their systems than did fertile men. At this point, this would strictly be a post-facto finding. It could be that fertile and infertile men also differ in other respects, such as their testosterone levels, and it's really this difference that causes fertile men to seek out foods that are rich in vitamin C. Then it might be noted that many infertile men had equal numbers of spermatozoa in their semen as did fertile men, but that the spermatozoa among the infertile men tended to clump together, or agglutinate. Laboratory test work might then show that this agglutination could be corrected in the test tube by the direct addition of vitamin C. Finally, an experiment could be designed in which matched pairs of infertile men are tested, one group receiving 500-mg capsules of vitamin C every 12 hours and the other group a simple placebo. If the wives of all the experimental subjects later became pregnant, while few if any of the wives of the men not given the vitamin C became pregnant, then a cause-and-effect conclusion would be justified.

Post-Facto Research and the Question of Ethics

It is also important to realize that for many kinds of research studies, post-facto techniques are the only ones which don't violate ethical principles. Post-facto techniques do allow a researcher to gather predictive evidence in areas which might be too sensitive, or possibly harmful to the subjects, to be handled experimentally. Suppose, for example, that a researcher is interested in discovering whether the heavy use of alcohol lowers academic achievement. To test this experimentally, one would have to select two groups of students randomly and then force one group to drink heavily on a daily basis, while preventing the other group from touching a drop. Then, if after a semester or two, grade-average differences were found between the two groups, it could be legitimately claimed that the use of alcohol lowered academic performance. But to isolate a causal factor, the experimental subjects may have suffered in more ways than just the low grades. Suppose that they developed liver problems, or delirium tremens, whatever. Should a researcher be allowed to expose subjects to possible long-term damage merely for the sake of nailing down the causal factor? Of course not; the experimental method should be used only when the risks are minimal compared to the potential benefit to humankind.

A study of this type could have been more ethically handled by the post-facto method. The researcher would simply identify students who are already heavy drinkers and then compare their grades with a group of non-drinkers. In this case, the subjects themselves have chosen whether or not to drink, and the researcher simply finds out whether the two groups also differ as to grade average.

Of course, no cause-and-effect statement is immediately possible, for even if a significant difference is found, one cannot unambiguously determine the direction of the relationship. Perhaps A (drinking) caused B (lower grades). Perhaps B (lower grades) caused A (drinking). Or perhaps X (unknown variable) caused both A and B. The X variable might be a depressed state of mind that caused the student to drink too much and also not do the work necessary for academic achievement.

ANALYSIS OF COVARIANCE: AN AFTER-THE-FACT CONTROL

A statistical technique called the **analysis of covariance (ANCOVA)** has been used in an attempt to create equivalent groups of subjects *after the fact,* that is, after the groups have been formed. The ANCOVA technique, in a sense, crosses the line between experimental and post-facto research techniques. It is not, strictly speaking, a substitute for experimental methodology in the pure sense, since the groups were not originally selected as equivalent. At the same time, it does allow for active manipulation of the independent variable. For example, assume that two groups of subjects are being compared on diastolic blood-pressure readings, one of the groups having been involved in an experimentally controlled exercise program and the other group not so involved. Later, it is wondered whether the groups may have originally differed on the basis of, say, weight. The ANCOVA technique can be used statistically to control any possible weight differences, the covariate in this case, that may have originally occurred between the groups. It will then assess the extent to which the blood-pressure differences may have been above and beyond the original weight differences.

ANCOVA may also be used to analyze data from straightforward post-facto studies, where no attempt to manipulate an I.V. has occurred. Suppose, for example, that one wished to test the hypothesis that persons with large bank accounts are more likely to vote for Republicans than are persons of less wealth. This is a post-facto study, since bank account size is obviously not a manipulated independent variable. Therefore, even if large and small bank account persons did differ significantly regarding voting preference, it may not have been the amount in the savings account which actually influenced the vote. Age, for example, might be a contributing factor, since older persons would have had more time to accumulate wealth. The covariance technique, then, would be applied as a statistical control on the age variable. The two groups, which were unequal on the age factor, would then be treated *as though* they were equal.

Although analysis of covariance does not completely isolate the causal factor, it does have the advantage of pointing a highly suspicious finger in that direction. Cause-and-effect, though not definitely proven, is at least strongly hinted at. This is a powerful technique, and is a long step past the naive post-facto analyses which have so often been done in the past. The analysis of covariance may not isolate the causal factor "beyond all reasonable doubt," but at least it does provide a case of "circumstantial evidence."

The Hypothesis of Difference in Experimental Research

In all experimental research, the hypothesis of difference must eventually be addressed. That is, after the groups of subjects are selected and exposed to different treatment conditions (I.V.), they are then measured (on the D.V.) to see if differences can be observed. In fact, the alternative hypothesis clearly states that differences will occur, or that $\mu_1 \neq \mu_2$. If these D.V. differences are found to be significant, not due to chance, then the researcher may conclude that they are caused by the differential treatment the subjects received.

Post-Facto Research and the Hypothesis of Difference

In post-facto research on differences, subjects are assigned to I.V. categories on the basis of various trait measures, such as socioeconomic class, sex, intelligence, and political affiliation. The subjects are then measured on some other variable (D.V.), and the differences obtained are tested for significance.

THE *t* RATIO AND RESEARCH METHODOLOGY

The *t* ratio can be used for the analysis of both experimental and post-facto data. Keep in mind that it is not the statistical test that determines if causation can be inferred, but rather the research methodology being employed. Concerning the *t* test, the logic applied to both methods of research is aimed at determining whether the two samples represent a single population.

The Experimental Method

With the experimental method, the two sample groups, experimental and control, are always chosen to represent a single population. This is the whole point of experimental design—the creation of equivalent groups. After the independent variable has been manipulated, if the two groups are found to differ significantly, they are then said to represent different populations. The change is caused by the action of the independent variable. For example, a certain drug is being tested as an antidepressant. Two groups of subjects are randomly selected from a population of depressed patients. At the outset, then, both groups represent a single population. Group A (experimental group) gets the drug, and Group B (control group) gets a placebo. If, after several weeks, the two groups are found to differ significantly when measured for depression, they do not represent a single population. The experimental group no longer represents the population it was originally selected from, but instead represents a nondepressed population.

The Post-Facto Method

As for the post-facto method, the groups obviously do not represent a single population at the outset. In fact, the subjects are assigned to groups *because they differ on at least one population characteristic.* The groups are then compared on some other measure to assess whether they represent different populations on this measurement as well. For example, a researcher is interested in whether business majors have higher grade-point averages than do liberal arts majors. A random sample of college seniors is selected and is assigned to one of two groups on the basis of college major. The groups representing two different populations are thereby created. The two groups are then compared regarding grade-point average. If the difference is statistically significant, they also represent different grade-point populations.

While the t test tells us whether or not two groups represent different populations, only the type of research methodology, experimental or post-facto, tells us whether a causal explanation can be used to account for the difference.

Post-Facto Research and the Hypothesis of Association

In some post-facto research, the goal is to establish not differences but associations. That is, a group of subjects might be measured on two or more response dimensions to determine whether any correlation exists. Or perhaps a correlation study is run on two or more external events, like the unemployment rate and the rate of inflation. In any event, all correlational research is post-facto, since no active manipulation of an independent variable is involved. (Numerous examples of the correlational technique are found in Chapter 10.)

Associational research uses only assigned independent variables, never manipulated ones. Once the assigned independent variable is determined, the subjects are measured in terms of this variable (age, height, weight, IQ, or whatever), and these measures are then compared with measures taken on the D.V. As we saw in Chapter 10, when using correlational research, the I.V. is always the variable that is being *predicted from,* and the D.V. is the variable being *predicted to.* Thus, if one were predicting reading speed on the basis of IQ, the IQ would be the I.V. and the reading speed the D.V.

When using post-facto research to test the hypothesis of association, one attempts to establish correlations among the various measures. The alternative hypothesis for the Pearson r, as you will recall, clearly states that $\rho \neq 0$, or that there really is *an association* between the variables existing out there in the population.

RESEARCH ERRORS

Although there is probably an endless list of ways in which the researcher can create uninterpretable results when conducting a study, some of the most common and most serious errors will be presented in this section. These will all be errors in research strategy, not statistical analysis. In fact, these are errors that lead to uninterpretable findings regardless of the power or elegance of the statistical analysis.

Again, the presentation to follow is not expressly designed just to create cynics among you. It is to aid you in becoming more sophisticated research consumers whose practiced eye will be able to see through the camouflage and identify the flaws. It is also important to learn to spot research errors in the headlines and stories appearing in popular magazines, Sunday supplements, and daily newspapers as well as in the scientific journals. Face it. For many of you, the reading of scientific journals may not be top priority after you graduate from college. The approach used in this section will be essentially that of the case study, an approach that will, it is hoped, teach by example. The major message, which you will see repeated, is aimed at alerting you to pay at least as much attention to the researcher's methodology as to the researcher's conclusions.

Confounding Variables—Secondary Variance

As you have learned, in experimental research cause-and-effect conclusions are only justified when, all other things being equal, the manipulation of an independent variable leads to a concomitant change in the dependent variable. The fact that other things must be kept equal means that the experimental and control groups must be as nearly identical as possible. Other possibly influencing variables must be held constant, so that the only thing that really varies among the groups is the independent variable. When other factors are inadvertently allowed to vary, these factors are called confounding variables and they often produce **secondary variance.** When an independent variable is manipulated, and its action leads to change, or difference, or *variation* in the dependent variable, this resulting D.V. variation is called **primary variance**—which, of course, is what the experimenter is trying to produce. When something else comes along, not under experimental control, which may also influence or produce changes in the D.V., then secondary variance has been allowed to occur. This, then, confounds or contaminates the independent variable in such a way that the pure effects of the I.V. are no longer available for analysis. When confounding variables are present, secondary variance is sure to follow, and we say that the research has been "flawed."

Some of this may seem to be just plain common sense, but as we shall see, even among some of the most famous of the social science researchers, this "sense" hasn't always been so common. *Confounding variables are any extraneous influences that may cause response differences above and beyond those possibly caused by the independent variable.*

Failure to Use an Adequate Control Group

Clearly, then, confounding variables occur when the researcher makes a major mistake—the mistake of not having an adequate control group. Without fully obtaining the key, experimental method ingredient—equivalent groups of subjects—free-floating confounding variables are bound to be present. Without an equivalent control group, secondary variance will surely be lurking on the sidelines, just waiting to pounce on the conclusions and stamp them "flawed."

When a design has been flawed, the experiment automatically loses its internal validity. That is, no matter how beautifully representative the sample is, the results of the experiment become impossible to evaluate when secondary variance is allowed to seep into the study.

EXPERIMENTAL ERROR: FAILURE TO USE ANY CONTROL GROUP

Case 1: The Hawthorne Effect

This first case goes way back to industrial psychology's earliest days when a pair of Harvard psychologists, Roethlisberger and Dickson, went out to the Hawthorne plant of the Western Electric Company in Homewood, Illinois.[4] Their job was to find out why productivity levels in this plant were so low. After a quick tour of the plant, they decided that maybe the problem was simply due to the low level of illumination inside the building. They did a preliminary study, first assessing productivity levels, then increasing the illumination, and then again measuring the levels of productivity. Just as they suspected, productivity among the workers went up. The problem here, of course, is the use of a before–after design with no separate control group.

These psychologists, however, became a little skeptical of their own findings and decided to repeat the study in another section of the plant by simply pretending to change the bulbs, but in reality leaving illumination the same. The result—productivity climbed. Finally, in still a third version of the study, they selected one group of workers and deliberately lowered the illumination to a level resembling soft moon light. Productivity sky rocketed! At this point they discovered an extremely important confounding variable. Worker productivity went up all right, but it had nothing whatever to do with the presumed I.V., illumination. The workers, instead, were simply reacting to the presence of these Harvard psychologists. That is, the seemingly dramatic results were really due to the very fact that the experiment was being conducted. This is now known as the **Hawthorne effect**—the fact that changes in the D.V. sometimes result from the increased motivation of the subjects, motivation probably instilled by the flattery and attention provided by the researcher. Clearly, if a control group isn't used, it becomes impossible to tell whether the subject's response improved because of the I.V. or because the subject somehow felt special and wanted to please the researcher.

Researchers must be especially careful of the Hawthorne effect when conducting studies of personal change, be it of personality change or intellectual change, whatever. One researcher in the field of learning disabilities has complained that, "any idea or finding which is unacceptable to anyone today can be explained away on the basis of the Hawthorne effect" (Kephart, 1971). In point of fact, the only time results can be "explained away" on the basis of the Hawthorne effect is when the researcher carelessly fails to use an adequate control group or, in the case of Kephart, any control group at all.

Case 2: Rogerian Therapy

Some time ago, a researcher attempted a study aimed at discovering the possible efficacy of the Rogerian, client-centered therapy technique in cases of reading retardation. A sample of 37 first- and second-grade children, all of whom had been diagnosed as retarded in reading, was selected on the basis of a standardized reading test *and* teacher evaluation.[5] These children were then placed in a special class, taught by a teacher trained in Rogerian, nondirective techniques. The atmosphere in this classroom was described as therapeutic, the children being encouraged to express their feelings openly in the presence of a warm, permissive, and accepting teacher. The Stanford–Binet IQs of the children ranged from 80 to 148. At the end of the school term, the children were again tested on the Gates Primary Reading Test, and, in general, their reading-age scores improved, in one instance by a phenomenal 17 months. These gains were interpreted by the researcher as resulting from the nondirective teaching technique, the assumed independent variable. However, the lack of a separate control group leaves this study seriously flawed. We cannot tell if the gains were really the effect of the teaching technique or of the several confounding variables, teacher personality, novelty effect of being placed in a special class, the growth changes that normally take place among children during a school term, and so on. Finally, in this particular study, no statistical tests were conducted leaving the reader unable to determine whether the reading gains were, in fact, simply a result of chance, or whether the high-IQ children profited more from the special class than did the low-IQ children, or, really, anything at all.

Case 3: Frustration-Aggression

Perhaps one of American psychology's most cherished theories of aggression was launched many years ago by those famous then-at-Yale psychologists, Neal Miller and John Dollard. They put forth the frustration-aggression hypothesis, which in its briefest form stated that frustration (in the form of goal blockage) always leads to aggression. They also believed that the aggression produced by frustration could sometimes be displaced onto minority groups, that is, could be exhibited in the form of racial prejudice. To test this, Neal Miller went to a government work camp, situated in the woods, yet close enough to civilization to allow the men to get to a nearby town about once a week. Needless to say, the men eagerly looked forward to these visits! On one occasion, the men were especially anxious. The town's movie theater had been holding a series of "bank night" drawings, and the previous week one of the workers at the camp had won a large sum of money. On the appointed night for the trip to town, the men were all brought together and given an attitude test toward Japanese and Mexicans. They were told that as soon as they finished filling out the questionnaires, the bus would be ready to take them to town. The men amiably filled out the questionnaires and then waited for the bus. It never came. Instead of going to town that night, they were told they had to stay in camp and take a series of difficult intelligence tests. No question about it, Miller knew how to manipulate the goal blockage I.V. After a lot of grumbling, the men settled down, took the tests, and then, to add insult to injury, were told they had to again take the Japanese–Mexican attitude test. Their scores, on the posttest, declined precipitously, indicating an increase in racial prejudice. This was interpreted by Miller as resulting from the action of the I.V.—not going to town led to outgroup aggression.[6]

As it stands, however, the problem with this study is that one cannot really be sure that the increase in prejudice resulted in not being allowed to go to town, the goal blockage. This was a one-group, before–after experimental design where the same group was used as its own control. In fact, a separate control group is desperately needed here, a group that was not told anything about going to town but was still forced to take the attitude test, the intelligence tests, and the attitude test a second time. Without this control, the researcher can't be sure whether the attitude change was the result of not going to town, or the result of the boredom generated by being forced to sit still for such a long period of time taking these tests, or the anger generated by having to take the attitude test a second time.

Case 4: Frustration-Regression

In another study in the area of frustration, the famous Gestalt psychologist Kurt Lewin hypothesized that frustration, again in the form of goal blockage, would cause psychological regression in young children.[7] In this study the independent variable was manipulated by allowing the children to view, but not reach, a glittering array of brand-new toys. Here's how it went. The children were all placed in a single room and were allowed to play with serviceable, but obviously well-used, toys. Observers followed the children around and rated each child as to the level of maturity displayed by the child's play behavior. Next, a curtain was drawn, revealing the beautiful, shiny new toys; but, alas, a wire screen prevented the children from getting at these toys. Then the children were again assessed as to maturity level of their play with the used toys, a level, incidentally that dropped considerably from the premeasure.

Here it is again, a before–after experimental design, with the same children serving as their own controls. Remember, by the time the children were measured for the second time, they had been cooped up in that room for some time. Could boredom have set in? Could attention spans have begun to diminish? We simply can't tell from this study. Only with a separate control group (another group of children playing with similar toys for the same length of time, but with no view of the unobtainable toys) can one hope to eliminate some of the most blatant of the confounding variables found in this study.

EXPERIMENTAL ERROR: FAILURE TO USE AN ADEQUATE CONTROL GROUP

Case 5: Play Therapy

An example of a study where a control group was used, but the control group was not adequate, comes from the clinical psychology literature. In this study, an attempt was made to test whether nondirective play therapy might improve personality adjustment scores among institutionalized children.[8]

All 46 residents of a children's home were tested on the Rogers Test of Personality Adjustment, and the 7 children with the lowest scores were selected for the special treatment. The other 39 children, thus, served as the separate control in this before–after design. However, the play therapy group was treated twice a week at a clinic that was located 10 miles from the institutional home. At the end of six weeks, both groups were again tested on the adjustment inventory. Two important confounding variables were allowed free rein in this study. First, the experimental group not only received the play therapy, but it also was treated to a 10-mile bus ride twice a week, and, hence, an opportunity to leave the possible boredom of the institution's confines. Second, the two groups were not equivalent to begin with. Now, although it may seem especially fair to select only the most maladjusted children for the special treatment condition, it might very well be that these were the very children who would change the most, regardless of the independent variable, simply because on the measurement scale they had the most upward room in which to change. In Chapter 14 we will discuss something called "regression toward the mean," but suffice it to say here that persons who score low on any test are more apt, if they change at all, to change upward than are persons who score high. Finally, this study is triply flawed in that 16 of the control group children left the institution during the six weeks the study was in progress and could therefore not be retested. This fact produces the nagging concern that the departure of these children might have resulted from their having been perceived as becoming adjusted enough to go home.

Case 6: College Teaching Techniques

In a study of college teaching techniques, three different introduction-to-psychology classes were taught by three different methods.[9] One class was taught in the traditional lecture format, the second used a "student-centered" fashion, and the third class alternated the two techniques at each meeting. The results showed that students in Groups 2 and 3 indicated higher interest and enjoyment in the course than did students in the traditional class. They also felt that the social-emotional value of the course was greater. On an objective test of the course's content, however, there were no significant differences among the three groups (although, interestingly, the students in the traditional class felt that they had gained more information and knowledge).

The problem with this study is twofold. First, this was a quasi-experiment and there very well could have been systematic differences among the students in the three classes to begin with. Perhaps students who elect afternoon courses are different from those who opt for morning classes. (Perhaps students who have part-time jobs, or students on athletic teams, are forced by their schedules to select morning classes.) More serious, however, was that the instructor, playing the different teaching roles, *was also the researcher*. It could very well be that this researcher's own convictions allowed him to carry out one of the roles more convincingly than the other. Also, the students must have been fully aware of who was in the experimental group and who was in the control group. This was, in short, definitely not a double-blind study, and because of this, the possibility of extraneous motivational variables could not have been controlled. The potential for confounding variables is far too great to make this a definitive test of teaching techniques.

Case 7: Group Decision

During World War II, Kurt Lewin did a series of studies on the topic of "group decision." The theoretical basis was that since a person's social attitudes are learned in a group situation (family, friends, school), then changing a person's attitude should also best take place in a group setting. In one study, a group of women was brought to a lecture, and, in the impersonal role of being part of an audience, listened to an impassioned speaker.[10] The speaker exhorted the women to use less expensive and more plentiful cuts of meat. They were told that it was their patriotic duty and also that it was more healthful for their families. Another group of women was also brought together, although in a very different group atmosphere. This group informally sat around a table and heard a group leader raise the same points as the "lecture" group had heard, but in this "discussion group" setting, members were encouraged to participate, offer suggestions, and become generally involved. Several weeks later both groups were checked at home to determine whether they were indeed using the meat cuts that had been urged. Only 3% of the "lecture" group members had the meat cuts, whereas 32% of the discussion group members had obliged. The difference was clearly significant, and it appeared as though the independent variable, lecture versus discussion group settings, did have an effect. However, Lewin used different people to lead the two groups. Perhaps the "discussion" leader had a more forceful personality, or was more believable, or whatever. The point is the independent variable was very much confounded.

In another variation of this study, Lewin again set up the same types of groups, one in the lecture and one in the discussion setting, but this time the same person, Dana Klisurich, conducted both groups.[11] The women were urged to give their children orange juice, and again, several weeks later, investigators called on the homes of all the women and checked their refrigerators for the juice. The women who had participated in the discussion group were far more apt to have the juice in their homes. However, this time Lewin told the discussion group members that they would be checked on, *but forgot to tell the lecture group!* This is confounding writ large. There is no way to tell whether the women had the juice because they had been in the discussion group or because they were told Lewin was going to check up on them. In fact, the children of the women in the experimental group may even have been deprived of getting any juice, since these mothers may have been saving it to show Kurt Lewin.

POST-FACTO ERRORS

Case 8: The Halo Effect

In many studies, especially correlational, the research becomes flawed due to the various evaluations of subjects being conducted by observers who know how the subjects scored on previous evaluations. Sometimes this knowledge is extremely intimate, since the same observer rated the subjects on several trait measures. In research, this error is called the **halo effect,** and it results from the very obvious fact that if an observer assigns a positive evaluation to a subject in one area, then, consciously or unconsciously, the observer tends to assign another positive evaluation when the subject is being measured in another area. In short, the subjects are having their trait measures generalized into a whole host of seemingly related areas. This is the reason why advertisers spend big money hiring well-known celebrities to do commercials. The viewer, it is hoped, will assume that because a certain individual is proficient with a tennis racket, he will also be an expert in determining which brand of razor to use.

The halo effect can be a hazard to both the researcher and the research consumer. The issue is especially acute in post-facto research testing the hypothesis of association. In one study, the investigators wished to study the impact of the halo effect on student grades.[12] The research was post-facto, that is, the students' measurements on one trait were compared with the measures the same observers assigned to the students on other traits. The researchers asked the classroom teacher to assign personality ratings to each student and then compared these ratings with the grades the students received from the same teacher. The correlation was high and positive, that is, the more favorable the personality rating, the higher the grade received. If this study had been designed to test the possibility of an independent relationship between personality and academic achievement, then the personality ratings should have been assigned by someone other than the person doing the grading. Also, this independent observer should not even be made aware of what those students' grades had actually been.

In another example, a study was conducted at the Institute for Child Behavior in San Diego in which subjects were asked to rate the 10 persons "whom you know best" on two variables—happiness and selfishness. The results showed an inverse relationship; the more a person was judged to be happy, the less that person was seen as being selfish. The conclusion implied a link between the two variables, or that being unselfish (helping others) tended to create a state of personal happiness in the helper.[13] However, since the evaluation of a person's selfishness and happiness was made by the same observer, these results might be more readily explained on the basis of the halo effect. When you like someone you may easily become convinced that that person abounds in a whole series of positive virtues, even in the face of contrary evidence.

Case 9: Smiling and Causation

In a study from the educational literature, the hypothesis was tested that students would achieve more academically when the classroom teacher spent more time smiling.[14] Observers visited a number of different classrooms and monitored the amount of time each teacher spent smiling. These results were then compared with the grades being received by the students in each of the classrooms. The results were significant—the more the teacher smiled, the higher were the student grades. This is, of course, post-facto research, since the teacher, not the experimenter, determined the amount of smile time. In fact, this would be a difficult study to conduct experimentally. To manipulate the I.V. actively, the researcher would have to perhaps sit in the back of the room and, at random time intervals, signal the teacher that it was time to smile. This might obviously lead to a rather bizarre scene, where the teacher, in the middle of a vigorous reprimand, would suddenly have to break out in a broad grin. In any case, though, the study as conducted was post-facto. Because of this, the results *cannot* tell us the direction of the relationship. It may well be, as the authors hypothesized, that smiling teachers produce achieving students. Or it may just as likely be that high-achieving students produce smiling teachers—teachers luxuriating in the fact that the student success rate is obvious proof of the teacher's own competence. Or it could also be that a third variable, say, the personality of the teacher, may have caused both the smiling and the high grades. Perhaps a smiling teacher is a happy optimist who always sees the best in everyone and is, therefore, more lenient when assigning grades.

METHODOLOGY AS A BASIS FOR MORE SOPHISTICATED TECHNIQUES

This quick foray into the world of research methodology has, at one level, jumped ahead of our statistical discussion. At a more important level, however, it has prepared you for what is to come. Statistical techniques and research methodology must necessarily be unfolded in this leapfrog fashion. In presenting the inferential techniques in the following chapters, we will often refer to the methodological issues outlined in this chapter. Even now, with a limited arsenal of inferential tests, t and r, you are still in a position to do and to analyze a variety of research studies.

Both t and r require interval data, but the majority of studies in the social sciences do, in fact, use interval data. The t ratio tests the hypothesis of difference, as do *all* studies based on the experimental method. Of the experimental designs presented in this chapter, the independent t test may be used with the two-group after-only design. In later chapters, tests for handling before–after and matched-subjects designs will be given. Also, many post-facto studies test the hypothesis of difference and when two groups are involved the independent t may be used. Those post-facto studies that test the hypothesis of association may be evaluated with the Pearson r, or if they employ ordinal data, r_s.

At this precise moment in time, you are not yet ready to conquer the world of research, but you can do something. You are now in a position to do more than sit back and admire. You are now in a position to get your hands dirty with the data.

SUMMARY

Some knowledge of methodology is necessary in order that the statistical analyses and conclusions drawn from research studies can be interpreted. Simply reading the conclusions of a research study puts the reader at the mercy of the researcher's own intentions and/or implicit assumptions.

There are two major research strategies, experimental and post-facto. When the independent variable is actively manipulated, the research is experimental and cause-and-effect inferences are possible. When the independent variable is based on a trait the subjects already possess, it is called a subject variable, a variable on which the subjects are then assigned to different categories. Although this method does allow for better than chance predictions, studies conducted by this method should not be interpreted as having proven the existence of causal factors.

There are three basic experimental designs, all aimed at creating that crucial experimental ingredient—equivalent groups of subjects.

1. *After-only (A/O):* where random selection of samples and random assignment of subjects to two or more groups are used to create equivalence. In the A/O design, the dependent variable is measured only after the independent variable has been manipulated.

2. *Before–after (B/A):* where a random sample is measured both before and after the introduction of the manipulated independent variable, and equivalence is achieved by using the same subjects in each treatment condition. Because of the possibility of confounding the independent variable (not being able to assess its pure effects), a refinement of the B/A design, called the before–after with separate control has been developed. The separate control group may either be independent or matched.

3. *Matched-subjects (M/S):* where subjects are selected for the groups on the basis of a subject-for-subject matching process. The subjects are equated and paired off on the basis of some relevant variable.

Experimental research always tests the hypothesis of difference, that is, that groups of subjects assumed to be originally equivalent now differ significantly on some measured trait. Since the subjects were the same before the independent variable was manipulated, if they now differ, the *cause* of this difference may be interpreted as resulting from the action of the independent variable.

Post-facto research sometimes tests the hypothesis of difference, that is, subjects are assigned to different groups on the basis of some original difference and then assessed for possible differences in some other area. Post-facto research may also test the hypothesis of association, that is, that a correlation exists among separate measures. In neither of these post-facto areas, however, should the results be interpreted as having isolated a causal factor.

Research errors may occur in both experimental and post-facto methodologies. In experimental research, a study will lack internal validity when the I.V.(s) cannot be unambiguously interpreted. Manipulated I.V.s become confounded, producing secondary variance, when their pure effects cannot be evaluated on a cause-and-effect basis. Failure to use an adequate control group is a major reason for the loss of internal validity. External validity results when a study's findings are generalizable to a real-life population in real-life conditions. When a laboratory study becomes so artificially contrived as to no longer reflect the real world, or when the subjects used in the study fail to represent the population, then a loss of external validity results.

Other major research errors are stressed. Among these are the halo effect and the Hawthorne effect. The halo effect occurs when a researcher measures a subject on one trait and is then influenced by that measure (either positively or negatively) when evaluating that same subject in a different area. This problem can be eliminated by using independent observers when measuring subjects on more than one trait. The Hawthorne effect occurs when a researcher, using a simple before–after design, assumes that a given difference is the result of the manipulated independent variable, when in fact the result may be due to the attention paid to the subjects by the experimenter. The judicious use of a separate control group can minimize this problem.

KEY TERMS

analysis of covariance (ANCOVA)
control group
counterbalancing
dependent variable
double-blind study
experimental design

experimental research
external validity
halo effect
Hawthorne effect
independent variable (subject or manipulated)

internal validity
post-facto research
primary variance
secondary variance

REFERENCES

1. KERLINGER, F. N.   (1986). *Foundations of behavioral research*. (3rd ed.). New York: Holt, Rinehart and Winston.
2. JENSEN, A. R.   (1969). How much can we boost IQ and scholastic achievement? *Harvard Educational Review, 39,* pp. 1–123.
3. BRUNER, J. S. & GOODMAN, C. C.   (1947). Value and need as organizing factors in perception. *Journal of Abnormal and Social Psychology, 42,* pp. 33–44.
4. ROETHLISBERGER, F. J., & DICKSON, W. J.   (1939). *Management and the worker.* Cambridge, MA: Harvard University Press.
5. AXLINE, V. M.   (1947). Nondirective therapy for poor readers. *Journal of Consulting Psychology, 1,* pp. 61–69.
6. MILLER, N. E., & BUGELSKI, R.   (1948). Minor studies of aggression. *Journal of Psychology, 25,* pp. 437–442.
7. BARKER, R. G., DEMBO, T., & LEWIN, K.   (1941). Frustration and regression. *University of Iowa Studies in Child Welfare,* 18.

8. FLEMING, L., & SNYDER, W. U.   (1947). Social and personal changes following nondirective group play therapy. *American Journal of Orthopsychiatry, 17,* pp. 101–116.
9. FAW, V. E.   (1949). A psychotherapeutic model of teaching psychology. *American Psychologist, 4,* pp. 104–109.
10. LEWIN, K.   (1952). Group decision and social change. In G. E. Swanson, T. M. Newcomb, & E. L. HARTLEY, EDS., *Readings in Social Psychology.* New York: Holt.
11. SHERIF, M., & SHERIF, C. W.   (1956). *An outline of social psychology.* New York: Harper.
12. RUSSELL, J. L., & THALMAN, W. A.   (1955). Personality: Does it influence teacher's marks? *Journal of Educational Research, 48,* pp. 561–564.
13. RIMLAND, B.   (1982). The altruism paradox. *Psychological Reports, 51,* pp. 521–522.
14. HARRINGTON, G. M.   (1955). Smiling as a measure of teacher effectiveness. *Journal of Educational Research, 49,* pp. 715–717.

PROBLEMS

Indicate both the type of research (experimental or post-facto) and the hypothesis being tested (difference or association) for problems 1 through 6.

1. A researcher wishes to find out if there is a difference in intelligence between men and women. A random sample of 10,000 adults is selected and divided

into two groups on the basis of gender. All subjects are given the Wechsler IQ test. The mean IQs for the two groups are compared, and the difference is found to be not significant.

2. A researcher wishes to test the hypothesis that fluoride reduces dental caries. A large random sample of college students is selected and divided into two groups. One group is given a year's supply of toothpaste containing fluoride; the other group is given a year's supply of seemingly identical toothpaste without fluoride. A year later all subjects are checked by dentists to establish the incidence of caries. The difference is found to be significant.

3. A researcher wishes to test the hypothesis that persons with high incomes are more apt to vote than are persons with low incomes. A random sample of 1500 registered voters is selected and divided into two groups on the basis of income level. After the election, voting lists are checked; it is found that significantly more persons in the high-income group went to the polls.

4. A researcher wishes to establish that special training in physical coordination increases reading ability. A group of forty 10-year-old children is randomly selected. Each child is given a standardized "Reading Achievement Test" and then placed in a specially designed program of physical coordination training. After six months of training, the "Reading Achievement Test" is again administered. The scores show significant improvement.

5. A researcher wishes to test the anxiety-prejudice hypothesis. A random sample of 200 white college students is selected and given the Taylor Manifest Anxiety Test. They are then given a test measuring prejudice toward minorities. The results show a significant positive correlation—the more anxiety, the more prejudice.

6. A researcher wishes to test the subliminal perception hypothesis. A movie theater is chosen, and the sales of popcorn and Coca-Cola are checked for a period of two weeks. For the next two weeks, during the showing of the feature film, two messages are flashed on the screen every 5 seconds, each lasting only 1/3000 of a second (a point far below the human visual threshold). The alternating messages are "Hungry? Eat popcorn." and "Thirsty? Have a Coke." Sales are checked again, and it is found that popcorn sales increased by 60% and Coke sales by 55%.
   a. What kind of research is this, post-facto or experimental?
   b. If experimental, what design has been used?
   c. What possible confounding variables might there be?

7. For problem 1, identify the independent variable and state whether it was a manipulated or an assigned-subject variable.

8. For problem 2, identify the independent variable and state whether it was a manipulated or an assigned-subject variable.

9. For problem 3, identify the independent variable and state whether it was a manipulated or an assigned-subject variable.

10. For problem 4, identify the independent variable and state whether it was a manipulated or an assigned-subject variable.

11. For problem 5, identify the independent variable and state whether it was a manipulated or an assigned-subject variable.

12. For problem 6, identify the independent variable and state whether it was a manipulated or an assigned-subject variable.

13. For problem 4, suggest possible confounding variables.
14. Of the six research examples, which seems most prone to the occurrence of the Hawthorne effect?
15. What is the major difficulty inherent in the use of the before–after experimental design?
16. What is the major difficulty inherent in the use of the matched-subjects design?
17. A researcher wishes to assess the possibility that a certain type of psychotherapy actually reduces neurotic symptoms. A random sample of subjects is selected from the population of names on a waiting list at a mental health clinic. For each subject selected, another subject is sought, also from the waiting list, who resembles the originally selected subject in a number of important respects, such as age, length of time on the waiting list, IQ, severity of symptoms, length of symptom duration, category of neurosis, and previous treatment history. By a flip of a coin, one group is chosen to receive six months of intensive psychotherapy, while the other group is told to continue waiting and not to seek treatment elsewhere. At the end of a six-month period, a panel of experts judges all the subjects with respect to the extent of symptomatology.
    a. What type of research is this, experimental or post-facto?
    b. What is the I.V. and what is the D.V.?
    c. What would you advise the researcher to do with respect to the judge's knowledge regarding which subject represented which group?
    d. What other variables might you wish to match on?

True or False—Indicate either T or F for problems 18 through 26.

18. When an independent variable is actively manipulated, the research method must be experimental.
19. If a researcher uses the post-facto method and establishes a significant relationship, the possibility of a cause-and-effect relationship must be ruled out.
20. If a researcher establishes a unidirectional relationship, the research method used must have been experimental.
21. If a researcher establishes a significant correlation in order to predict college grades on the basis of high school grades, the high school grades are the independent variable.
22. The main purpose of the various experimental designs is to establish equivalent groups of subjects.
23. In experimental research, the independent variable always defines the differences in the conditions to which the subjects are exposed.
24. A study designed to test the effect of economic inflation on personal income should establish personal income as the dependent variable.
25. In correlational research, the independent variable is always manipulated.
26. Correlational research must always be post-facto research.

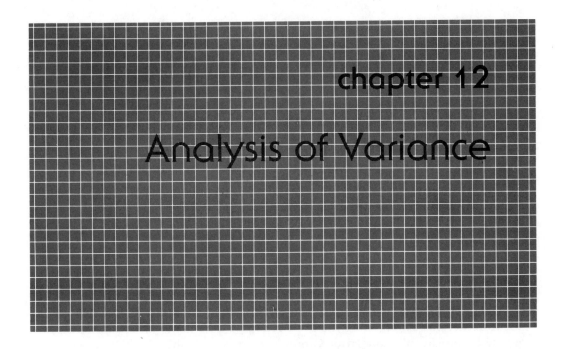

chapter 12

Analysis of Variance

Since the time around the turn of the century that William Sealy Gossett first created the *t* test, another statistical test of the hypothesis of difference has been developed. During the 1920s, **Sir Ronald Fisher** introduced a technique called the **analysis of variance,** and the resulting statistic, the *F* ratio, was named in his honor. The procedure is now known by the acronym **ANOVA,** for ANalysis Of VAriance. ANOVA is a powerful and versatile statistical technique. It allows us to do virtually everything we use the *t* test for, plus a lot more. Why, then, did we bother with the *t* test? Why not skip directly to *F*? There are two reasons: first, the concepts on which the *t* ratio is built provide an excellent background for appreciating the overall logic of statistical reasoning, from samples to sampling error and on to parameter predictions; second, one form of the *t* test, the one tail *t*, allows the researcher to predict the direction of a difference, which is a feature ANOVA simply does not possess.

THE ADVANTAGES OF ANOVA

The main advantage of ANOVA is that it allows the researcher to compare differences among many sample groups. Whereas *t* is "for two," the *F* ratio can theoretically handle any number of group comparisons. This is a big plus; it means that we can design experiments in which the independent variable is manipulated through a whole range of values. Analysis using the *t* test means that the independent variable can have only two levels, one for the experimental group and one for the control group. With ANOVA, we may set up a number of experimental groups to compare to the control group.

For example, in problem 3 at the end of Chapter 9, a t test was performed for a study in which increased illumination was hypothesized to cause an increase in worker productivity. Two groups of subjects were selected, one working under normal light conditions (control group) and the other working under an illumination level that was increased by 50% (experimental group). The independent variable (illumination) was manipulated at two levels. With ANOVA, we can, by adding more experimental groups, manipulate the independent variable over a range of lighting conditions, for instance, as in the following:

Control Group A: Normal illumination.
Experimental Group B: Illumination increased by 50%.
Experimental Group C: Illumination increased by 75%.
Experimental Group D: Illumination increased by 100%.

Drawbacks of Doing Successive t Tests.   It may immediately be obvious that comparisons among more than two groups can be made by doing successive t tests, for example, comparing Group A with Group B, then A with C, A with D, B with C, B with D, and finally C with D. That means calculating 6 t tests for only a 4-group experimental design. For a 6-group design, it would take 15 t tests (see the box shown below), and for a 10-group design, it would take 45 t tests). One reason, then, for not performing successive t tests is the enormous amount of work involved to calculate them. However, a far more important reason concerns the alpha error, the probability of being wrong when the null hypothesis is rejected. If H_0 is rejected several times, as it might when performing successive t tests, the alpha errors for each decision combine to produce a dangerously high level. Table 12.1 illustrates how the alpha error originally set at the .05 level inflates as the number of decisions to reject increases.

Table 12.1   Alpha error levels for successive rejections of H_0.*

Number of decisions to reject	Alpha error level
1	.05
2	.10
3	.14
4	.19
5	.23
6	.26
7	.30
8	.34
9	.37
10	.40

*Constructed from the general formula P of $\alpha = 1 - (1 - \alpha)^d$, where d equals the number of decisions. Thus, with alpha set at .05 and three decisions to reject, P of .05 $= 1 - (1 - .05)^3 = 1 - .95^3 = 1 - .86 = .14$.

Setting the alpha error level at .05 and making only one decision to reject H_0 keeps the alpha error precisely at .05. But by the tenth decision to reject, the alpha error has skyrocketed to .40. Were the alpha error to have a value of .50, which it would reach after 13 decisions, we may as well not have done the experiment, because the probability of significant results would be equal merely to flipping a coin.

ANOVA and the Null Hypothesis. Using ANOVA, the overall hypothesis of difference among more than two groups can be tested while making *only one statistical decision.* This means that if the alpha error is originally set at .05, it remains .05, no matter how many groups are being compared. For a four-group experimental design, for example, the null hypothesis states that $\mu_1 = \mu_2 = \mu_3 = \mu_4$, that is, that the means of the populations being represented by the four sample groups are all identical (or that the four samples represent the same population). If the statistical decision for the study is to reject H_0, then H_a, the alternative hypothesis (that the same groups do not all represent the same population), is accepted. Thus, a *single* decision to reject encompasses *all* sample groups. In short, then, if we begin an ANOVA with the alpha error level set at .05 and reject H_0, the alpha error level stays at .05, no matter how many sample groups are involved.

ANALYZING THE VARIANCE

In Chapter 3, we saw that variance, like the standard deviation, was a measure of the amount that all scores in a distribution vary from the mean of that distribution. A large variance indicates a large spread in the distribution—many of the scores are deviating widely from the mean. A large variance, like a large standard deviation, is characteristic of a platykurtic distribution. A small variance indicates that there is little dispersion of the scores from the mean and that the distribution of scores is leptokurtic. The basic fact to remember is that when scores are very similar to one another, or homogeneous, the variance is small, but when scores are dissimilar, or heterogeneous, the variance is large.

When several groups of scores are being compared, however, the question arises as to which mean a given score should be compared to, the mean of its own sample group or the mean of all of the scores from all of the sample groups. Furthermore, the means of the separate sample groups are almost certain to differ; this could be yet another source of variability. Sir Ronald Fisher's solution to this apparent dilemma is, in fact, rather simple—do it all! Analyze (sort out) the different sources of variance into separate components, and then compare these variance components; hence, his term analysis of variance.

The Sum of Squares

Before calculating the actual variance components, we must look at a necessary ANOVA concept called the **sum of squares,** symbolized by SS. The sum of squares is our route to the calculation of the variance. It is equal to the sum of all of the squared deviations of scores from the mean.

$$SS = \Sigma x^2$$
$$x = X - \bar{X}$$

Therefore, for the sets of scores given in Table 12.2 the sum of squares equals 80 for Sample A and only 10 for Sample B. These two samples demonstrate that the sum of squares, *like the standard deviation and the variance,* is large when the scores are spread out and small when the scores are bunched together. The sum of squares is, in fact, another measure of variability. Note, therefore, that the sum of squares can *never be negative.* The smallest possible sum of squares is zero, and that value only occurs when every score in a distribution is the same.

The procedure in Table 12.2, although it nicely conveys the relationship between the definition of the sum of squares and its calculation, can be exceedingly cumbersome in practice. The samples given there have scores that are nice, whole numbers, and the means come out as nice, whole numbers. As we know by now, that is not the way it usually happens out there in research land. To lighten the mathematical burden, another equation has been derived for the sum of squares.

The Computational Method for Calculating the Sum of Squares. The equation for the computational method for calculating the sum of squares is

$$SS = \Sigma X^2 - \frac{(\Sigma X)^2}{N}$$

That is, the sum of squares is equal to the summation of the squared raw scores minus the squared summation of raw scores divided by the number of scores.

Table 12.2 The sum of squares for two samples.

Sample A			Sample B		
X	x	x_A^2	X	x	x_B^2
14	+6	36	10	2	4
10	+2	4	9	1	1
8	0	0	8	0	0
6	−2	4	7	−1	1
2	−6	36	6	−2	4
$\Sigma X_A = 40$		$\Sigma x_A^2 = 80$	$\Sigma X_B = 40$		$\Sigma x_B^2 = 10$

$$\bar{X}_A = \frac{\Sigma X_A}{N} = \frac{40}{5} = 8 \qquad\qquad \bar{X}_B = \frac{\Sigma X_B}{N} = \frac{40}{5} = 8$$

As we saw in Chapter 7, those parentheses are important. ΣX^2 does not equal $(\Sigma X)^2$. ΣX^2 tells us to square the scores and then add, whereas $(\Sigma X)^2$ tells us to add the scores and then square.

The value $(\Sigma X)^2/N$ is also labeled C, a correction factor for using raw scores rather than deviation scores.

$$C = \frac{(\Sigma X)^2}{N}$$

When this value is found, we circle it. It will be used more than once in the ANOVA process. Table 12.3 shows the computational method for calculating the sum of squares using the same data as in Table 12.2.

This method results in the same sum of squares values, 80 and 10. It is important to realize that with this computational method the mean is never calculated directly. However, the resulting sum of squares still affords information regarding the amount by which the scores vary around the mean. Thus, we gain information about the mean without having calculated it.

The Components of Variability.   Depending on which mean the scores are compared with, three different sums of squares can be calculated.

1. *The total sum of squares.* When several groups are being compared, the scores from all the groups can be added to compute a total mean. A total sum of squares can then be calculated; it is based on how far each score in each group differs from this total mean. SS_t is therefore composed of all values of $X - \bar{X}_t$, the difference between each score and the total mean. Figure 12.1 illustrates one such value in a comparison of four sample groups.

Table 12.3   The computational method for calculating the sum of squares.

Sample A		Sample B	
X_A	X_A^2	X_B	X_B^2
14	196	10	100
10	100	9	81
8	64	8	64
6	36	7	49
2	4	6	36
$\Sigma X_A = 40$	$\Sigma X_A^2 = 400$	$\Sigma X_B = 40$	$\Sigma X_B^2 = 330$

$$SS_A = \Sigma X_A^2 - \frac{(\Sigma X_A)^2}{N} = \Sigma X_A^2 - C \qquad SS_B = \Sigma X_B^2 - \frac{(\Sigma X_B)^2}{N} = \Sigma X_B^2 - C$$

$$SS_A = 400 - \frac{40^2}{N} = 400 - \frac{1600}{5} \qquad SS_B = 330 - \frac{40^2}{N} = 330 - \frac{1600}{5}$$

$$SS_A = 400 - \boxed{320} = 80 \qquad SS_B = 330 - \boxed{320} = 10$$

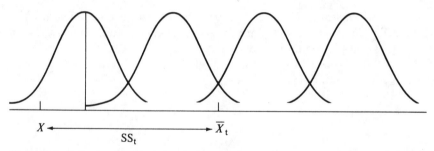

Figure 12.1 Illustration of the total sum of squares.

2. *The variability between groups.* When several groups are being compared, the differ-ence between each group mean and the total mean can be calculated. This is called the sum of squares between groups, SS_b. Figure 12.2 illustrates this value again as part of a comparison of four groups.

3. *The variability within groups.* Finally, when several groups are being compared, the difference between each individual score and the mean of the group from which that score comes can be calculated. This is called the sum of squares within groups, SS_w. Figure 12.3 is an illustration of this value.

Now, by combining each of the three sum of squares components on a single graph, we get the relationship shown in Fig. 12.4. This shows that the total sum of squares is equal to the sum of squares between groups *plus* the sum of squares within groups, or $SS_t = SS_b + SS_w$. Knowing any two of these three values, then, automatically gives us the third. Since

$$SS_t = SS_b + SS_w$$

then,

$$SS_b = SS_t - SS_w$$

or

$$SS_w = SS_t - SS_b$$

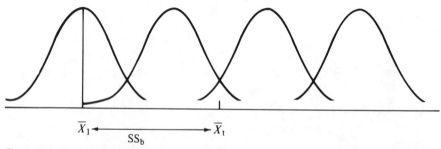

Figure 12.2 Illustration of the sum of squares between groups.

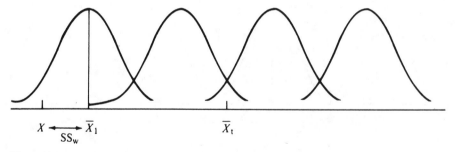

Figure 12.3   Illustration of the sum of squares within a group.

The total sum of squares can, therefore, be partitioned into two components, the sum of squares between groups and the sum of squares within groups. In other words, the total variability is composed of between-group variability and within-group variability.

Interpreting the Variability Components.   A close look at these variability components should now shed some light on the reason for this type of analysis.

The sum of squares between groups gives us a measure of how far each group mean is from the total mean *and of how far the group means are from each other.* Figure 12.5a shows a situation in which the SS_b is relatively large, meaning that the sample groups are spread apart from one another. In this case, the spread is so great that there is no overlap among the distributions.

Figure 12.5b, on the other hand, illustrates a smaller SS_b, with much less spread among the samples, in fact, considerable overlap. Intuitively, it seems likely that the sample groups in Fig. 12.5a represent different populations and those in Fig. 12.5b represent the same population. But what about the sum of squares within groups? Both parts of Fig. 12.5 have the same SS_w.

The value of the SS_w reveals how far each individual score varies from the mean of its own sample group. The more leptokurtic the sample distribution is, the smaller the SS_w is; the more platykurtic the sample distribution, the larger the

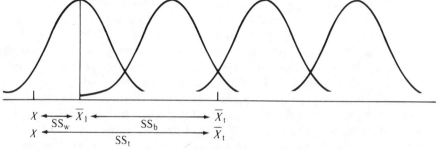

Figure 12.4   Illustration showing the relationship between total sum of squares, sum of squares between groups, and sum of squares within groups.

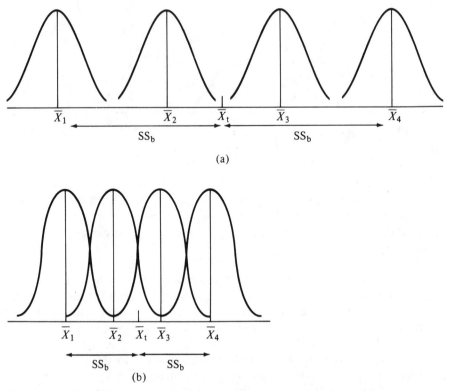

Figure 12.5   Difference in the values of the sum of squares between groups (a) when the distributions are spread out and (b) when the distributions overlap.

SS_w. Figure 12.6 illustrates this difference in the size of the SS_w, holding SS_b constant. Figure 12.6a shows four very leptokurtic distributions, samples whose SS_w is quite small. Figure 12.6b, however, illustrates a larger value of SS_w, hence the platykurtic shape of the four distributions. Since leptokurtic distributions are less apt to overlap, *the smaller the SS_w, the more likely it is that the samples are representing different populations.*

Putting together what we now know of the two variability components, SS_b and SS_w, sample groups are most likely to reflect different populations when SS_b is relatively large and SS_w relatively small. Figure 12.7 illustrates this relationship. In this illustration, the sample means are spread apart, meaning a large SS_b, but each sample distribution is itself tightly leptokurtic, meaning a small SS_w. Analysis of these data would undoubtedly lead to a reject of H_0 and a conclusion that the samples represent different populations. Figure 12.8, on the other hand, shows four groups whose means lie close to each other, meaning a small SS_b, but whose distributions are platykurtic, meaning a large SS_w. Analysis of these data would lead to an acceptance of H_0 and a conclusion that the sample groups represent the same population.

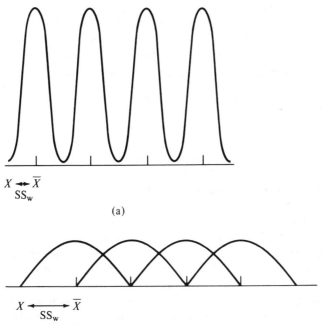

(a)

Figure 12.6   Difference in the value of the sum of squares within a group (a) when the distribution of sample scores is leptokurtic and (b) when the distribution of sample scores is platykurtic.

Therefore, when comparing the variability components, if the SS_b is large relative to the SS_w, we reject the null hypothesis. The less the sample groups overlap each other, the more sure we are that they represent different populations. The F ratio will tell us precisely how much overlap can be tolerated while remaining confident that the samples represent different populations.

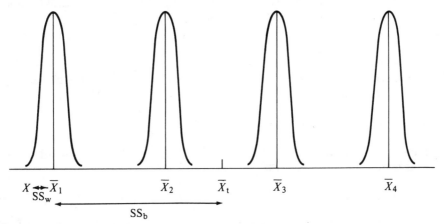

Figure 12.7   Sample groups that represent different populations; SS_b is relatively large and SS_w is relatively small.

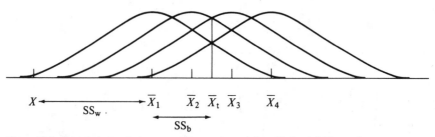

Figure 12.8   Sample groups that represent the same population; SS_b is relatively small and SS_w is relatively large.

Converting Sums of Squares to Variance Estimates

Using generalities, such as "the SS_b is large relative to the SS_w," is fine for openers, but a far more precise method must be used to calculate a testable inferential statistic. The sums of squares, total, between groups, and within groups, must be converted into variance estimates. The reason for this is that the sum of squares is greatly affected by the sample size. Adding scores to a distribution increases the value of the sum of squares, unless the additional scores happen to fall right on the mean. Therefore, to make accurate variability comparisons regardless of sample size, a method must be used that will, in effect, "average out" the variability. (We place "average out" in quotation marks because the method used does not achieve an exact arithmetic average, but it does come very close to it.) The sums of squares must be divided by some value that reflects the sample size. The resulting value is called the variance, or the mean square.

Degrees of Freedom.   The conversion of a sum of squares to a variance is simple and direct—we divide the sum of squares by the appropriate degrees of freedom. The total degrees of freedom are equal to the total number of scores in all the groups combined, N, minus one.

$$df_t = N - 1$$

The between degrees of freedom are equal to the actual number of sample groups, k, minus one.

$$df_b = k - 1$$

Finally, the within degrees of freedom are equal to the total number of scores, N, minus the number of sample groups, k.

$$df_w = N - k$$

Suppose that we use a three-group design: Groups A, B, and C have three subjects each. The df_t equals 8 (the 9 subjects less 1). The df_b equals 2 (the 3 groups less 1). The df_w equals 6 (the 9 subjects less the 3 groups). As a check, df_b and df_w must add up to the df_t. In this case, df_b plus df_w equals df_t, or $2 + 6 = 8$.

*The **F** Ratio.* After the sums of squares are converted into variance components, the F ratio can be computed; the F ratio is equal to the variance between groups, V_b, divided by the variance within groups, V_w.

$$F = \frac{V_b}{V_w}$$

If the value of the F ratio is large, it tells us that the V_b is larger than the V_w; that is, the variability between the groups is larger than the variability occurring within the groups. Thus, the larger the F ratio, the more likely it is that the null hypothesis will be rejected. An F ratio of 12, for example, indicates that the variance between groups is precisely 12 times greater than the variance within groups. The smaller the F ratio, on the other hand, the greater is the likelihood of accepting the null hypothesis. An F ratio of .50, for example, indicates that the variance between the groups is only half as large as the variance within groups.

The Mean Square. When the variance is calculated by using the sum of squares, it is called the mean square, or MS. The mean square is the mean of the squared deviations, *not* the square of the mean. Throughout the following calculations, then, variance will be symbolized as MS, and

$$F = \frac{MS_b}{MS_w}$$

Calculation of the One-Way *F* Ratio: Between Subjects

The following procedure is for a *one-way* ANOVA, that is, where there is only one independent variable set at various levels and where each of the sample groups is independent of the others. To use this analysis, each level of the independent variable must consist of a different group of subjects. A given subject, thus, may *serve in only one condition,* and each score represents a separate subject. Hence, the number of scores will always equal the number of subjects.

Using the following data, we will now calculate an F ratio.

	X_1	X_1^2	X_2	X_2^2	X_3	X_3^2
	1	1	2	4	3	9
	2	4	3	9	4	16
	3	9	4	16	5	25
$\Sigma =$	6	14	9	29	12	50

1. We find the total sum of squares. We square all the scores for each group and add the squares. We calculate C, the correction factor: add all scores in all groups combined, square the total, and divide by the total number of scores. Then we subtract C from the total of the squares. The value for C will be used again in the next step.

$$SS_t = \Sigma X^2 - \frac{(\Sigma X)^2}{N} = \Sigma X^2 - C$$

$$SS_t = 14 + 29 + 50 - \frac{(6 + 9 + 12)^2}{9} = 93 - \frac{27^2}{9}$$

$$SS_t = 93 - \frac{729}{9} = 93 - \boxed{81} = 12$$

2. We find the sum of squares between groups. We add the scores for each group, square each total, and divide it by the number of subjects *within each group*, or n_1, n_2, and so on. We add the resulting values, and again subtract the $(\Sigma X)^2/N$ term, that is, C.

$$SS_b = \frac{(\Sigma X_1)^2}{n_1} + \frac{(\Sigma X_2)^2}{n_2} + \frac{(\Sigma X_3)^2}{n_3} - \frac{(\Sigma X)^2}{N}$$

$$SS_b = \frac{6^2}{3} + \frac{9^2}{3} + \frac{12^2}{3} - \boxed{81}$$

$$SS_b = \frac{36}{3} + \frac{81}{3} + \frac{144}{3} - 81$$

$$SS_b = 12 + 27 + 48 - 81 = 87 - 81 = 6$$

3. We find the sum of squares within groups. From the SS_t value, we simply subtract the SS_b value. Remember that the total sum of squares is composed of the sum of squares between groups plus the sum of squares within groups.

$$SS_w = SS_t - SS_b$$

$$SS_w = 12 - 6 = 6$$

4. We find the mean square between groups by dividing the sum of squares between groups by its degrees of freedom ($df_b = K - 1$).

$$MS_b = \frac{SS_b}{df_b} = \frac{6}{2} = 3.00$$

5. We find the mean square within groups by dividing the sum of squares within groups by its degrees of freedom ($df_w = N - k$).

$$MS_w = \frac{SS_w}{df_w} = \frac{6}{6} = 1.00$$

6. To obtain the F ratio, we divide the mean square between groups by the mean square within groups.

$$F = \frac{MS_b}{MS_w} = \frac{3}{1} = 3.00$$

Summarizing the Results.   Statisticians always set up a summary ANOVA table of the results of the F ratio calculation, with the between-group values in the first row and the within-group values just below.

SOURCE OF VARIANCE	SS	df	MS	F
Between groups	6	2	3	3.00
Within groups	6	6	1	

The last column of the table contains the numerator and denominator of the F ratio, in this case 3 and 1, F equals 3/1, or 3.00. This technique should be followed, not only because it's standard operating procedure, but also because it provides a convenient summary of the ANOVA results.

Using the* F *Table.   With a three-group design, the H_0 being tested is that $\mu_1 = \mu_2 = \mu_3$. We turn now to the table of critical values of F, Appendix Table G, and compare our obtained value of F with the critical value of F for the appropriate degrees of freedom. The row across the top indicates the between degrees of freedom, and the column on the far left indicates the within degrees of freedom. For the calculations just done, then, the column for 2 df between and the row for 6 df within intersect on Table G at two F values: 5.14 for an alpha error level of .05 and, in boldface, 10.92, for an alpha error level of .01. As with the t test, the null hypothesis is rejected when the obtained value of F is equal to, or greater than, the critical, or table, value of F. Thus,

$$F_{.05(2,6)} = 5.14$$
$$F = 3.00 \qquad \text{Accept } H_0; \text{ the difference is not significant.}$$

If our obtained F had been equal to, say, 6.25, then

$$F_{05(2,6)} = 5.14$$
$$F = 6.25 \qquad \text{Reject } H_0; \text{ significant at } P < .05.$$

Or, if our obtained F had been equal to, say, 12.00, then

$$F_{.01(2,6)} = 10.92$$
$$F = 12.00 \qquad \text{Reject } H_0; \text{ significant at } P < .01.$$

Example. A researcher wishes to study the effects of weight training on muscle size. A total sample of 16 males is randomly selected from a population of 20-year-olds. Each subject is randomly assigned to one of four different exercise groups that vary from no weight training to heavy weight training. The subjects are closely supervised, fed the same diet, allowed the same amount of sleep, and generally treated similarly. At the end of three months, flexed bicep measures are taken of each subject. The results, in inches, are as follows:

No Training		Light Training		Medium Training		Heavy Training	
X_1	X_1^2	X_2	X_2^2	X_3	X_3^2	X_4	X_4^2
11	121	13	169	16	256	17	289
12	144	13	169	16	256	21	441
11	121	14	196	15	225	19	361
10	100	12	144	17	289	19	361
$\Sigma = 44$	486	52	678	64	1026	76	1452

1. $SS_t = \Sigma X^2 - \dfrac{(\Sigma X)^2}{N} = \Sigma X^2 - C$

$$SS_t = (486 + 678 + 1026 + 1452) - \frac{(44 + 52 + 64 + 76)^2}{16}$$

$$SS_t = 3642 - \frac{236^2}{16} = 3642 - \frac{55{,}696}{16}$$

$$SS_t = 3642 - \boxed{3481} = 161$$

2. $SS_b = \dfrac{(\Sigma X_1)^2}{n_1} + \dfrac{(\Sigma X_2)^2}{n_2} + \dfrac{(\Sigma X_3)^2}{n_3} + \dfrac{(\Sigma X_4)^2}{n_4} - C$

$$SS_b = \frac{44^2}{4} + \frac{52^2}{4} + \frac{64^2}{4} + \frac{76^2}{4} - \boxed{3481}$$

$$SS_b = \frac{1936}{4} + \frac{2704}{4} + \frac{4096}{4} + \frac{5776}{4} - \boxed{3481}$$

$$SS_b = 484 + 676 + 1024 + 1444 - \boxed{3481}$$

$$SS_b = 3628 - 3481 = 147$$

3. $SS_w = SS_t - SS_b = 161 - 147 = 14$

SOURCE OF VARIANCE	SS	df	MS	F
Between groups	147	3	49	41.88
Within groups	14	12	1.17	

$F_{.01(3,12)} = 5.95$

$F = 41.88$ Reject H_0; significant $P < .01$.

The calculated F ratio of 41.88 leads to a clear rejection of the null hypothesis, the probability of alpha error being less than .01. Since this research utilizes the experimental method (weight training being a manipulated independent variable), the significant difference in bicep size can be attributed to the independent variable. This is an after-only design; group equivalence is created by random assignment, and the dependent variable is not measured until after the manipulation of the independent variable.

Requirements for Using the* F *Ratio.   To use the F ratio, the following requirements must be met.

1. The sample groups have been randomly and independently selected.
2. There is a normal distribution in the population from which the samples are selected.
3. The data are in interval form (or, of course, ratio).
4. The within-group variances of the samples should be fairly similar. This is called **homogeneity of variance** and it simply means that ANOVA demands sample groups that do not differ too much with regard to their internal variabilities. For example, you cannot do an ANOVA on three sample groups, one of which has a leptokurtic shape, the second a mesokurtic shape, and the third a platykurtic shape.

Tukey's HSD: A Post Hoc Multiple Comparison Test

If the F ratio *has been found to be significant,* it is important, after the fact (or post hoc), to ferret out precisely where the sample differences occurred. The reason for this step is that a significant F ratio can be the result of different patterns of group differences.

For example, the results of a three-group design graphed in Fig. 12.9 produce a significant F ratio, but the actual pattern of the group differences makes it clear that the effects are not spread out evenly. The between-group variance is certainly large, but most of it is due to the fact that Group C is so far away from both Groups A and B. In fact, perhaps Groups A and B do not differ significantly from each other. The alternative hypothesis for the F ratio is not, therefore, a

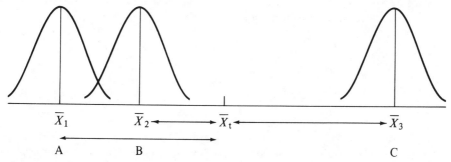

Figure 12.9   One possible pattern for the between-group differences with three sample groups.

straightforward $\mu_1 \neq \mu_2 \neq \mu_3 \neq \mu_4$, and so on, but instead is any one of a series of choices, $\mu_1 \neq \mu_2 = \mu_3 = \mu_4$ or $\mu_1 \neq \mu_2 \neq \mu_3 = \mu_4$ or any other such combination.

To determine exactly where the differences are, multiple comparison tests have been developed. Although several statistical tests may be used for this analysis, one of the most popular is **Tukey's HSD,** or Honestly Significant Difference Test. This test should only be performed when the overall F ratio has been shown to be significant, meaning that the use of HSD for these internal comparisons does not inflate the alpha error (as doing successive t tests would).*

Calculating Tukey's HSD. The first step in calculating Tukey's HSD is to set the alpha error level, typically .05, and then to turn to Appendix Table H. This table has the within degrees of freedom in the left column and k, the number of sample groups, across the top row. Since the F ratio obtained for our weight training research study was found to be significant, we will use those data to illustrate the HSD calculation.

$$HSD = \alpha_{.05} \sqrt{\frac{V_w}{n}} = \alpha_{.05} \sqrt{\frac{MS_w}{n}}$$

The data are from a four-group design, so we look across the top row of Table H until we find 4. Then we follow straight down that column until it intersects the row led by 12, which is our within degrees of freedom. The value found there, in this case 4.20 (or 5.50 if the alpha error level had been set at .01), is then plugged into the equation in place of $\alpha_{.05}$. Then we simply plug in the mean square within groups, 1.17 and divide by the number of subjects in any single sample, 4.

$$HSD = 4.20 \sqrt{\frac{1.17}{4}}$$

$$HSD = 4.20\sqrt{.29} = (4.20)(.54) = 2.27$$

Interpreting HSD Values. The calculated value for HSD just obtained signifies that in order to be statistically significant at a probability of .05, the difference between any pair of means must be at least 2.27. From our weight training data, we calculate the following sample means:

1. $\bar{X}_1 = \dfrac{\Sigma X_1}{n_1} = \dfrac{44}{4} = 11$

2. $\bar{X}_2 = \dfrac{\Sigma X_2}{n_2} = \dfrac{52}{4} = 13$

---

*Tukey's HSD is considered to be one of the best, although conservative, multiple comparison tests, especially when the homogeneity of variance assumption is met.

3. $\bar{X}_3 = \dfrac{\Sigma X_3}{n_3} = \dfrac{64}{4} = 16$

4. $\bar{X}_4 = \dfrac{\Sigma X_4}{n_4} = \dfrac{76}{4} = 19$

All the mean differences equal to or greater than the calculated HSD of 2.27 are significant at the .05 level. Placing the calculated mean differences in matrix form, we mark the significant differences with asterisks.

	$\bar{X}_1 = 11$	$\bar{X}_2 = 13$	$\bar{X}_3 = 16$	$\bar{X}_4 = 19$
$\bar{X}_1 = 11$	—	2	5*	8*
$\bar{X}_2 = 13$	2	—	3*	6*
$\bar{X}_3 = 16$	5*	3*	—	3*
$\bar{X}_4 = 19$	8*	6*	3*	—

We therefore conclude that weight training does indeed increase bicep muscle size among 20-year-old males in all training conditions, except for the no exercise-light training comparison.

Applications of Tukey's HSD.   The procedure just outlined can be used only when comparing pairs of sample means taken from sample groups of equal size. Tukey's HSD may also be used on data from samples of unequal sizes, and even for comparing clusters of means, rather than individual pairs. These procedures are covered in more advanced statistics texts.

APPLICATIONS OF ANOVA

Obtaining the Value of **t.**   ANOVA sometimes is used for testing the hypothesis of difference between two independent sample means. In this situation, the between degree of freedom, $k - 1$, equal $2 - 1$, or 1. The resulting F ratio is equal to t^2.

$$F = t^2 \qquad t = \sqrt{F}$$

The interpretation of a t ratio calculated in this way must be consistent with the two-tail t format, since the alternative hypothesis for F can never be directional. That is, if t is computed by taking the square root of F, then its significance must be evaluated against the critical values in the two-tail t table.

ANOVA can be used to test the hypothesis of difference in both experimental and post-facto research.

Analysis of Experimental Research. The action of the independent variable in experimental research is always reflected in the variability between groups, while the variability within groups simply reflects random error and/or individual differences among the subjects. (Between-group variability also reflects individual differences to a small degree, since individuals do differ in their receptivity to the independent variable.) The effects of the independent variable tend to separate the groups, to cause the means of the samples to fall farther and farther apart. This produces an increase in the size of the variability between the groups and, therefore, an increased probability of rejecting the null hypothesis. When the independent variable "takes," that is, has an effect, then the following are observed:

1. Between-group variance is increased.
2. The F ratio is larger.
3. The probability of rejecting the null hypothesis is increased.

Analysis of Post-Facto Research. In the analysis of post-facto differences, the F ratio, of course, is not used for establishing whether or not an independent variable has any effect, since in such studies the independent variable is assigned, not manipulated. Therefore, in post-facto research, an increase in the variance between the groups relative to the variance within groups reflects the fact that the samples represent different populations. Why this difference exists cannot be determined. But since it does exist, predictions can be made using it.

THE FACTORIAL ANOVA

All the experimental research studies covered so far in this book have been interested in the influence one independent variable might have on a dependent variable, with all other variables held constant. That is, we have created equivalent groups of subjects and exposed the groups to different levels of the independent variable. However, in all other ways the groups have been treated equally (for example, testing all groups at the same time of day, under identical conditions of illumination, etc.). Sometimes, however, researchers are interested in the effects of more than one independent variable. In these situations, the sample groups must be treated differently on a number of different dimensions.

For example, suppose that we are interested in conditions that might affect the height to which grass grows. One research technique would be to manipulate several independent variables separately, holding all other factors constant. We could treat two lawn patches with different amounts of nitrogen, but keep the patches equal with regard to amount of sunlight, moisture, lime, temperature, and so on. Then, in another study, the amount of moisture could be manipulated, and nitrogen content (as well as all other factors) would be held constant. However, studies of this nature are often somewhat misleading, because the different conditions involved often have an interactive effect. Perhaps nitrogen is most effective under certain moisture conditions and is even damaging under others. The point is that when several independent variables are manipulated *simultaneously,* it often leads to the discovery of cumulative effects acting above and beyond the effects of any of the independent variables working separately. We all know that phenobarbitol has a depressing effect on the human physiology, as does alcohol. But put the two together and you get that big dependent variable in the sky.

Factors

Any variation in an independent variable can also be called a *factor.* For example, if groups of subjects are treated differently with regard to, say, amount of illumination, that would be one factor. Similarly, if the groups are also treated differently on something else, such as noise level, that would be another factor. Whenever a study focuses on more than one factor, the data analysis is called factorial. If the data scale is at least interval, the statistic used for this analysis is called the **factorial ANOVA.**

Between-Group Variability

We know that the between-group variability is that portion of the total variability separating the sample groups and that it results from the action of the independent variable. When more than one independent variable is involved, we must factor out the portion of the between-group variability that is attributable to each independent variable. Furthermore, since the independent variables may interact to produce a cumulative effect, we also need to know what part of the between-group variability is being created or increased by such interaction among the independent variables. Just as we have previously taken the total variability and partitioned out its between-group and within-group components, we can now take the between-group variability and separate its components.

The actions of the independent variables taken separately are called the **main effects,** as opposed to their combination or **interaction effects.**

Calculating the Two-Way ANOVA

In the two-independent variable research design, which calls for the two-way ANOVA (rest assured, this is as far as we will go in this chapter), one of the independent variables is set up in columns and the other in rows to form a block of cells. A cell is a combination of treatment conditions that is unique to one group of subjects, or *where a row intersects a column.* In the following example, we examine a four-cell, completely randomized design. Two independent variables, diet and exercise, are each being manipulated at two levels.

Example. A researcher is interested in the effects of both diet and exercise on percentage of body fat. A sample of 20 male, high school sophomores is randomly selected. The subjects are randomly assigned to four different treatment conditions. Since the two independent variables, diet and exercise, are each manipulated in two ways, the four cells shown represent all possible combinations of treatment conditions.

Factor A

	DIET BELOW 2000 CALORIES	DIET ABOVE 2000 CALORIES
Factor B EXERCISE	a	b
NO EXERCISE	c	d

Cell *a* contains only those subjects who follow the exercise program and are also on the low-calorie diet. Cell *b* contains only those subjects who follow the exercise program and are on the high-calorie diet. Cell *c* contains only those subjects who follow the no-exercise program and are on the low-calorie diet. Cell *d* contains those subjects who follow the no-exercise program and are on the high-calorie diet.

After a three-month period, each subject is measured for percentage of body fat compared to total weight, and their scores are recorded as follows:

	DIET BELOW 2000 CALORIES		DIET ABOVE 2000 CALORIES		
	a		b		
EXERCISE	X_a	X_a^2	X_b	X_b^2	
	10	100	12	144	
	12	144	14	196	
	14	196	15	225	
	12	144	16	256	
	10	100	14	196	$\Sigma X_a + \Sigma X_b = 129$
	$\Sigma X_a = 58$	$\Sigma X_a^2 = 684$	$\Sigma X_b = 71$	$\Sigma X_b^2 = 1017$	
	c		d		
NO EXERCISE	X_c	X_c^2	X_d	X_d^2	
	16	256	22	484	
	18	324	24	576	
	20	400	24	576	
	22	484	22	484	
	20	400	24	576	
	$\Sigma X_c = 96$	$\Sigma X_c^2 = 1864$	$\Sigma X_d = 116$	$\Sigma X_d^2 = 2696$	$\Sigma X_c + \Sigma X_d = 212$

$\Sigma X_a + \Sigma X_c = 154$ $\Sigma X_b + \Sigma X_d = 187$ $\Sigma X = 341$

To perform a two-way ANOVA on these data, we go through the following steps:

1. We calculate the total sum of squares, just as we did earlier in this chapter for a one-way ANOVA.

$$SS_t = \Sigma X^2 - \frac{(\Sigma X)^2}{N} = \Sigma X^2 - C$$

$$SS_t = (684 + 1017 + 1864 + 2696) - \frac{341^2}{20} = 6261 - \frac{116,281}{20}$$

$$SS_t = 6261 - 5814.05 = 446.95$$

2. We calculate the sum of squares between groups, again as we did for a one-way ANOVA.

$$SS_b = \frac{(\Sigma X_a)^2}{n_a} + \frac{(\Sigma X_b)^2}{n_b} + \frac{(\Sigma X_c)^2}{n_c} + \frac{(\Sigma X_d)^2}{n_d} - C$$

$$SS_b = \frac{58^2}{5} + \frac{71^2}{5} + \frac{96^2}{5} + \frac{116^2}{5} - 5814.05$$

$$SS_b = \frac{3364}{5} + \frac{5041}{5} + \frac{9216}{5} + \frac{13,456}{5} - 5814.05$$

$$SS_b = 672.80 + 1008.20 + 1843.20 + 2691.20 - 5814.05$$

$$SS_b = 6215.40 - 5814.05 = 401.35$$

3. We calculate the sum of squares within groups from our values for SS_t and SS_b.

$$SS_w = SS_t - SS_b$$

$$SS_w = 446.95 - 401.35 = 45.60$$

4. We calculate the sum of squares for rows, SS_{row} for the first main effect. We add all the scores for subjects on the exercise program, square the total, and divide by the number of scores in that row. We do the same for the row of subjects on the no-exercise program. Then we add the values for the two rows and subtract the correction factor.

$$SS_{row} = \frac{(\Sigma X_a + \Sigma X_b)^2}{n_a + n_b} + \frac{(\Sigma X_c + \Sigma X_d)^2}{n_c + n_d} - C$$

$$SS_{row} = \frac{(58 + 71)^2}{5 + 5} + \frac{(96 + 116)^2}{5 + 5} - 5814.05$$

$$SS_{row} = \frac{129^2}{10} + \frac{212^2}{10} - 5814.05$$

$$SS_{row} = \frac{16,641}{10} + \frac{44,944}{10} - 5814.05$$

$$SS_{row} = 1664.10 + 4494.40 - 5814.05 = 344.45$$

5. We calculate the sum of squares for columns, SS_{col} for the second main effect. We add all the scores for subjects on the diet below 2000 calories, square the total, and divide it by the number of scores in that column. We do the same for the

column of subjects on the diet above 2000 calories. We then add the values for the two columns and subtract the correction factor.

$$SS_{col} = \frac{(\Sigma X_a + \Sigma X_c)^2}{n_a + n_c} + \frac{(\Sigma X_b + \Sigma X_d)^2}{n_b + n_d} - C$$

$$SS_{col} = \frac{(58 + 96)^2}{5 + 5} + \frac{(71 + 116)^2}{5 + 5} - 5814.05$$

$$SS_{col} = \frac{154^2}{10} + \frac{187^2}{10} - 5814.05$$

$$SS_{col} = \frac{23{,}716}{10} + \frac{34{,}969}{10} - 5814.05$$

$$SS_{col} = 2371.60 + 3496.90 - 5814.05 = 54.45$$

6. We calculate the interaction sum of squares, $SS_{r \times c}$. Since the sum of squares between groups is composed of the sum of squares for columns plus the sum of squares for rows plus the interaction sum of squares, we can get the interaction value by finding the difference between values we already have.

$$SS_{r \times c} = SS_b - SS_{col} - SS_{row}$$

$$SS_{r \times c} = 401.35 - 54.45 - 344.45 = 2.45$$

7. We set up a summary table of our calculations for the factorial (two-way) ANOVA. The degrees of freedom for the table are allocated as follows. The degrees of freedom for rows equal the number of rows minus one.

$$df_{row} = n_{row} - 1$$

$$df_{row} = 2 - 1 = 1$$

The degrees of freedom for columns equal the number of columns minus one.

$$df_{col} = n_{col} - 1$$

$$df_{col} = 2 - 1 = 1$$

The interaction degrees of freedom equal the degrees of freedom for rows multiplied by the degrees of freedom for columns.

$$df_{r \times c} = (df_{row})(df_{col}) = (n_{row} - 1)(n_{col} - 1)$$

$$df_{r \times c} = (1)(1) = 1$$

The degrees of freedom within groups are equal to the total sample size, N, minus the number of sample groups, k.

$$df_w = N - k$$

$$df_w = 20 - 4 = 16$$

As a check, we can add the various degrees of freedom, $1 + 1 + 1 + 16 = 19$. This should equal $N - 1$; $20 - 1 = 19$.

SOURCE OF VARIABILITY	SS	df	MS	F
Rows (exercise)	344.45	1	344.45	120.86
Columns (diet)	54.45	1	54.45	19.11
r × c (interaction)	2.45	1	2.45	.86
Within groups	45.60	16	2.85	

The mean square values for the table are calculated as for the one-way ANOVA; that is, the mean square equals the sum of squares divided by the appropriate degrees of freedom.

$$MS_{row} = \frac{SS_{row}}{df_{row}} = \frac{344.45}{1} = 344.45$$

$$MS_{col} = \frac{SS_{col}}{df_{col}} = \frac{54.45}{1} = 54.45$$

$$MS_{r\times c} = \frac{SS_{r\times c}}{df_{r\times c}} = \frac{2.45}{1} = 2.45$$

$$MS_w = \frac{SS_w}{df_w} = \frac{45.60}{16} = 2.85$$

The F ratios are determined exactly as in a one-way ANOVA. There we divided the mean square between groups by the mean square within groups to get the value of F. Here the mean square between groups is broken down into three values, so we get three F ratios.

$$F_{row} = \frac{MS_{row}}{MS_w} = \frac{344.45}{2.85} = 120.86$$

$$F_{col} = \frac{MS_{col}}{MS_w} = \frac{54.45}{2.85} = 19.11$$

$$F_{r\times c} = \frac{M_{r\times c}}{MS_w} = \frac{2.45}{2.85} = .86$$

8. Each of the three F ratios obtained is now compared with the critical value of F in Table G. The degrees of freedom are 1 in the numerator and 16 in the denominator (each of the mean squares between groups has 1 df and the mean square within groups has 16 df). Again, the rule is that the null hypothesis is rejected whenever the calculated value of F equals or exceeds the table value.

$$F_{.01(1,16)} = 8.53$$

$F_{row} = 120.86$ Reject H_0; significant at $P < .01$.

$F_{col} = 19.11$ Reject H_0; significant at $P < .01$.

$F_{r\times c} = .86$ Accept H_0; not significant.

We conclude that each of the independent variables, diet (F_{col}) and exercise (F_{row}), caused a significant difference in the amount of body fat. However, in interaction ($F_{r \times c}$), the two independent variables did not combine to produce any greater effect. Any interaction that occurred is accounted for by chance.

The reason for the lack of an interaction effect in this study might be due to the chosen levels for the independent variables. It is possible that if diet had been manipulated at a third level of 1000 calories, an interaction effect would have occurred.

Because each independent variable in this problem had only two levels, Tukey's HSD is not needed.

The Theory of the Two-Way ANOVA

Whenever there are significant differences among several sample groups, these differences are related to the between-group variability. When only one independent variable is involved, we have no problem identifying the source of these differences, so only one F ratio is calculated (a one-way ANOVA). But when significant differences result from the effects of several independent variables, a number of explanations of the sources of difference are possible; each of these possible accountings requires a separate F ratio. Since the action of the independent variables is reflected in the between-group variability, it is this *between-group variability* that must be broken up into its various parts—one part for each independent variable and the remaining part for the interaction.*

To accomplish the two-way ANOVA, each component of the between-group variability is analyzed separately. For one of the independent variables, we look only at the column scores, holding the row comparisons in abeyance. This is why in our example we added the scores of all of the subjects on the low-calorie diet, regardless of whether they exercised or not. Next, we look only at the row scores, holding column variability constant. In our example, that meant adding all the scores for an exercise level, regardless of which diet the subjects followed. Finally, by subtraction, we factor out the interaction effect, that is, whether the independent variables are in any way acting in concert to produce still another effect. The effects of each independent variable separately are called *main effects*, as opposed to the interaction effect produced by the combination of variables.

Limitation on the Use of ANOVA Techniques. All the ANOVA techniques outlined in this chapter can only be applied if the groups of subjects are randomly selected and are *independent* of each other. Among the various experimental designs, then, our ANOVAs are restricted to after-only studies, (or studies based on between-group comparisons, where each subject has been randomly assigned to a

---

*Note that if there are three or more independent variables, then more than just one possible interaction effect must be included in the analysis.

specific combination of treatment conditions). Later, in Chapter 15, other ANOVA techniques will be presented that do allow for the analysis of repeated-measure and matched-subject designs.

SUMMARY

ANOVA, an acronym for ANalysis Of VAriance, is used on interval data when more than two sample means are to be compared for differences. The resulting statistic, the F ratio, determines the ratio between the variability occurring between the sample groups and the variability occurring within each of the sample groups. The higher the F ratio, the greater is the likelihood that the samples represent different populations; a high F ratio indicates that there is a great deal of between-group variability and little within-group variability occurring (the sample distributions show little or no overlap). The lower the F ratio, the greater is the chance that the samples represent a single population; a low F ratio indicates that there is little between-group variability compared to the amount of within-group variability occurring (the sample distributions show a great deal of overlap).

Variability can be determined through the use of the sum of squares, the sum of the squared deviations of all the scores from the mean. The variance, V, or mean square, MS, is calculated by dividing the sum of squares by the appropriate degrees of freedom, df. This is a necessary step; although adding new scores to a distribution has a dramatic effect on the sum of squares, this effect is minimized through the averaging process used to obtain the mean square. The F ratio, then, is equal to MS_b/MS_w, the mean square between the groups divided by the mean square within the groups.

The calculated F ratio must then be checked for significance by comparing it against the critical values of F for the appropriate degrees of freedom. When the calculated F is equal to or greater than the table value, the null hypothesis is rejected. As with the t test, the alpha error should be kept at a probability level of .05 or less.

Although a significant F ratio tells us that there are significant differences among the several group means, it does not, in and of itself, specify precisely where those differences are occurring. For this analysis, Tukey's HSD (Honestly Significant Difference) is required.

The F ratio so far described is called a one-way ANOVA, in that, although several sample groups are being compared, only one independent variable is used. When the experimental design calls for more than one independent variable, the factorial ANOVA must be used. This technique allows us not only to discover whether the independent variables taken separately are having any effect, but *also* whether the independent variables are interacting and producing a cumulative effect. The actions of the separate independent variables are called main effects, and the cumulative effects are called interaction effects (produced by the combination of independent variables).

KEY TERMS AND NAMES

analysis of variance (ANOVA)
factorial ANOVA
Fisher, Sir Ronald
homogeneity of variance

interaction effect
main effect
sum of squares
Tukey's HSD

REFERENCE

1. STOLINE, M. R. (1981). The status of multiple comparisons: Simultaneous estimation of all pairwise comparisons in one-way ANOVA designs. *The American Statistician, 35,* pp. 134–141.

PROBLEMS

1. Calculate an F ratio for the following five-group design. Do you accept or reject the null hypothesis?

A	B	C	D	E
3	6	7	9	12
4	7	8	9	10
5	6	7	8	9

2. If the results in problem 1 make it appropriate to do so, compare the five sample means using Tukey's HSD.
3. For the following two-group design, calculate both a t and an F ratio. Do you accept or reject the null hypothesis? (Be sure that $t = \sqrt{F}$ and $F = t^2$, within rounding error.)

A	B
12	8
10	6
8	4
6	2
4	0

4. A researcher is interested in determining the possible effect of shock on memory of verbal material. Subjects are randomly selected and then randomly assigned to one of three treatment conditions. Group A receives no shock, Group B receives medium shock, and Group C gets high-intensity shock levels. The various shock levels are introduced during the learning period. All subjects then take a verbal retention test, with high scores indicating more retention. Their scores are as follows:

A	B	C
24	20	16
30	18	14
25	22	12
24	20	15
20	17	16

a. Calculate the F ratio.

b. Do you accept or reject the null hypothesis?

c. If appropriate, calculate Tukey's HSD.

5. A researcher is interested in proving that ingestion of the drug magnesium pemoline (MgPe) increases retention of learned material. A group of 16 subjects is randomly selected from the population of students at a large university. The subjects are then randomly assigned to one of four conditions: A receives placebo, B receives 10 cc of MgPe, C receives 20 cc of MgPe, and D receives 30 cc of MgPe. All subjects are then given some material to read and four hours later are tested for retention (high scores indicating high retention).

A	B	C	D
8	10	11	10
6	7	6	8
7	8	8	7
5	6	9	9

a. Calculate the F ratio.

b. Do you accept or reject the null hypothesis?

c. If appropriate calculate Tukey's HSD.

d. State the kind of research this represents, experimental or post-facto.

e. Identify the independent and dependent variables.

f. What conclusions can you legitimately draw from this study?

6. Hypothesis: Increasing movie violence increases viewer aggression. Three groups of subjects are randomly selected. All groups view a 30-minute motion picture, but in each version of the movie there are a different number of violent scenes. Following the movie, all subjects are given the TAT (Thematic Apperception Test) and scored on the amount of fantasy aggression displayed, high scores indicating higher levels of need aggression. The data follow:

GROUP A (NO VIOLENT SCENES), X_A	GROUP B (5 VIOLENT SCENES), X_B	GROUP C (10 VIOLENT SCENES), X_C
1	2	3
2	3	4
1	2	3

a. Calculate the F ratio.
b. Should H_0 be accepted or rejected?
c. If appropriate, calculate Tukey's HSD.

7. Hypothesis: Increasing temperature and physical exercise increases the body's skin conductivity (as measured by a galvanometer). A sample of 24 males, all 21 years old, is randomly selected, and the subjects are randomly assigned to four treatment conditions. Group A is subjected to high temperature and physical exercise (jogging in place for 10 minutes). Their G.S.R. (galvanic skin response) scores in decivolts are 9, 8, 8, 10, 7, and 8. Group B is subjected to low temperature and physical exercise (again, jogging in place for 10 minutes). Their G.S.R. scores are 7, 6, 5, 5, 4, and 5. Group C is subjected to high temperature without exercise. Their G.S.R. scores are 5, 3, 3, 4, 3, and 4. Group D is subjected to low temperature without exercise. Their G.S.R. scores are 3, 2, 2, 1, 2, and 1. (High scores indicate greater skin conductivity.) Test the hypothesis.

8. A psychologist is interested in discovering the possible effect of both a certain drug and psychotherapy on the anxiety levels of anxious patients. Six groups of anxious patients, 10 per group, were randomly selected and independently assigned to the various treatment conditions.

| | | *Factor A* | |
	PLACEBO A1	DRUG (LOW DOSE) A2	DRUG (HIGH DOSE) A3
B1 PSYCHO-THERAPY	9 7 8 4 7 9 10 5 6 9	6 4 3 6 5 6 2 4 8 2	7 1 1 2 4 5 4 2 1 1
Factor B **B2** NO PSYCHO-THERAPY	9 6 9 10 14 6 8 5 4 5	7 5 7 6 5 3 2 4 5 9	6 1 1 5 4 3 2 1 2 3

(Each of the scores represents the measure of anxiety taken on the patient after the treatments were completed, with higher scores indicating higher levels of anxiety).

a. Was there a significant effect for the drug?
b. Was there a significant effect for the psychotherapy?

c. Was there a significant interaction between the drug and the psychotherapy?

Fill in the blanks in problems 9 through 18.

9. Total variability results from the accumulated differences between each individual score and _____.
10. Between-group variability results from the accumulated differences between each sample mean and _____.
11. Within-group variability results from the accumulated differences between each individual score and _____.
12. When a calculated F ratio has a large value, it indicates that the variability between groups is _____ (larger or smaller) than the variability within groups.
13. The variance, or mean square, results from dividing the sum of squares by _____.
14. The total sum of squares is made up of two major components, the _____ and the _____.
15. When an ANOVA results in the rejection of the null hypothesis, then the _____ variability must be larger than the _____ variability.
16. If the within-group variability is small, then the separate sample groups are most likely to have _____ (platykurtic or leptokurtic) distributions.
17. A five-group research design with six subjects in each group has _____ between degrees of freedom and _____ within degrees of freedom.
18. The greater the spread among the various sample means, the larger is the _____ (between or within) variability.

True or False—Indicate either T or F for problems 19 through 27.

19. The F ratio is a nondirectional, two-tail test of differences among sample groups used whenever the data are in interval form.
20. ANOVA demands that *at least* four sample groups must be compared.
21. On a four-group design, the between degrees of freedom for a one-way ANOVA must equal 4.
22. ANOVA assumes that the data are at least interval.
23. An F ratio of 5.00 indicates that the variance between groups is five times greater than the variance within groups.
24. The use of the factorial ANOVA is required whenever there is more than one independent variable and the data are in interval form.
25. On a factorial ANOVA, the interaction effect will always be significant if the main effects are themselves significant.
26. To do a factorial ANOVA, there must be a minimum of at least four different treatment conditions.
27. When the obtained value of F is larger than the table value of F for a given number of degrees of freedom, the null hypothesis must be accepted.

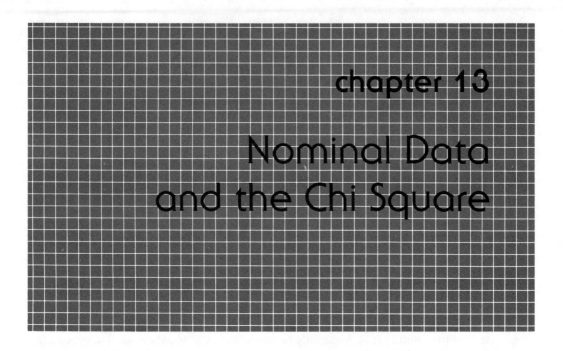

chapter 13

Nominal Data
and the Chi Square

By far the most popular test for **nominal data** is the chi square. Although nominal data were described in Chapter 1, it is important to repeat here that the nominal case only gives us information regarding the frequency of occurrence within categories. From the nominal scale we get nose-counting data. It tells us in how *many* cases a certain trait occurs. From nominal data we get no information as to how much of the given trait any individual possesses, only information as to whether that individual has the trait at all. A nominal scale can be constructed for political party affiliation by grouping individuals according to party loyalties—so many Democrats, so many Republicans, so many Independents, and so on. We cannot characterize the strength of any individual's party affiliation from this nominal scale; we can only know what that affiliation is. With nominal data, there are no shades of gray; an observation either has the trait or not. In short, then, nominal data are generated by sorting and counting—sorting the data into discrete, mutually exclusive categories and then counting the frequency of occurrence within each category.

THE CHI SQUARE AND INDEPENDENT SAMPLES

When we are setting up a nominal scale of measurement, the categories must be totally independent of each other. That is, an individual that is counted as a Democrat cannot also be counted as a Republican. All the cases in any given category must share a common trait, but they cannot share the same trait with cases from any other category. This is what is meant by the equality versus nonequality rule for the nominal scale that was mentioned in Chapter 1. All observations in a given category are equal to all other observations in that category and not equal to observations in any other category.

The 1 × k Chi Square (Goodness of Fit)

Suppose that a market researcher wishes to find out if one radio station is more popular than others among teen-agers. A random sample of 100 teen-agers is selected, and they are categorized on the basis of their radio station preference. The data are as follows: Station A, 40; Station B, 30; Station C, 20; and Station D, 10. These data—40, 30, 20, and 10—make up the frequency *observed*, or f_o, which is to be compared to the frequency expected, f_e. The frequency expected can be calculated either on the basis of pure chance or on the basis of some a priori hypothesis. (An a priori hypothesis is one that is either known or is presumed to be true, either because of previous research findings or, more commonly, because it is consistent with some scientific theory.) We then compare the observed data with an expected data set to determine how well these observations "fit" the expectations.

The Frequency Expected Due to Chance. We first consider the effect of chance. If 100 teen-agers are selected and sorted into four categories, and if only chance determines the sorting, then the result should be a frequency of 25 individuals per category. The frequency expected due to chance, then, is the total number in our sample, N, divided by the number of categories, k.

$$\text{chance } f_e = \frac{N}{k}$$

Calculating the Chi Square Value. The value of **chi square** is obtained from the following equation:

$$\chi^2 = \Sigma \frac{(f_o - f_e)^2}{f_e}$$

To illustrate the method for calculating chi square with this equation, we can use the data on radio station preference among teen-agers. We construct a table of four columns (one for each category, that is, station preference) and five rows (see Table 13.1). We fill in this table according to the following steps:

1. In the first row, f_o, we place our observed data on radio station preference, that is, 40 for Station A, 30 for Station B, 20 for Station C, and 10 for Station D.
2. Our calculated value for frequency expected, f_e, is placed in each category in the second row. For the radio station survey, this value is 25.

Table 13.1 Chi square table of radio station preference among 100 teenagers.

	STATION A	STATION B	STATION C	STATION D	
f_o	40	30	20	10	
f_e	25	25	25	25	
$f_o - f_e$	15	5	−5	−15	
$(f_o - f_e)^2$	225	25	25	225	$\Sigma \dfrac{(f_o - f_e)^2}{f_e} = 20$
$(f_o - f_e)^2/f_e$	9	1	1	9	

3. The third row of the table represents the difference between the observed and expected frequency in each category, that is, $f_o - f_e$. Subtracting in each column of Table 13.1, we get 15, 5, −5, and −15 across the third row.

4. We square each difference and place these values in the fourth row. The squared differences are 225, 25, 25, and 225.

5. For the fifth row, each squared difference is divided by the value for f_e for that category. For Table 13.1, this step yields 9, 1, 1, and 9.

6. Adding the values across the fifth row gives us the summation result for chi square. Our chi square value for the radio station survey is, therefore, 20.

Interpreting the Chi Square Value. The chi square value tells us whether or not the frequency observed differs significantly from that expected. The null, or chance, hypothesis, of course, says that there is no difference.

$$H_0: f_o = f_e$$

The alternative hypothesis says there is a difference.

$$H_a: f_o \neq f_e$$

To make the statistical decision, we compare our obtained value with the critical chi square values in Appendix Table I. The table is set up with the degrees of freedom in the column at the far left and the critical chi square values for the .05 and .01 alpha error levels in the next two columns. Note that in this table as the degrees of freedom increase, the critical chi square values also increase. Unlike the t or r tables, here, as the degrees of freedom increase, it becomes more, rather than less, difficult to reject the null hypothesis.

The degrees of freedom for $1 \times k$ chi square are based on $k - 1$, the number of categories minus one. Note that for the chi square the degrees of freedom are *not* based on the size of the sample.

Degrees of Freedom for the 1 × k Chi Square (1 by k)

For the $1 \times k$ chi square, degrees of freedom are found on the basis of how many cell entries, not number of subjects, are free to vary after Σf_o or N is fixed. For example, if we were to place a total of 100 people into four categories, once the first three categories were filled

	A	B	C	D	
f_o	30	20	40	?	$\Sigma f_o = N = 100$

the fourth now must contain 10 people. The value allowed in that fourth category is, therefore, *not free to vary*. With four categories, then, only three of the values are free to vary, giving us $k - 1$, or $4 - 1 = 3$ degrees of freedom. This remains true whether N is equal to 100, 1000, or whatever.

For the chi square, as for both t and r, if our obtained value is equal to or greater than the table value, the null hypothesis is rejected. For our radio station survey the degrees of freedom are equal to $4 - 1$, or 3. Our results for this chi square are written as follows:

$$\chi^2_{.01(3)} = 11.34$$

$$\chi^2 = 20.00 \quad \text{Reject } H_0; \text{ significant at } P < .01.$$

Therefore, we reject the null hypothesis and conclude that Station A is indeed more popular among the population of teen-agers from which the sample was selected. The observed frequency does differ significantly from that predicted by chance. In this case, the chi square value is large enough that the probability of alpha error is less than one in a hundred.

The Frequency Expected Due to an a Priori Hypothesis.   Sometimes a certain theory, or perhaps an existing piece of research, is used to generate the values for the expected frequency of occurrence of a trait being studied. For example, geneticists know that of the genes controlling eye color, those for brown eyes are dominant over those for blue. This means that blue-eyed parents will only have children with blue eyes. It also means that a brown-eyed parent who carries *only* genes for brown eyes will only have children with brown eyes, even if the other parent's eyes are blue. However, if a known hybrid, that is, a man with brown eyes who carries a gene for blue eyes because one of his parents had blue eyes, has children with a brown-eyed hybrid woman, the chances are 3 to 1 that each resulting child will have brown eyes. This 3-to-1 ratio, generated from Mendel's genetic theory of dominants and recessives, can be used as an a priori hypothesis.

To test this a priori hypothesis, a researcher might select a sample of parents and observe the eye color of their children. These observations could be compared with the theoretical frequencies in an attempt to assess how well the observed frequencies fit with those expected—in short, a goodness-of-fit test.

Example.   A sample of 40 couples is selected. In each couple both of the parents are brown-eyed hybrids. The eye color of each couple's first-born child is

noted. The data are as follows: brown-eyed children, 23; blue-eyed children, 17. These values for observed frequency are then compared with the frequency of eye colors expected on the basis of the a priori, or 3-to-1, hypothesis. Of the total of 40, then, the frequency expected is that 30 of the children will have brown eyes and 10 will have blue. Now we have values for f_o and f_e and can complete the chi square table as follows:

	BROWN EYES	BLUE EYES	
f_o	23	17	
f_e	30	10	
$f_o - f_e$	−7	7	
$(f_o - f_e)^2$	49	49	$\Sigma \dfrac{(f_o - f_e)^2}{f_e} = 6.53$
$(f_o - f_e)^2/f_e$	1.63	4.90	

$$\chi^2_{.05(1)} = 3.84$$

$$\chi^2 = 6.53 \quad \text{Reject } H_0; \text{ significant at } P < .05.$$

The frequency observed, therefore, deviates significantly from that expected on the basis of the a priori hypothesis.

Do the results in this example disprove the genetic theory? They certainly do not validate it, but neither do they refute it. Remember, this was post-facto research. The subjects are assigned to categories on the basis of a trait that they already possessed. It may be that the theory is wrong, or perhaps people do not always know who their real fathers are.

The $r \times k$ Chi Square (r by k)

The chi squares shown so far are called $1 \times k$ chi squares, meaning that one sample group of subjects has been assigned to any number, k, of categories. Sometimes, however, a researcher wishes to select more than one group, for example, an experimental and a control group, and then to compare these groups with respect to some observed frequency. For this, the $r \times k$ chi square is used.

Although the basic equation remains the same, there are two differences in the steps required to complete the $r \times k$ chi square. First, the frequency expected on the basis of chance cannot be computed in the same way. Second, the degrees of freedom are assigned in a slightly different manner.

A researcher is interested in discovering whether vitamin C aids in the prevention of influenza. Two groups are randomly selected, with 30 subjects in each group. Group A is given 250 mg of vitamin C daily for a period of three months, while Group B is given a placebo. In Group A, 10 subjects report having caught influenza during that time, while in Group B, 15 subjects report having flu.

The Contingency Table. The data are set up in what is called a contingency table, shown in Table 13.2. The cells in the contingency table are lettered a, b, c, and d. The cells represent specific and unique categories. For example, cell a contains only those subjects who took vitamin C and also caught influenza, cell b contains only those who took vitamin C and did not catch influenza, and so on. Note that this particular contingency table has two rows and two columns; it is called a 2 × 2 contingency table. Also note that the two groups define the rows, whereas the categories on which the subjects are being measured (flu versus no-flu) are heading the columns. Another way to remember this is that the independent variable (whether or not the subjects took the vitamin) is set in the rows and the dependent variable (whether or not they caught the flu) in the columns.

To the right of each row, and at the bottom of each column, we place the marginal totals: 30 and 30 for the rows and 25 and 35 for the columns. This is an important step; these marginal totals are used for calculating the values of the frequency expected. Also, as a check, we verify that the row total (30 + 30) equals the column total (25 + 35) and that *each adds up to the total N*, in this case 60.

Calculating the Chi Square Value. As we did for the 1 × k, we set up a table. Here, each column is headed by a particular cell. As we fill this table in, we must keep the contingency table (Table 13.2) clearly in view, because some of the values are taken from it.

The value of the r by k chi square is obtained by the following steps (see Table 13.3):

1. We take the values for f_o directly from the contingency table and put them in the first row: 10 for cell a, 20 for cell b, and so on.
2. To fill the second row, we calculate the values for f_e for all cells of the contingency table. For a given cell, f_e equals the product of its column total and its row total divided by N, the total number of cases. Cell a is in the column whose total is 25 and in the row whose total is 30. Therefore, f_e for cell a equals (25 × 30) ÷ 60, or 12.50. By this same formula, we obtain values of f_e for b, c, and d of 17.50, 12.50, and 17.50, respectively.
3. In the third row, we place the difference between f_o and f_e. For cell a, this value is −2.50, from 10 − 12.50, and so on.

Table 13.2 Contingency table for vitamin C/influenza data.

	HAD INFLUENZA	DID NOT HAVE INFLUENZA	
GROUP A: VITAMIN C	a 10	b 20	30 = a + b
GROUP B: NO VITAMIN C	c 15	d 15	30 = c + d
	a + c = 25	b + d = 35	60 = a + b + c + d = N

Table 13.3 An r × k chi square with data from Table 13.2.

	a	b	c	d	
f_o	10	20	15	15	
f_e	12.50	17.50	12.50	17.50	
$f_o - f_e$	−2.50	2.50	2.50	−2.50	
$(f_o - f_e)^2$	6.25	6.25	6.25	6.25	
$(f_o - f_e)^2/f_e$	.50	.36	.50	.36	$\chi^2 = 1.72$

4. We square the values in the third row and place the squared differences in the fourth row; that is, 6.25 for all cells.

5. We divide each squared difference by the f_e for that cell and put the resulting values in the fifth row: .50, .36, .50, and .36.

6. We add across all the values in the fifth row to obtain the value for chi square, that is, $\chi^2 = 1.72$.

Interpreting the Chi Square Value. The degrees of freedom for the r × k chi square are found by the following equation: df $= (r - 1)(k - 1)$. The numbers of rows and columns are taken *from the contingency table*. Thus, in the vitamin C study, we had two rows (vitamin C versus no vitamin C) and two columns (influenza versus no influenza). The degrees of freedom, then, are equal to $(2 - 1)$ $(2 - 1)$, or 1. Then,

$$\chi^2_{.05(1)} = 3.84$$

$$\chi^2 = 1.72 \quad \text{Accept } H_0; \text{ not significant.}$$

Checking our obtained chi square value of 1.72 against the critical value of 3.84 for an alpha error level of .05 shows that the statistical decision must be an acceptance of the null hypothesis. There are no significant differences between the two groups with respect to their frequency of catching flu. Since this was experimental methodology (after-only design), we conclude that the independent variable (vitamin C) did not affect the dependent variable (influenza).

Degrees of Freedom and the r × k Chi Square

The r × k chi square allots degrees of freedom on the basis of how many cell entries are free to vary, *once the marginal totals are fixed*. For example, on a 2 × 2 chi square, with marginal totals of 30, 70, 60, and 40,

a 10	b ?	40
c ?	d ?	60
30	70	100

Table 13.4 Contingency table for a 3 × 2 chi square.

	HAD INFLUENZA	DID NOT HAVE INFLUENZA
GROUP A: PLACEBO	a	b
GROUP B: VITAMIN C	c	d
GROUP C: VITAMIN C	e	f

once the value of cell a is entered, in this case 10, all the other cell entries are fixed. Cell b can now only have a value of 40 − 10, or 30. Similarly, cell c now must contain 20 and cell d, 40. Thus, with a 2 × 2 chi square, only the value for one cell is free to vary—producing, of course, one degree of freedom.

Variations of the Chi Square Design. The $r \times k$ chi square just shown, because the contingency table has two rows and two columns, is called a 2 × 2 chi square. Without question, this is a very popular chi square design. However, many other variations are possible. For example, in the vitamin C study, the researcher might have wished to compare three sample groups with respect to influenza. The contingency table then would appear as in Table 13.4. This is a 3 × 2 chi square, with degrees of freedom equal to (3 − 1)(2 − 1), or 2. The researcher might even have wished to make a finer discrimination on the dependent variable by setting up a 3 × 3 chi square, as shown in Table 13.5. In this case, the degrees of freedom are equal to (3 − 1)(3 − 1), or 4.

Checking the Calculated Degrees of Freedom. To be sure that the degrees of freedom for a chi square have been calculated correctly, we can perform a simple manual check using the contingency table. We merely cross out one row and one column of the contingency table and then count the cells remaining. A contingency table for a 3 × 3 chi square yields a value of 4 by this process, as shown.

Table 13.5 Contingency table for a 3 × 3 chi square.

	HAD SEVERE INFLUENZA	HAD MILD INFLUENZA	DID NOT HAVE INFLUENZA
GROUP A: PLACEBO	a	b	c
GROUP B: VITAMIN C	d	e	f
GROUP C: VITAMIN C	g	h	i

A contingency table for a 2 × 2 chi square handled in the same manner yields a value of 1 for the degrees of freedom.

Yates Correction for the 2 × 2 Chi Square

Whenever the expected frequency within a given cell is small, that is, less than 10, the 2 × 2 chi square value may become slightly inflated. Since higher chi square values make it easier to reject the null hypothesis, then any such artificial increase in chi square may increase the alpha error. To prevent this, a correction formula has been developed that slightly lowers the chi square value. The **Yates correction** for continuity is simply this: whenever any *expected frequency* is less than 10, subtract .50 from the absolute size of all the $|f_o - f_e|$ values. Even if only one value of f_e is less than 10, all of the differences must be reduced. This correction is only necessary when df = 1, as for a 2 × 2 chi square. Although there has been some controversy regarding the worth of this adjustment, there is no question as to its conservatism.[1]

Example. Hypothesis: Persons with high incomes are more apt to watch the 11 P.M. TV news than the 6 P.M. news. A random sample of 40 subjects is selected. The subjects are separated on the basis of whether or not they earn $40,000 per year; 23 report income below that level and 17 above. Of the high-income subjects, 7 indicate watching the 6 P.M. news, whereas 10 watch the 11 P.M. news. Of the low-income subjects, 15 watch the news at 6 P.M. and 8 at 11 P.M. We set up the 2 × 2 contingency table as follows:

	6 P.M. NEWS	11 P.M. NEWS	
HIGH INCOME	a 7	b 10	17 = a + b
LOW INCOME	c 15	d 8	23 = c + d
	a + c = 22	b + d = 18	40 = a + b + c + d = N

Then we calculate the chi square value, using a table format as before. The circled values of f_e trigger the Yates correction. The Yates correction stipulates the subtraction of the value .50 from the *absolute* value of each difference between f_o and f_e. Thus, when the Yates correction is applied, all these absolute difference values are *decreased*.

	a	b	c	d	
f_o	7	10	15	8	
f_e	9.35	7.65	12.65	10.35	
$f_o - f_e$	−2.35	2.35	2.35	−2.35	
|Yates correction| − .50	1.85	1.85	1.85	1.85	
$(f_o - f_e)^2$	3.42	3.42	3.42	3.42	
$(f_o - f_e)^2/f_e$	.37	.45	.27	.33	$\chi^2 = 1.42$

We compare our obtained value for chi square with the critical value from Table I:

$$\chi^2_{.05(1)} = 3.84$$

$$\chi^2 = 1.42 \quad \text{Accept } H_0; \text{ not significant.}$$

Therefore, the null hypothesis is accepted. The two groups, with high and low incomes, do not differ significantly with respect to TV news. This, of course, is post-facto research; the independent variable (income) is a subject variable and is assigned rather than manipulated.

Some Checkpoints When Calculating a Chi Square Value

1. The cell entries must be independent of each other.
2. The sum total of the values for f_o must equal N, the total number of observations ($\Sigma f_o = N$).
3. All the values of f_o must be whole numbers, as opposed to fractions or decimals. (With nominal data, either an event is placed in a given category or it is not. No observation can be halfway in or three-quarters of the way in a given category.)

LOCATING THE DIFFERENCE

Problem: A social psychologist is interested in testing the "mere presence" phenomenon, that is, that just having another person around you can affect your performance—even if that other person is neither watching nor judging you. Hiding a movie camera beside the university's jogging path, the psychologist filmed passing male joggers under three conditions: (1) a woman, seated on the grass, and facing the joggers, (2) a woman, seated on the grass, but with her back to the joggers; and (3) no observer at all. A sample of 300 joggers was filmed, 100 under each condition. The joggers were then categorized as to whether they (a) increased their pace, (b) remained at the same place, or (c) decreased their pace.

	INCREASE	SAME	DECREASE	
WOMAN FACING	a 80	b 10	c 10	100
WOMAN'S BACK	d 30	e 40	f 30	100
NO OBSERVER	g 33	h 33	i 34	100
	143	83	74	300

	a	b	c	d	e	f	g	h	i
f_o	80	10	10	30	40	30	33	33	34
f_e	47.67	27.67	24.67	47.67	27.67	24.67	47.67	27.67	24.67
$f_o - f_e$	32.33	−17.67	−14.67	−17.67	12.33	5.33	−14.67	5.33	9.33
$(f_o - f_e)^2$	1045.23	312.23	215.21	312.23	152.03	28.41	215.21	28.41	87.05
$\dfrac{(f_o - f_e)^2}{f_e}$	21.93	11.28	8.72	6.55	5.49	1.15	4.51	1.03	3.53

$$\chi^2 = \Sigma \frac{(f_o - f_e)^2}{f_e} = 64.19$$

$$\chi^2_{.01(4)} = 13.28$$

$\chi^2 = 64.19$ Reject H_0; significant at $P < .01$.

The decision to reject the null hypothesis means that the observed frequencies found among the three categories did not occur by chance under the assumption that all frequencies were the result of independent selection from the same population. On the basis of the chi square value, we have decided that significant differences did occur, *but the chi square value itself has not defined their precise location.* We must, therefore, go back and inspect the contingency table. We are then able to find that the major source of the difference occurs in the first row, that is, among joggers being watched. In the other two conditions, the frequencies, roughly a third in each category, are not showing divergence from the chance hypothesis. We may conclude, then, that male joggers do pick up the pace when being observed (by a woman) but that the "mere presence" of another person seems to have no apparent effect—a finding that challenges the "mere presence" hypothesis.

The 2 × 2 Chi Square: A Special Equation

A special equation has been derived *to use for the 2 × 2 chi square.* With this equation the tedious process of calculating values of f_e is avoided.

$$\chi^2 = \frac{N(ad - bc)^2}{(a + b)(c + d)(a + c)(b + d)}$$

Using the data from the vitamin C study, we have

	HAD INFLUENZA	DID NOT HAVE INFLUENZA	
GROUP A: VITAMIN C	a 10	b 20	$30 = a + b$
GROUP B: PLACEBO	c 15	d 15	$30 = c + d$
	$a + c = 25$	$b + d = 35$	$60 = a + b + c + d = N$

$$\chi^2 = \frac{(60)[(10)(15) - (15)(20)]^2}{(30)(30)(25)(35)*}$$

$$\chi^2 = \frac{(60)(150 - 300)^2}{787,500} = \frac{(60)(-150)^2}{787,500} = \frac{(60)(22,500)}{787,500}$$

$$\chi^2 = \frac{1,350,000}{787,500} = 1.71$$

The difference between this answer, 1.71, and the result of 1.72 found previously using the long method is due to rounding. Since the long method requires many more calculations and roundings, this answer is probably more accurate. In any case, the two equations for chi square are algebraically identical, and the small difference has no effect on the statistical decision.

We can make the Yates correction with a form of this equation. We correct by subtracting $N/2$ from the *absolute* value of the difference between ad and bc.

$$\chi^2 = \frac{N\left(\left|ad - bc\right| - \dfrac{N}{2}\right)^2}{(a + b)(c + d)(a + c)(b + d)}$$

Using the data from the TV news study earlier in this chapter, we have

	6 P.M. NEWS	11 P.M. NEWS	
HIGH INCOME	a 7	b 10	$17 = a + b$
LOW INCOME	c 15	d 8	$23 = c + d$
	$a + c = 22$	$b + d = 18$	40

*To obtain the denominator value, use your calculator to chain multiply, that is, key in 30 times 30 times 25 times 35 and then hit the "equals" key.

$$\chi^2 = \frac{(40)\left[\left|(7)(8) - (10)(15)\right| - \frac{40}{2}\right]^2}{(17)(23)(22)(18)}$$

$$\chi^2 = \frac{(40)(|56 - 150| - 20)^2}{154{,}836} = \frac{(40)(94 - 20)^2}{154{,}836}$$

$$\chi^2 = \frac{(40)(74)^2}{154{,}836} = \frac{(40)(5476)}{154{,}836} = \frac{219{,}040}{154{,}836} = 1.41$$

Chi Square: A Test of Independence

Since the $r \times k$ tests whether or not the variables in the contingency table are independent, it can be used for testing the hypothesis of difference. Therefore, chi square can be applied to both experimental data, which is always aimed at the hypothesis of difference, and post-facto data, which sometimes tests for difference.

CHI SQUARE AND PERCENTAGES

If the chi square has been calculated on the basis of percentages, it must be corrected to reflect the sample size. The same percentages taken from different sample sizes yield very different chi square values; the larger the sample, the larger the chi square.

To make this correction, you must take the percentage-based chi square and multiply it by $N/100$, where N is equal to the sample size. (A percentage-based chi square, of course, may have cell entries which are not whole numbers.)

For example, suppose that a $1 \times k$ chi square has been calculated based on the following percentages:

	a	b	c	d	
f_o	10%	20%	30%	40%	
f_e	25	25	25	25	
$f_o - f_e$	−15	−5	5	15	
$(f_o - f_e)^2$	225	25	25	225	
$\dfrac{(f_o - f_e)^2}{f_e}$	9	1	1	1	$\chi^2 = 20$

This value must then *be corrected for sample size before being checked for significance.* If the sample size had been 200, then

$$\chi^2 = \frac{20(200)}{(100)} = 40$$

Or, if the sample size had been 1000, then

$$\chi^2 = \frac{20(1000)}{(100)} = 200$$

The same procedure is used on the $r \times k$ chi square. Assume that a 2×2 chi square has been calculated on the following percentages:

a	b	
5%	45%	50%
c	d	
15%	35%	50%
20%	80%	100%

	a	b	c	d
f_o	5	45	15	35
f_e	10	40	10	40
$f_o - f_e$	-5	5	5	-5
$(f_o - f_e)^2$	25	25	25	25
$(f_o - f_e)^2$	2.50	.625	2.50	.625
f_e				$\chi^2 = 6.25$

Had N been equal to 2000, then

$$\chi^2 = \frac{6.25(2000)}{(100)} = 125.00$$

Or, if the sample size had been 200, then

$$\chi^2 = \frac{6.25(200)}{(100)} = 12.50$$

Both of these values clearly lead to a reject of the null hypothesis, since $\chi^2_{.05(1)} = 3.84$. But if the same percentages had been obtained when N was equal to 60 (still a fairly large sample), then

$$\chi^2 = \frac{6.25(60)}{(100)} = 3.75$$

which, of course, would then lead to an accept of the null hypothesis. With this test, then, although sample size does not enter into the allocation of the degrees of freedom, it is absolutely crucial in determining the actual chi square value.

CHI SQUARE AND z SCORES

The chi square test is another Karl Pearson creation. Notice that the chi square significance values, Appendix Table I, for one degree of freedom are based on the squares of the z distribution. At the .05 significance level, the chi square value of 3.84 is equal to 1.96 squared, ± 1.96 being the z score that excludes the extreme 5% of the normal curve. Also, at the .01 level, the chi square value of 6.66 is equal to the square of 2.58 and, again, z's of ± 2.58 exclude the extreme 1%.

THE CHI SQUARE AND DEPENDENT SAMPLES

The chi square test has very few requirements; therefore, it is safe and extremely versatile. However, the *chi square does demand independent cell entries.* There can be no exceptions to this fundamental requirement. It would seem that this might be a rather severe limitation and that chi square might not be so versatile after all. For example, how can we possibly analyze nominal data when the sample groups are correlated, as they are when the same group of subjects is measured twice or when different groups are matched?

    Suppose that a researcher is interested in determining whether a direct-mail campaign sells a certain product. A sample of 100 persons is selected, and they are asked whether or not they have the product in their homes—30 have the product and 70 do not. The sample is then subjected to a heavy direct-mail campaign extolling the virtues of this product. Then the subjects are checked again. This time 70 have the product and only 30 do not. A four-cell contingency table for this is as follows:

	HAD PRODUCT	DID NOT HAVE PRODUCT
BEFORE CAMPAIGN	a 30	b 70
AFTER CAMPAIGN	c 70	d 30

    200

    At this point, the researcher realizes that something has gone awry. Adding all values for f_o should yield the total number of subjects, $N = 100$. Instead it yields 200. What went wrong? The basic independence rule for the chi square has been violated, and the data are simply not testable in this form. However, there is a solution to this dilemma—set up the categories on the basis of change scores. That is, instead of categorizing the same group of subjects, before and after, on whether or not they have the product, the subjects should be categorized on the basis of whether or not they change from not having the product to later having the product, or vice versa. The *change scores are independent of each other*, and the sum of the values of f_o now equal N, the sample size.

The McNemar Test for Dependent Samples

The originator of this type of chi square analysis for dependent samples is Quinn McNemar, and the procedure is known as the **McNemar test.** The equation for the McNemar test is as follows:

$$\chi^2 = \frac{|a - d|^2}{a + d}$$

The data must be set up in a 2×2 contingency table as follows:

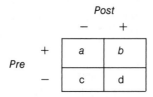

N of possible changes

Here, cell a contains *only* those subjects who changed from a + precondition to a − postcondition. Cell b contains only those subjects who did not change, remaining + both pre and post. Cell c contains only those subjects who did not change, remaining − both pre and post, and cell d contains only those subjects who were − in the precondition and changed to + in the post. The equation for the analysis, as we can see, concerns only the change cells, a and d.

Reworking the results of the direct-mail campaign study, we find that of the 30 who had the product before the campaign, 20 still had the product after the campaign, leaving 10 who did not. Also, of the 70 who did not have the product before the campaign, 50 had the product after the campaign, leaving 20 who did not.

		Post		
		NO PRODUCT (−)	HAD PRODUCT (+)	
Pre	HAD PRODUCT (+)	a 10	b 20	30
	NO PRODUCT (−)	c 20	d 50	70
		30	70	100

$$\chi^2 = \frac{|a - d|^2}{a + d} = \frac{|10 - 50|^2}{10 + 50} = \frac{40^2}{60} = \frac{1600}{60}$$

$$\chi^2 = 26.67$$

Checking our result for significance (Table I), we find that

$$\chi^2_{.01(1)} = 6.64$$

$$\chi^2 = 26.67 \quad \text{Reject } H_0; \text{ significant at } P < .01.$$

Therefore, the difference in change scores is significant between the pre- and postconditions. The manipulated independent variable (direct-mail campaign) did cause a change in the direction of obtaining the product. The null hypothesis is rejected at an alpha probability level of .01.

The McNemar Test and Matched-Subjects Designs

The same type of analysis can be done on matched-subjects designs. The cells of the contingency table in this case indicate the agreement ($+$) or the disagreement ($-$) between the nominal scores of the matched pairs. The agreement–disagreement matchups are independent of each other.

Group 2

	DIFFERENT ($-$)	SAME ($+$)
SAME ($+$)	a	b
DIFFERENT ($-$)	c	d

Group 1

N of possible changes

The Yates Correction for the McNemar Test

Like any 2×2 chi square, the McNemar test may be adjusted for continuity when dealing with small cell entries. With one degree of freedom (as is the case for all 2×2 chi squares), the McNemar equation becomes

$$\chi^2 = \frac{(|a - d| - 1)^2}{a + d}$$

Remember that the vertical bars in this equation indicate that the value for the difference between a and d is inserted without regard to its sign. We subtract one from the absolute value of the difference between a and d, the resulting value is squared, and then it is divided by the sum of a and d.

Correlation and Nominal Data: The Coefficient of Contingency

There are a variety of nominal tests for correlation, but the most popular among social researchers is the **coefficient of contingency,** symbolized by C. It is a versatile test that can be used with any size of contingency table, 2×2, 2×3, 4×4, whatever. Also, C is very easy to calculate. When the chi square value is known, C is only a few seconds away via this equation:

$$C = \sqrt{\frac{\chi^2}{N + \chi^2}}$$

Finally, C is easy to check for significance. If the calculated chi square is significant, so too is the resulting value of C (Table I).

Example. Hypothesis: There is a significant correlation among male children between watching violent TV shows and actual overt aggression on the part of the viewer. A random sample of 100 fifth-grade, male children is selected from a certain school, and the children are asked to indicate their favorite TV show. Observers then visit the children during recess and record, for each child, whether or not overtly aggressive acts are exhibited. Of the 65 children who prefer violent TV shows, 50 are recorded as being overtly aggressive. Of the 35 children who prefer nonviolent TV, 10 are recorded as overtly aggressive.

	DID COMMIT AGGRESSIVE ACTS	DID NOT COMMIT AGGRESSIVE ACTS	
PREFERRED VIOLENT TV	a 50	b 15	$65 = a + b$
PREFERRED NONVIOLENT TV	c 10	d 25	$35 = c + d$
	$a + c = 60$	$b + d = 40$	$100 = a + b + c + d = N$

We calculate the value of chi square.

	a	b	c	d	
f_o	50	15	10	25	
f_e	39	26	21	14	
$f_o - f_e$	11	−11	−11	11	
$(f_o - f_e)^2$	121	121	121	121	
$(f_o - f_e)^2/f_e$	3.10	4.65	5.76	8.64	$\chi^2 = 22.15$

We compare our calculated value with the critical value from Table I:

$$\chi^2_{.01(1)} = 6.64$$

$$\chi^2 = 22.15 \quad \text{Reject } H_0; \text{ significant at } P < .01.$$

Since the chi square value is significant, we can use that value in the C equation. No matter what value we now get for C, it must also be significant.

$$C = \sqrt{\frac{\chi^2}{N + \chi^2}}$$

$$C = \sqrt{\frac{22.15}{122.15}} = \sqrt{.18}$$

$C = .42$ Reject H_0; significant at $P < .01$.

Therefore, there is a significant correlation between watching violent TV and actual overt aggression on the part of the young male viewer.

This is an example of correlational, post-facto research. No conclusions as to causation are permissible from the results of this study. Perhaps watching violent TV does cause aggressive behavior, or perhaps aggressive behavior causes one to watch violence on TV. Finally, there may be a third factor, variable X, that causes both overt aggression and preference for TV violence.

Limitations of the Coefficient of Contingency. The resulting correlation of .42 in the last example may seem to be somewhat low, especially in light of the rather dramatic differences between the two groups of TV viewers. The value of C was low, far lower than a Pearson r would have been had we been able to obtain interval measures on the subjects. The fact is that the C correlation calculated for a 2×2 chi square can only, at best, reach a value of .87, not 1.00. Because it is a nonparametric method of determining correlation, C is a much less powerful test than the parametric test, r. All nonparametric tests underestimate the degree of differences and correlations, and the C test is no exception.

REQUIREMENTS FOR USING THE CHI SQUARE

The requirements for using the chi square are as follows:

1. The samples must have been randomly selected.
2. The data must be in nominal form.
3. There must be independent cell entries.
4. No value for expected frequency should be less than 5.

The last requirement is only of interest when the contingency table has just four cells. With larger tables, as in a 4×5 table, a few values of less than 5 are permissible.* Some statisticians even argue that the *5-Rule* may be too stringent in smaller contingency tables.[2]

*For a more thorough discussion of this exception, see S. Siegel, *Nonparametric Statistics,* 2nd ed. (New York: McGraw-Hill, 1988).

There's Always Chi Square.   Although as a nonparametric test, chi square is not the world's most powerful statistical test, it certainly is one of the most popular. It is an extremely safe test to use because it requires so few assumptions. It really is a workhorse test for researchers. Distributions can be skewed, variability can be dramatically different for two samples, the intervals between successive scale points can vary, and the researcher can relax—because if nothing else works, one can usually use chi square.

SUMMARY

Nominal data are those that have been identified on the basis of equality versus nonequality. Observations are sorted into mutually exclusive categories and then counted to get the frequency of occurrence within each category. The categories must be completely independent of each other, that is, an observation sorted into one category cannot also be placed in another category.

The statistical test spotlighted in this chapter is the chi square, a test of whether or not the observed frequency of occurrence differs significantly from the frequency expected on the basis of chance or on the basis of some a priori hypothesis. The $1 \times k$ chi square is applicable if one sample group is sorted into any number of categories. The $r \times k$ chi square is applied when two or more sample groups are sorted into any number of categories.

An extremely common form of the $r \times k$ is the 2×2 chi square (two sample groups sorted into two categories), and a special computational equation has been derived for this particular situation. Whenever any of the values of f_e for the 2×2 chi square are less than 10, the Yates correction is typically applied to the basic chi square equation.

Since one of the limitations of the chi square is that it only can be used when all cell entries are independent of each other, the testing of correlated samples requires a special technique. This is the McNemar test, and to apply it only the change scores are used in the analysis.

Although the chi square itself only tests the hypothesis of difference, variations on the chi square theme have been developed for assessing the possibility of a correlation. The test of correlation covered in this chapter is the coefficient of contingency, C, which can be applied to any number of independent cells, 2×2, 3×4, and so on.

Chi square, since it makes no assumptions regarding either the population mean, μ, or the shape of the underlying distribution, is called a nonparametric test. Also, for calculating the degrees of freedom for the chi square, the size of the sample is irrelevant; the only information that is required is the number of sample groups and the number of categories.

KEY TERMS AND NAMES

chi square
coefficient of contingency
McNemar test

nominal data
Yates correction

REFERENCES

1. CONOVER, W. J. (1974). Some reasons for not using the Yates continuity correction on 2 × 2 contingency tables. *Journal of the American Statistical Association*, 69, pp. 374–376.

2. EVERITT, B. S. (1977). *The analysis of contingency tables*. London: Chapman & Hall.

PROBLEMS

1. A researcher is interested in whether or not a significant trend exists regarding the popularity of certain work shifts among police officers. A random sample of 60 uniformed police officers is selected from a large metropolitan police force. The officers are asked to indicate which of three work shifts they preferred. The resulting data show that 40 officers prefer the first shift, 10 the second shift, and 10 the third. Do the results deviate significantly from what would be expected due to chance?

2. A sample of 48 college students is randomly selected, and the students are asked to indicate their attitudes toward the statement: "The United States military should invade the island of Bermuda." Of the students polled, 12 agreed with the statement, 12 had no opinion, and 24 disagreed. Do these results deviate significantly from what would be expected due to chance?

3. A group of 90 economists are randomly chosen from members of the profession in this country. They are asked to predict whether the prime lending rate (the interest charged by major banks to their best customers) six months from now will be higher, the same, or lower than it is today. Of 90 economists, 30 said "higher," 35 said "the same," and 25 said "lower." Do these results deviate significantly from what would be expected due to chance?

4. A dog-racing enthusiast wonders if there is any systematic bias connected with the lane in which the dogs run. Results of the first race on a given night are checked at 108 dog tracks selected randomly from throughout the country. The following data indicates that the number of winning dogs is 26 for lane one, 17 for lane two, 16 for lane three, 15 for lane four, 17 for lane five, and 17 for lane six. Check the hypothesis that dog-track wins are a function of the lane.

5. Two random samples have been selected. One sample consists of 220 Catholics and the other contains 200 Protestants. All subjects were then asked to indicate their attitudes toward voluntary birth control. The data are as follows:

	FAVOR	OPPOSE
Catholics	70	150
Protestants	120	80

Do the two groups differ regarding their attitudes toward birth control?

6. A random sample of 150 persons was selected from the membership rolls of the National Rifle Association and another random sample of 150 persons was selected from the list of contributors to the "Save the Whales" foundation. The two populations do not overlap. All subjects were asked to indicate their attitudes toward hand-gun control. The data are as follows:

	FAVOR	OPPOSE
NRA group	90	60
"Save the Whales" group	50	100

Do the two groups differ regarding their attitudes toward hand-gun control?

7. Hypothesis: Union membership differs on the basis of sex. A random sample of 100 adult workers, 60 men and 40 women, is selected in a large midwestern city. Of the group, 42 men and 17 women are union members.

	UNION MEMBERS	NOT UNION MEMBERS	
MEN	42	18	60
WOMEN	17	23	40
			100

   a. Check the hypothesis.
   b. Indicate the type of research (experimental or post-facto).
   c. State the independent and dependent variables.

8. A researcher suspects there are differences in reading problems as a function of gender among first-grade children. A random sample of 100 first-graders is selected—50 boys and 50 girls. Of the boys, 25 are found to have reading problems and 25 are found to have no reading problems. Of the girls, 10 are found to have reading problems and 40 to have no reading problems. The contingency table is as follows:

	READING PROBLEMS	NO READING PROBLEMS	
BOYS	a 25	b 25	50
GIRLS	c 10	d 40	50
	35	65	100

Test the hypothesis.

9. A researcher hypothesizes that there are differences in personality type as a function of body type among adult males. A random sample of 200 men is selected and categorized as to body type. Of the sample selected, 50 are found to be ectomorphs (underweight and small-boned), 60 are found to be mesomorphs (muscular and normal weight), and 90 are found to be endomorphs

(overweight). Of the ectomorphs, 20 are diagnosed as having extraverted personalities and 30 as introverted. Of the mesomorphs, 32 are diagnosed as extraverted and 28 as introverted. Of the endomorphs, 70 are diagnosed as extraverted and 20 as introverted. The contingency table is as follows:

	EXTRAVERT	INTROVERT	
ECTOMORPH	a 20	b 30	50
MESOMORPH	c 32	d 28	60
ENDOMORPH	e 70	f 20	90
	122	78	200

Test the hypothesis.

10. A major pharmaceutical company is testing the possible effectiveness of a certain new antiarthritic drug. A sample of 50 arthritic patients is randomly selected and divided into two groups of 25 each. Group A is treated with pills containing the new drug, and Group B is given a placebo (a pill that appears identical to the drug but is nonactive). After three months of treatment, the two groups are compared. Of the patients in Group A, 15 report symptom "improvement," and 10 report "no improvement." Of the patients in Group B, 12 report "improvement," and 13 report "no improvement." Test the hypothesis that the new drug produces significantly higher rates of reported symptom improvement.

11. Hypothesis: More Republicans than either Democrats or Independents own their own homes. A random sample of 71 city residents is selected in a large eastern city. Of these, 23 indicated that they are Republicans, 25 that they are Independents, and 23 that they are Democrats. Of the Republicans, 15 are homeowners and 8 are not. Of the Independents, 10 are homeowners and 15 are not. Of the Democrats, 8 are homeowners and 15 are not. Test the hypothesis.

12. Hypothesis: There is a significant difference between students living in dorms and students who commute regarding whether or not they receive financial aid. A random sample of 55 university students is selected—15 commuters and 40 dorm students. Of the commuters, 5 are receiving university aid and 10 are not. Of the dorm students, 30 are receiving financial aid, and 10 are not. Test the hypothesis.

13. Hypothesis: A weekend seminar in assertiveness training increases the assertiveness of the female participants. A random sample of 90 female university students is selected. Each woman is first evaluated for assertiveness by specially trained interviewers. In that pretest, 25 are judged to be assertive and 65 nonassertive. All the women then attend a weekend seminar in assertiveness training. At the end of the weekend, each woman is again interviewed and evaluated. In the postcondition, 60 of the women are judged to be assertive and 30 nonassertive. The data table, with marginal totals, follows:

Post

	NONASSERTIVE	ASSERTIVE	
Pre ASSERTIVE	10	15	25
NONASSERTIVE	20	45	65
	30	60	90

Test the hypothesis.

14. A researcher is interested in establishing whether or not a correlation exists between a student's choice of college major and qualification for the academic honor roll. A random sample of 80 university seniors is selected and categorized on the basis of college major. Of the students selected, 30 are majoring in liberal arts, 30 are in business, and 20 in education. Of the liberal arts majors, 10 are on the honor roll; of the business majors, 15 are on the honor roll; of the education majors, 10 are on the honor roll. The data table follows:

	HONORS	NO HONORS	
LIBERAL ARTS	10	20	30
BUSINESS	15	15	30
EDUCATION	10	10	20
			80

    a. Calculate the chi square.
    b. If appropriate, calculate the correlation.

15. A researcher is examining the possibility of a correlation existing between participation in varsity athletics and the incidence of personal and/or emotional problems among male university students. Random samples of 50 varsity athletes and 50 nonathletes are selected from among the male population of several large midwestern universities. The incidence of personal problems is gauged by asking each student whether or not he had ever visited the university counseling center. Of the athletes, 10 acknowledged that they had paid at least one visit to the counseling center. Among the nonathletes, 22 said they had gone to the counseling center at least once. Test the correlation.

16. When data are in the form of frequencies of occurrence within mutually exclusive categories, which scale of measurement is being used?

17. How is the null hypothesis for the chi square test stated?

18. Under what condition(s) should the Yates correction be used?

19. For which experimental design(s) is the McNemar test an appropriate analysis of nominal data?

True or False—Indicate T or F for problems 20 through 24.

20. Chi square may only be used with nominal data.

21. Chi square can only be used if the frequency expected value of all cell entries is at least 5.

22. An a priori hypothesis always states that the frequency expected must be the same in every cell.
23. Chi square is a nonparametric statistical test.
24. To test for "goodness of fit," a $1 \times k$ chi square may be performed.

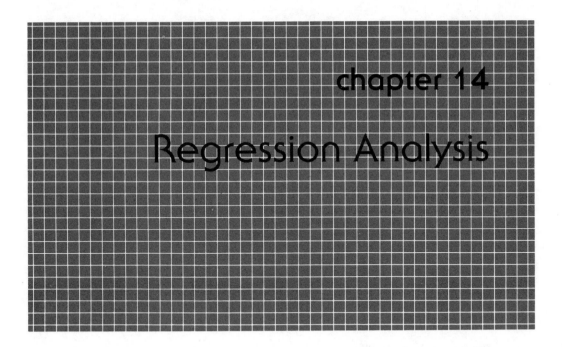

chapter 14

Regression Analysis

One of the major themes of this book, recurring again and again like the cadence of a marching drumbeat, is that correlation in and of itself cannot be used to assign causation. Correlation does not rule out causation; however, it cannot guarantee it. What a correlational technique can do, however, is to allow the researcher to make better than chance predictions. Regression analysis provides one method for making such predictions and even, at times, pointing a finger of suspicion in the direction of possible causation.

When using the correlation for purposes of prediction, the goal is to estimate one variable on the basis of the information contained in one or more other variables. The variable being predicted is called the criterion variable, and the variable(s) from which the prediction is being made is (are) called the predictor variable(s).

THE REGRESSION OF Y ON X

Throughout this chapter the focus is on predicting a value of Y having been given the value of X. This is called the regression of Y on X, and it attempts to utilize the correlation between X and Y to make very specific predictions of the Y value. The correlation between X and Y specifies how much information about Y is contained in X. If the correlation is a perfect $+1.00$, then X contains all there is to know about Y. A correlation of zero, on the other hand, indicates that X tells us

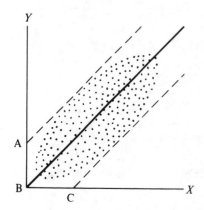

Figure 14.1
A bivariate scatter plot.

nothing at all about Y. Thus, the higher the correlation, the more information about Y is contained in the X score. The variable whose value is known is called the independent variable, whereas the variable whose value is to be determined from the formula is called the dependent variable.

The Bivariate Scatter Plot

In Chapter 10, a discussion of the scatter plot was presented. We learned then that a scatter plot is a graphic format in which *each point represents a pair of scores*—a measurement on X as well as a measurement on Y. A bivariate, or two-variable, scatter plot is shown in Fig. 14.1. The points in the scatter plot are arrayed from lower left to upper right, indicating a positive correlation. Also, the points form the elliptical shape that is a result of the central tendency occurring within the two distributions of X and Y scores.

The Regression Line. Three straight lines are drawn on the plot in Fig. 14.1. Line A is close to the points along the top of the array of points, but is relatively far from those along the bottom. Similarly, line C, although near the points along the bottom, is a considerable distance from those along the top. However, line B is fairly close to *all* the points in the scatter plot. For this reason, line B is called the prediction line or, more often, the **regression line** of Y on X. The regression line is, therefore, the single straight line that lies closest to all the points in the scatter plot. This regression line allows us to make the aforementioned correlational predictions.

To make predictions, three important facts about the regression line must be known: (1) the extent of the scatter around the line, (2) the slope of the line, and (3) the point where the line crosses the Y axis.

The Extent of the Scatter Around the Regression Line. The closer the points on the scatter plot cluster around the regression line, the higher is the resulting correlation between X and Y, *and the more accurate is the resulting prediction.*

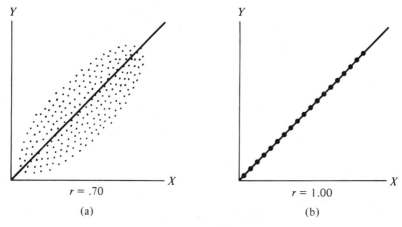

Figure 14.2 Scatter plots showing positive correlations.

Figure 14.2 shows two scatter plots, both of which illustrate positive correlations. Figure 14.2b, however, portrays a much higher correlation—the points do not deviate at all from the regression line. In this scatter plot, the correlation is a perfect 1.00; each point that has a higher value on the X variable is also higher on Y. With a correlation of 1.00, there are no exceptions, that is, no reversals. Figure 14.2a, on the other hand is more typical of real research in the social sciences. In this scatter plot, although there is still a positive trend to the correlation, the scattering of points around the regression line shows that the correlation is not perfect. Here, while in general the high scores match with high scores and the low scores with low scores, there are numerous exceptions. In short, then, the higher the correlation, the closer the scatter points cluster around the regression line. Also, the higher the correlation, the more accurate is the prediction, because more information about Y is being carried in X.

The Slope of the Regression Line. The manner in which the regression line tips, or slopes, greatly affects the prediction. For example, the regression lines in Fig. 14.3a (going from lower left to upper right) and in Fig. 14.3b (going from upper left to lower right) demonstrate entirely different correlations. The line in Fig. 14.3a portrays a positive correlation. Here, for a high score on X, we would predict a similarly high score on Y. The line in Fig. 14.3b, however, shows a negative correlation. A high score on X would lead to a prediction of a low score on Y. Therefore, the line in Fig. 14.3a has a positive slope and the line in Fig. 14.3b a negative slope.

Knowing the sign of the slope, however, is not enough. Figure 14.4 shows three regression lines with positive slopes. These would yield very different predictions. The degree of the slope is determined by the amount of change in Y that accompanies a given unit change in X. Figure 14.4a shows a line with a slope of 1.00; Y increases by one unit for each single unit increase in X. The line in Fig.

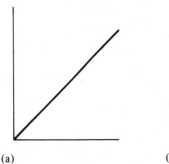 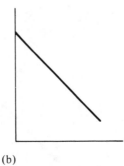

Figure 14.3
Regression lines showing (a)
positive and (b) negative
correlations.

(a)   (b)

14.4b is much steeper. The slope is 2.00; that is, Y increases by two units for each single unit increase in X. Finally, the line in Fig. 14.4c has a slope of one-half, or .50. Here, Y only increases by half of a unit for each full unit increase in X.

      The Point Where the Regression Line Crosses the Y Axis. Finally, to make a prediction we must know precisely where the regression line crosses the ordinate. That is, we must establish the value of Y when X is equal to zero. This is called the **point of intercept,** or the Y intercept. Figure 14.5 shows two possible situations.

      In Fig. 14.5a, the regression line meets the ordinate at zero. In this situation, when X equals zero, so does Y. In Fig. 14.5b, however, the regression line intercepts the ordinate at a value of 2 (Y equals 2, when X equals 0). Note, in these two examples, that both lines have the same slope, despite their different points of intercept.

The Regression Equation

To draw the graph of the regression line, the following equation is used:

$$Y_{pred} = bX + a$$

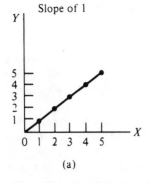 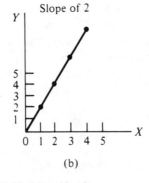 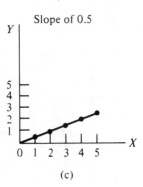

(a)   (b)   (c)

Figure 14.4 Regression lines with different positive slopes.

Slope of 1

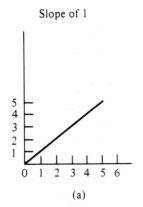

(a)

Slope of 1

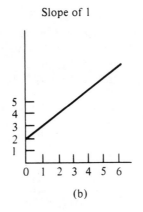

(b)

Figure 14.5
Regression lines with the same positive slope but different Y intercepts.

In this equation, Y_{pred} is the predicted value of Y; b is the slope of the regression line; X is a known value, from which Y_{pred} is to be calculated; and a is the point of intercept.

The equation for the b term (slope) is

$$b = \frac{rSD_y}{SD_x}$$

The equation for the a term (intercept) is

$$a = \bar{Y} - b\bar{X}$$

Thus, since

$$Y_{pred} = bX + a$$

then

$$Y_{pred} = \frac{rSD_y}{SD_x} X + \bar{Y} - \frac{rSD_y}{SD_x} \bar{X}$$

Rearranging terms, we write

$$Y_{pred} = \frac{rSD_y}{SD_x} X - \frac{rSD_y}{SD_x} \bar{X} + \bar{Y}$$

This equation is much easier to work out than it may appear. For one thing, the rSD_y/SD_x, or slope, term appears twice, but we only have to calculate it once.

Example. A correlation between hours per week spent studying and grade-point average was found in Chapter 10 to be .98. The mean hours of study time is 22.86, with a standard deviation of 12.20. The mean grade-point average is 2.71, with a standard deviation of .69. Predict the grade-point averages of students who spend the following weekly study hours: 37, 22, and 8.

1. Set up the problem by clearly separating the X and Y values. Remember, we are given the hours of study time, making study time the X distribution, and we are asked to predict grade-point average, making grade-point average the Y distribution.

STUDY HOURS, X	GRADE AVERAGE, Y
$\bar{X} = 22.86$	$\bar{Y} = 2.71$
$SD_x = 12.20$	$SD_y = .69$
$X_1 = 37$	$Y_{pred} = ?$
$X_2 = 22$	$Y_{pred} = ?$
$X_3 = 8$	$Y_{pred} = ?$

2. We plug all values, except the specific values of X into the regression equation.

$$Y_{pred} = \frac{rSD_y}{SD_x} X - \frac{rSD_y}{SD_x} \bar{X} + \bar{Y}$$

$$Y_{pred} = \frac{(.98)(.69)}{12.20} X - \frac{(.98)(.69)}{12.20} (22.86) + 2.71$$

$$Y_{pred} = \frac{.68}{12.20} X - \frac{.68}{12.20} (22.86) + 2.71$$

$$Y_{pred} = .06X - (.06)(22.86) + 2.71$$

$$Y_{pred} = .06X - 1.37 + 2.71$$

To add two numbers with unlike signs, take the difference between the two quantities and affix the sign of the larger quantity.

$$Y_{pred} = .06X + 1.34$$

Slope $= .06$

Y intercept $= 1.34$

3. Using the elements of the regression equation, we set up a regression table, including the specific values of X. We multiply the X values by the slope (in this case, .06) and then add the intercept (in this case, +1.34).

X	.06X	.06X + 1.34 = Y_{pred}
37	2.22	3.56
22	1.32	2.66
8	.48	1.82

The predicted grade-point averages are 3.56 for 37 hours of study time, 2.66 for 22 hours, and 1.82 for 8 hours.

We hold the X values out of the regression equation initially, so that when we set up the regression table, we can use a long list of X values to predict the corresponding Y values very quickly, without having to resolve the equation each time. These predicted Y values are point estimates of population parameters, since they're based on sample data.

There are, in fact, two regression lines in any bivariate analysis. By simply reversing the procedure, that is, predicting X from a knowledge of Y, a second regression line may be obtained.

When both the X and Y values are known, we can easily check how much discrepancy there is between the actual values of Y and their predicted values. This deviation, between the known values of Y and the regression line, is known in correlational analysis as the residual error.* If the reader prefers to use s, the unbiased estimate of the population's standard deviation, the regression equation remains exactly the same:

$$Y_{pred} = \frac{rs_y}{s_x} X - \frac{rs_y}{s_x} \bar{X} + \bar{Y}$$

Although with small samples these estimated standard deviations have slightly larger values than do the actual standard deviations of the sample, the ratios remain the same. Thus, within rounding errors, both methods produce identical answers.

Using the Regression When $\rho = 0$. We can now use the regression equation to predict how future members of a population might score; but, to do this, the *correlation must be significant.* That is, the null hypothesis ($\rho = 0$) must be rejected. If the correlation in the population between X and Y is zero, then the only prediction we can make about the Y variable is to predict the *mean* of the Y distribution. This is because of the effect of central tendency, or the fact that most scores do cluster around the mean. Therefore, with a correlation of zero (the null hypothesis accepted) or with no knowledge of the correlation value, the best prediction we can make is simply the mean of the Y distribution.

*The term "residual error" is also used in the analysis of variance (see page 361) and defines the denominator in the within-subjects F ratio.

This is clear if we look again at the regression equation. If r is equal to zero, then all the terms, since they are products, also become zero, except for $\bar{Y}$:

$$Y_{\text{pred}} = \frac{(0)(SD_y)}{SD_x} X - \frac{(0)(SD_y)}{SD_x} \bar{X} + \bar{Y}$$

$$Y_{\text{pred}} = 0 - 0 + \bar{Y}$$

$$Y_{\text{pred}} = \bar{Y}$$

When r assumes a significant value, however, the resulting predictions of Y may deviate from the mean of the Y distribution. The higher the correlation, then, the more the predicted Y values may vary from $\bar{Y}$, and the lower the correlation, the more the predictions regress toward the mean. Thus, if you are told that the average rainfall in India is 58 inches, and then are asked to predict the height of a specific male college student, your best estimate would be 5 feet 9 inches, the mean of the male height distribution.

The Beta Coefficient

The equation for the slope of the regression line is called the **beta coefficient.** As we have seen, this b value indicates the amount of change in Y (a rise or drop in the ordinate) for a given change in X (a run on the abscissa). If the correlation between X and Y is perfect ($r = 1$), then the slope is the direct relationship of S_y over S_x. When $r = 1.00$, then

$$b = \frac{(1.00)SD_y}{SD_x} = \frac{SD_y}{SD_x}$$

When r is less than 1.00, however, then the SD_y over SD_x relationship is a weighted proportion. An r of .50 yields half as much of a Y-on-X increase as does an r of 1.00. For example, if we assume $SD_y = 2$ and $SD_x = 1$, then with an r of 1.00,

$$b = \frac{(1.00)(2)}{1} = 2.00$$

That is, with a beta coefficient of 2, a rather steep slope, there will be a two-unit increase in Y for each unit increase in X. But with an r of .50,

$$b = \frac{.50(2)}{1} = 1.00$$

Here, Y increases only one unit for each unit increase in X. Thus, with an r of .50, the increase in Y is half of what it is if r is equal to 1.00.

Note that when the standard deviations for both the X and Y distributions are equal to 1.00, as when comparing two distributions of z scores, then b (the slope) becomes identical with the Pearson r.

$$b = \frac{rSD_y}{SD_x} = \frac{r(1)}{(1)} = r$$

The Theory of Regression

It was **Sir Francis Galton** (1822–1911), friend and mentor of Karl Pearson, who first began talking about regression. This is how it happened. Galton had noted that there was a significant correlation between the height of fathers and the adult height of their sons. However, when attempting to predict a son's height, an interesting phenomenon occurred. If the father was extremely tall, the son would attain an above-average height, but would typically not be as tall as his father. Similarly, although the sons of very short fathers did attain below-average heights, they were still taller than their fathers. The predicted values kept drifting or *regressing* toward the mean. Unless the correlation between X and Y is 1.00, the predicted Y value will always be closer to the mean of the Y distribution than the original X value is to the mean of the X distribution.

Galton's quest—a method for quantifying relationship. (Historical Pictures Service, Inc., Chicago.)

Example.   The correlation between the heights of fathers and their adult sons is .55.[1] The mean of both height distributions is 68 inches, and the standard deviations are both 3 inches. Given two fathers with heights of 78 inches and 58 inches, predict the heights of their sons.

FATHERS, X	SONS, Y
$\bar{X} = 68$	$\bar{Y} = 68$
$SD_x = 3.00$	$SD_y = 3.00$

$$Y_{\text{pred}} = \frac{rSD_y}{SD_x} X - \frac{rSD_y}{SD_x} \bar{X} + \bar{Y}$$

$$Y_{\text{pred}} = \frac{(.55)(3)}{3} X - \frac{(.55)(3)}{3} (68) + 68$$

$$Y_{\text{pred}} = .55X - (.55)(68) + 68$$

$$Y_{\text{pred}} = .55X - 37.40 + 68$$

$$Y_{\text{pred}} = .55X + 30.60$$

FATHER'S HEIGHT, X	.55X	PREDICTED SON'S HEIGHT, .55X + 30.60
78	42.90	73.50
58	31.90	62.50

Note here that the beta coefficient of +.55 is telling us that for each full inch of change on the X distribution (father's height), we can expect a corresponding change of .55 inches in the Y distribution (son's height). The sons are changing about a half inch for each full-inch change among the fathers.

Secular Trend Analysis

Instead of predicting the adult height of a son as a function of the father's height, as Galton did, what if we wanted to predict the son's height or IQ at various ages? To do this, we can use the same regression techniques for predicting trends over a period of time. The term "secular trend" refers to data that are spread over time and are not overly cyclic in nature. Thus, any time we use past and current data to predict future results, we may use a time series called **secular trend analysis.** To do this type of analysis the time period should be as long as possible.

Example.   The mean height of the U.S. adult male has increased by a full inch during the years 1955 to 1985.

YEAR	MEAN U.S. ADULT MALE HEIGHT (INCHES)
1955	68.00
1960	68.20
1965	68.50
1970	68.40
1975	68.60
1980	68.80
1985	69.00

How tall will the average U.S. adult male be in 1995? In the year 2005? In the year 2020?

1. We code the years, using equal time intervals, by converting them into sequential digits beginning with 1, so that 1955 = 1, 1960 = 2, and so on. (When using the computer program PH-STAT, you can put the years in directly.) We then set up the coded years as the X distribution and the height measures as the Y distribution, and calculate the Pearson r.

YEARS	X	X²	Y	Y²	XY
1955	1	1	68.00	4,624.00	68.00
1960	2	4	68.20	4,651.24	136.40
1965	3	9	68.50	4,692.25	205.50
1970	4	16	68.40	4,678.56	273.60
1975	5	25	68.60	4,705.96	343.00
1980	6	36	68.80	4,733.44	412.80
1985	7	49	69.00	4,761.00	483.00

$\Sigma X = 28 \quad \Sigma X^2 = 140 \quad \Sigma Y = 479.50 \quad \Sigma Y^2 = 32,846.45 \quad \Sigma XY = 1922.30$

$N = 7$

$$\bar{X} = \frac{\Sigma X}{N} = \frac{28}{7} = 4.00 \qquad \bar{Y} = \frac{\Sigma Y}{N} = \frac{479.50}{7} = 68.50$$

$$SD_x = \sqrt{\frac{\Sigma X^2}{N} - \bar{X}^2} = \sqrt{\frac{140}{7} - 4.00^2}$$

$$= \sqrt{20 - 16} = \sqrt{4.00} = 2.00$$

$$SD_y = \sqrt{\frac{\Sigma Y^2}{N} - \bar{Y}^2}$$

$$= \sqrt{\frac{32,846.45}{7} - 68.50^2}$$

$$= \sqrt{4692.35 - 4692.25}$$

$$= \sqrt{.10} = .32$$

$$r = \frac{\Sigma XY/N - (\bar{X})(\bar{Y})}{SD_x SD_y}$$

$$= \frac{1922.30/7 - (4.00)(68.50)}{(2.00)(.32)} = \frac{.61}{.64} = .95$$

$$r_{.01(5)} = .87 \qquad \text{Reject } H_0; \text{ significant at } P < .01.$$

2. The correlation is significant. We next calculate the regression equation.

$$Y_{\text{pred}} = \frac{rSD_y}{SD_x} X - \frac{rSD_y}{SD_x} \bar{X} + \bar{Y}$$

$$= \frac{(.95)(.32)}{2.00} X - \frac{(.95)(.32)}{2.00} (4.00) + 68.50$$

$$= .15X - (.15)(4.00) + 68.50$$

$$= .15X - .60 + 68.50$$

$$= .15X + 67.90$$

3. Now we set up the regression table, plugging in the X value for 1995 as 9, for 2005 as 11, and for 2020 as 14.

YEAR	X	.15X	.15X + 67.90 = Y_{pred}
1995	9	1.35	69.25
2005	11	1.65	69.55
2020	14	2.10	70.00

Thus, we predict a mean height of 69.25 in 1995, 69.55 in the year 2005, and 70.00 inches in 2020.

If the reader prefers the estimated standard deviation, then,

$$s_x = \sqrt{\frac{\Sigma X^2 - (\Sigma X)^2/N}{N-1}} = \sqrt{\frac{140 - 28^2/7}{6}} \qquad s_y = \sqrt{\frac{\Sigma Y^2 - (\Sigma Y)^2/N}{N-1}} = \sqrt{\frac{32{,}846.45 - 479.50^2/7}{6}}$$

$$s_x = \sqrt{\frac{140 - 112}{6}} \qquad\qquad s_y = \sqrt{\frac{32{,}846.45 - 32{,}845.75}{6}}$$

$$s_x = \sqrt{\frac{28}{6}} = \sqrt{4.67} = 2.16 \qquad s_y = \sqrt{\frac{.70}{6}} = \sqrt{.12} = .35$$

$$r = \frac{\dfrac{\Sigma XY - [(\Sigma X)(\Sigma Y)]/N}{N-1}}{s_x s_y} = \frac{\dfrac{1922.30 - [(28)(479.50)]/7}{6}}{(2.16)(.35)} = \frac{.72}{.76} = .95$$

YEAR, X	HEIGHT, Y
$\bar{X} = 4.00$	$\bar{Y} = 68.50$
$s_x = 2.16$	$s_y = .35$

$$Y_{\text{pred}} = \frac{rs_y}{s_x} X - \frac{rs_y}{s_x} \bar{X} + \bar{Y}$$

$$Y_{\text{pred}} = \frac{(.95)(.35)}{2.16} X - \frac{(.95)(.35)}{2.16} (4.00) + 68.50$$

$$Y_{\text{pred}} = \frac{.33}{2.16} X - \frac{.33}{2.16} (4.00) + 68.50$$

$$Y_{\text{pred}} = .15X - (.15)(4.00) + 68.50$$

$$Y_{\text{pred}} = .15X - .60 + 68.50$$

$$Y_{\text{pred}} = .15X + 67.90$$

The Hazards of Trend Analysis. Trend analysis is an extremely useful and powerful technique. A certain exercise of caution, however, is advisable whenever future trends are being extrapolated. Trend analysis always assumes that the present trend will continue, that other things will remain equal. Although a trend that has existed up to the present time is certainly a fact, the future trend must remain a probability estimate. The farther into the future the prediction is made, the more chance there is for other things not to remain equal.

Another problem results from the fact that each year's value is not really independent of the adjoining year, but may instead be correlated to some extent. This is called an autocorrelation. If the autocorrelation is extremely high, the accuracy of the forecast is reduced. (Tests for evaluating the extent of the autocorrelations are available in more advanced texts.*) Still another issue occurs when the two series are increasing each year, as they were in the preceding problem. This tends to produce a Pearson r which may overestimate the degree of correlation, since both of the values are smaller in the early years and larger in the later year.

Secular trend analysis is rearview-mirror research—one is always looking behind to predict what is ahead. A brief look back can sometimes be dangerous as it may lead one to make blind extrapolations.

Also, if one stops looking back and thereby misses the new images appearing in the rearview mirror, major accidents may result. New data must constantly be inserted as soon as it becomes available. In one method, called the "moving average," the addition of the latest data entry is accompanied by the deletion of the oldest entry, which of course moves or changes the average. Thus, the secular trend line is always being updated.

Despite all these warnings, the regression line can be an extremely useful forecasting tool. Extrapolations are always dangerous to some extent, but in the area of forecasting they are, alas, necessary.

*Spurr, W. A., & Bonini, C. P. (1989). *Statistical analysis for business decisions.* (Revised ed.). Homewood, Illinois: Richard D. Irwin.

THE STANDARD ERROR OF ESTIMATE

Values predicted from X, the independent variable, yield the most probable values of the dependent variable. However, "most probable" does not always produce a practical or useful prediction. As we have seen, with a correlation of zero, the most probable Y value for any given X is simply the mean of Y. Thus, we must determine how accurate the predicted or most probable value is.

In Chapter 10, the coefficient of determination, r^2, was introduced. It was pointed out that r^2 provides a value for determining what proportion of the information about Y was contained in X. By utilizing this r^2 value, a new statistic, the **standard error of estimate,** can be generated that will allow us to predict the accuracy level of our predicted Y value. That is, after the data have been cranked through the regression equation, a technique is needed for establishing the accuracy of each of the predicted Y values. The standard error of estimate is that technique.

Calculating the Standard Error of Estimate

One formula for the standard error of estimate, and rest assured the easiest to calculate, is

$$\text{SE}_{\text{est}} = SD_y\sqrt{1 - r^2}$$

where $SD_y =$ the true standard deviation of the Y sample.

Before using this equation in a research situation, we need to take some time to look it over. Although there are two variables in the equation, SD_y and r, the key value is r. For example, if $r = 1$, then, regardless of the value of SD_y, SE_{est} must equal zero.

$$\text{SE}_{\text{est}} = SD_y\sqrt{1 - r^2} = SD_y\sqrt{1 - 1} = SD_y\sqrt{0} = 0$$

Thus, when the correlation between X and Y is perfect, there *is no error in estimating from X*. Every prediction, in this happy situation, will be right on the nose.

However, if $r = 0$, then

$$\text{SE}_{\text{est}} = SD_y\sqrt{1 - r^2} = SD_y\sqrt{1 - 0^2} = SD_y\sqrt{1}$$

$$\text{SE}_{\text{est}} = SD_y$$

Thus, when the correlation is zero (or unknown), the standard error of estimate is identical to the standard deviation of the Y distribution.

We have just determined the range of potential values for the standard error of estimate. The greatest it can ever be (the most error) is when it equals SD_y, and that only occurs when the correlation is zero. The smallest it can ever be is zero, and that only occurs when the correlation is perfect. (Like any variability estimate, SE_{est} can obviously never be less than zero; that is, it can never take on a negative value). Most of the time, SE_{est} lies somewhere between these extremes. When calculating SE_{est}, then, we must be sure that the obtained value never falls outside the range between SD_y and zero. If it does, we should check our calculations.

Establishing the Confidence Interval

Assume that we have selected a random sample of 1000 college students and discovered a significant correlation of .85 between scores on an introversion-extraversion personality test and hours per week spent in cocurricular activities. The mean personality score was 50, with a standard deviation of 7.50 (higher scores indicate more extraversion). The mean hours-per-week value was 20 with a standard deviation of 3.10. If we were to select a student whose personality score was 55 (slight extraversion), how many weekly cocurricular hours should be predicted? First, solve the regression equation for Y on X, where Y equals hours per week and X equals personality scores.

$$Y_{pred} = \frac{(.85)(3.10)}{7.50} X - \frac{(.85)(3.10)}{7.50} (50) + 20$$

$$Y = .35X - (.35)(50) + 20$$

$$= .35X - 17.50 + 20$$

$$= .35X + 2.50$$

Then, using the results of this equation, set up the regression table.

X	$.35X$	$.35X + 2.50 = Y_{pred}$
55	19.25	21.75

We are, thus, predicting that a student whose personality score is 55 should spend 21.75 hours per week on cocurricular activities.

Next, we find the standard error of estimate.

$$SE_{est} = SD_y\sqrt{1 - r^2} = 3.10\sqrt{1 - .85^2} = 3.10\sqrt{1 - .72} = 3.10\sqrt{.28}$$

$$= (3.10)(.53) = 1.64$$

Is it a reasonable value? Yes, since the value of 1.64 falls within its limits of zero on the low side and the standard deviation of Y, or 3.10, on the high side.

We will now use this statistic to generate a range of possible Y values within which we can feel confidence that the parameter Y value could likely fall.

Since the predicted Y value is a point estimate of the true parameter, we are again looking for a range of values that may lie around this estimate. In this example, we will use the z distribution because

1. The Pearson r is significant.
2. The Pearson r assumes a normal distribution in the underlying population.
3. The sample is of sufficient size.*

THE CONFIDENCE INTERVAL EQUATION

In order to establish the confidence interval, we will use a variation of the basic equation

$$X = zSD + \bar{X}$$

that we learned to use in Chapter 5 to find the raw score when both the mean and standard deviation were known. The only modifications in this z equation are that now we will use SE_{est} in place of the standard deviation and the predicted Y value in place of the mean. That is, the confidence interval equals the z score times the standard error of estimate plus the predicted value of Y for a given X.

Finally, with samples large enough to assume normality, creating a confidence interval of .95 demands z scores of ± 1.96 (since 95% of the cases under the normal curve are included between z's of $+1.96$ and -1.96). Similarly, for a .99 confidence interval, we must have z's of ± 2.58:

$$CI = (z)(SE_{est}) + Y_{pred}$$

For a .95 level of confidence, the limits are

$$CI_{.95} = (\pm 1.96)(SE_{est}) + Y_{pred}$$

And for a .99 level of confidence, the limits are defined by

$$CI_{.99} = (\pm 2.58)(SE_{est}) + Y_{pred}$$

Using the previous data on personality and hours per week, the student whose personality score was 55 yielded a Y_{pred} value of 21.75 hours. For .95 confidence,

---

*With small samples, typically less than 50, further corrections of the standard error of estimate may be made. Such corrections, derived from the analysis of variance, can be found in advanced texts. One such formula, however, is $SE_{est} = SD_y \sqrt{(1 - r^2)(N/N - 2)}$.

$$CI = (\pm 1.96)(1.64) + 21.75$$

$$\left. \begin{array}{l} = +3.21 + 21.75 = 24.96 \\ = -3.21 + 21.75 = 18.54 \end{array} \right\} \text{confidence interval for } P = .95$$

To be .99 confident,

$$CI = (\pm 2.58)(1.64) + 21.75$$

$$\left. \begin{array}{l} = +4.23 + 21.75 = 25.98 \\ = -4.23 + 21.75 = 17.52 \end{array} \right\} \text{confidence interval for } P = .99$$

Again, it must be recalled that since the parameter value is unknown, we are not able to determine whether any particular interval contains it.

The procedures outlined thus far in this chapter are based on three important assumptions: (1) linearity, (2) normality, and (3) **homoscedasticity.**

1. *Linearity.* The two variables being correlated must bear a straight-line relationship to one another. The regression line, therefore, must maintain a single direction. It cannot start in an upward direction, then turn and continue in a downward one.
2. *Normality.* The two sets of sample scores must represent populations whose distributions are normal, or at least nearly so. These regression procedures will not work on badly skewed distributions.
3. *Homoscedasticity.* All sample scores on the Y axis of the scatter plot must have similar amounts of variability. There cannot be dramatic differences in the variability of the Y scores from column to column. For example, if we were predicting grade-point averages, we could not use the standard error of estimate if all of the GPAs above 2.50 had large amounts of variability, whereas those below 2.50 had little or none.

THE MULTIPLE *R*

When we first discussed correlation in Chapter 10, it was defined as a numerical statement as to the relationship among two or more variables. Up to now, however, the only procedures shown have been for two-variable, or bivariate, problems. We are now going to confront multivariate problems, those in which several variables are intercorrelated with one another.

Instead of simply attempting to find out how much information about Y is being contained in X (the goal of the Pearson r), a multiple correlation could establish how much information about Y is contained in X_1, X_2, X_3, and so on. Suppose, for example, that your job is to predict the height of the very next person to enter your room. Knowing that you're to predict the height of a *person* helps you somewhat—you wouldn't guess 3 inches or 60 feet. Next, we add some information and tell you that the person in question is an adult male. You know that gender correlates to some extent with height and that the adult male height distribution tends to have a higher average value than does the female distribution. You might now guess a value of 5 feet 9 inches, the actual mean of the male distribution. Next, we provide even more information and tell you that the person weighs 230 pounds. Again, since you know that height and weight tend to correlate positively, you might now increase your estimate to 6 feet or 6 feet 1 inch. Finally, we give one last bit of information and tell you that the person plays center on a professional basketball team. You now jump your estimate to 6 feet 10 inches or 7 feet. The point is, your prediction becomes increasingly more accurate as more and more predictor variables are supplied. Now, although the example has been based partly on nominal variables (male versus female), the same sort of results can be obtained when all the correlated measures are in the form of interval/ratio data, and the appropriate statistical test then becomes the multiple R.

Although the technique we use applies to virtually any number of variables, our discussion is restricted to the three-variable correlation. This is really all that is needed by a student of elementary statistics. It is also, because of the limitations of inexpensive calculators, about all that you can realistically be asked to calculate. In any event, the reason for doing all multivariate analyses is the same—to increase predictive efficiency. After all, when it comes to correlation, prediction is the basic goal.

Recently, there has been a controversy regarding the use of Scholastic Aptitude Test (SAT) scores as predictors of college success. Critics of the SAT point to the fact that high school grades are better than SAT scores as predictors of college performance. So why bother with the SAT? Defenders of the SAT, on the other hand, argue that since SAT scores do correlate significantly, though admittedly not perfectly, with college performance, why throw them out? Doesn't every little bit help?

The nationwide correlation between high school grades and college grades is .52, whereas the correlation between SAT scores (combined verbal and math) and college grades is only .41. (Also, the correlation between high school grades and SAT scores is .31.) There is no question about it, then, high school grades do predict college performance better than SAT scores do.[2] But what happens when we combine them? Does combining high school grades and SAT scores increase their predictive efficiency above and beyond the performance of either factor alone? To answer this question, the **multiple R** can be used.

The Multiple R Equation

Before using the multiple R equation, the following designations must be made. The subscript y always refers to the criterion variable, the variable we are trying to predict. The subscripts 1, 2, 3, 4, and so on, always refer to the predictor variables, the variables we are predicting from. The multiple R equation is as follows:

$$R_{y1,2} = \sqrt{\frac{r_{y,1}^2 + r_{y,2}^2 - 2r_{y,1}\,r_{y,2}\,r_{1,2}}{1 - r_{1,2}^2}}$$

Identify the y Variable

Before plugging the values into the equation, take a moment to sort out the various components. The first step, *and the most important,* is to identify and *clearly label* the y (or criterion) variable, that is, the variable whose value is being predicted. In this example, then, y must be identified as representing college grades. Once this is done, it makes no difference which variable is called "1" and which "2." For this problem, we happened to choose high school grades as the "1" variable (the first predictor) and the SAT score as the "2" variable (the second predictor).

Label the components as follows:

y = criterion: college grades
1 = first predictor: high school grades, or X_1
2 = second predictor: SAT scores, or X_2

The second step requires that you go back to the problem and pull out the correlations as they apply to your labeled components.

$r_{y,1}$ = .52 (the correlation between college grades and high school grades)
$r_{y,2}$ = .41 (the correlation between college grades and SAT scores)
$r_{1,2}$ = .31 (the correlation between high school grades and SAT scores)

Finally, and this is really the easy part, simply plug these values into the multiple R equation.

$$R_{y \cdot 1,2} = \sqrt{\frac{.52^2 + .41^2 - 2(.52)(.41)(.31)}{1 - .31^2}}$$

1. Square $r_{y,1}$.

.52² = .27

Then square $r_{y,2}$.

.41² = .17

Plug both values into the numerator of the equation.
2. Chain multiply the term $2r_{y,1}r_{y,2}r_{1,2}$. That is, let your calculator be a computer. Enter 2, times .52, times .41, times .31, and then press the equals button. The result should equal .13. Plug this value into the numerator of the equation.
3. Square $r_{1,2}$.

.31² = .10

Subtract it from 1. This should equal .90. Plug this value into the denominator.

4. Complete the solution as follows:

$$R_{y \cdot 1,2} = \sqrt{\frac{.27 + .17 - .13}{.90}} = \sqrt{\frac{.44 - .13}{.90}} = \sqrt{\frac{.31}{.90}} = \sqrt{.34}$$

$$R_{y \cdot 1,2} = .58$$

Thus, combining high school grades with SAT scores yields a multiple correlation of .58, which is higher than either separate correlation. With a correlation of .58, will predictions ever be wrong? Of course they will, sometimes. But it does beat flipping a coin!

The value of the multiple R should always be compared only with those internal correlations that associate with the y (or the criterion) variable. The multiple R should then be used for making predictions only if it is higher than all the internal correlations with y. For example, suppose that $r_{y1} = .70$, $r_{y2} = .80$, and the multiple $R = .60$. Then, of course, go back and use r_{y2} for making the prediction and in this case disregard the multiple R.

The Multiple R and Regression

To make predictions to the criterion variable, having been given performance measures on the predictor variables, a **multiple regression** equation must be used. The use of this equation, like that of the regression equation for Y on X, requires the calculation of a, the point of intercept. Unlike the regression of Y on X, however, the multiple regression equation also requires the calculation of at least two slopes, b_1 and b_2. The value for b_1 tells us how much change in Y will occur for a given unit change in the first predictor, *when the effects of the second predictor have been held constant*. The b_2 value tells us, on the other hand, how much change in Y will occur for a given unit change in the second predictor, *when the effects of the first predictor have been ruled out*. The multiple regression equation takes the following form, with X_1 and X_2 being specific values from the two predictor distributions.

$$Y_{M \text{ pred}} = b_1 X_1 + b_2 X_2 + a$$

The equation for b_1 is as follows:

$$b_1 = \left(\frac{SD_y}{SD_1} \right) \left(\frac{r_{y,1} - r_{y,2}\, r_{1,2}}{1 - r_{1,2}^2} \right)$$

and for b_2:

$$b_2 = \left(\frac{SD_y}{SD_2} \right) \left(\frac{r_{y,2} - r_{y,1}\, r_{1,2}}{1 - r_{1,2}^2} \right)$$

Finally, for *a* we have the following:

$$a = \bar{Y} - b_1 \bar{X}_1 - b_2 \bar{X}_2$$

Example. With the multiple *R* we established a correlation of .58 between the criterion (college GPA) and the two predictors (high school grades and SAT scores). We can now take the data and set up a multiple regression. A college GPA will be predicted for two students, one with a high school grade average of 80 and an SAT of 500, the other with a high school grade average of 70 and an SAT score of 400. Let us assume (1) a mean college GPA of 2.71 with a standard deviation of .69, (2) a mean high school grade of 80 with a standard deviation of 6.00, and (3) a mean SAT score of 450 with a standard deviation of 69.00.

COLLEGE GPA, Y	HIGH SCHOOL GRADES, 1	SAT, 2
$\bar{Y}$ = 2.71	$\bar{X}_1$ = 80.00	$\bar{X}_2$ = 450
SD_y = .69	SD_1 = 6.00	SD_2 = 69

The correlations, as shown previously, are as follows:

$$r_{y,1} = .52$$
$$r_{y,2} = .41$$
$$r_{1,2} = .31$$

"Things just aren't so elementary anymore, Watson."

Multiple regression may not be elementary, but by following the step-by-step program, the procedure is not difficult. (© Cartoon Features Syndicate.)

1. We plug in the appropriate values to solve for b_1.

$$b_1 = \frac{SD_y}{SD_1}\left(\frac{r_{y,1} - r_{y,2}\, r_{1,2}}{1 - r_{1,2}^2}\right) = \left(\frac{.69}{6}\right)\left(\frac{.52 - (.41)(.31)}{1 - .31^2}\right)$$

$$b_1 = (.12)\left(\frac{.52 - .13}{1 - .10}\right) = (.12)\left(\frac{.39}{.90}\right) = (.12)(.43)$$

$$b_1 = .05$$

2. We plug in the appropriate values to solve for b_2.*

$$b_2 = \left(\frac{SD_y}{SD_2}\right)\left(\frac{r_{y,2} - r_{y,1}\, r_{1,2}}{1 - r_{1,2}^2}\right) = \left(\frac{.69}{69}\right)\left(\frac{.41 - (.52)(.31)}{1 - (.31)^2}\right)$$

$$b_2 = (.01)\left(\frac{.41 - .16}{1 - .10}\right) = (.01)\left(\frac{.25}{.90}\right) = (.01)(.28)$$

$$b_2 = .0028 = .003$$

3. We plug in the appropriate values to solve for a.

$$a = \bar{Y} - b_1\bar{X}_1 - b_2\bar{X}_2$$

$$a = 2.71 - (.05)(80) - (.003)(450)$$

$$a = 2.71 - 4.00 - 1.35$$

$$a = -2.64$$

4. Using the values obtained, we set up a multiple regression table, including the specific X_1 and X_2 values for each student.

STUDENT	HIGH SCHOOL GRADE AVERAGE, X_1	$.05X_1$	SAT, X_2	$.003X_2$	$.05X_1 + .003X_2$	COLLEGE GPA, $Y_{M\,pred} = .05X_1 + .003X_2 - 2.64$
A	80	4.00	500	1.50	5.50	2.86
B	70	3.50	400	1.20	4.70	2.06

Thus, for Student A, whose high school grade average was 80 and whose SAT score was 500, the predicted college GPA is 2.86. For Student B, the predicted GPA is 2.06.

*In this instance, we must modify our rounding rule. If rounding to two places leaves only zero, then we round to three places.

The Standard Error of Multiple Estimate

We can check the accuracy of a multiple prediction in essentially the same way we earlier evaluated two-variable predictions. The only difference is the substitution of the multiple R for the previously used Pearson r. The **standard error of multiple estimate** equation is as follows:

$$\text{SE}_{\text{M est}} = SD_y\sqrt{1 - R^2_{y\cdot 1,2}}$$

For the preceding example, then, we can calculate a value as follows:

$$\text{SE}_{\text{M est}} = .69\sqrt{1 - (.58)^2} = .69\sqrt{1 - .34}$$
$$\text{SE}_{\text{M est}} = .69\sqrt{.66} = .69(.81) = .56$$

And, checking the estimated GPAs for accuracy (exactly the same as for the Y_{pred}), for a .95 level of confidence, we get for Student A

$$\text{CI}_{.95} = (\pm 1.96)(\text{SE}_{\text{M est}}) + Y_{\text{M pred}}$$
$$\text{CI}_{.95} = (\pm 1.96)(.56) + 2.86$$

or,
$$\left. \begin{aligned} &= \quad 1.10 + 2.86 = 3.96 \\ &= -1.10 + 2.86 = 1.76 \end{aligned} \right\} \text{confidence interval for } P = .95$$

And for Student B, we get

$$\text{CI}_{.95} = (\pm 1.96)(.56) + 2.06$$

or,
$$\left. \begin{aligned} &= \quad 1.10 + 2.06 = 3.16 \\ &= -1.10 + 2.06 = \quad .96 \end{aligned} \right\} \text{confidence interval for } P = .95$$

PATH ANALYSIS, THE MULTIPLE R, AND CAUSATION

Multiple regression analysis is also being used in a causal modeling technique called **path analysis.** Although path analysis is definitely a correlation technique, it is being used to establish the possibility of cause-and-effect relationships. The multiple regression is used to determine the interrelationships among a series of variables which are logically ordered on the basis of time.[3] Since, logically, a causal variable must precede (in time) a variable it is supposed to influence, the multiple regression analyzes a whole series of variables, each presumed to show a causal ordering. The attempt is made to find out whether a given variable is being influenced by the variables that precede it and then, in turn, influencing the variables that follow it. A "path" diagram is drawn that indicates the direction of the various relationships. Although not as definitive a proof of causation as when the independent variable is experimentally manipulated, proponents of path analysis tell us that their technique has taken us a long step forward from the naive extrapolations of causation which at one time were taken from simple bivariate correlations.* Path analysis, however, is not without its share of critics. Paul Games, for example, insists that the use of correlational data as a vehicle for causative assumptions is still fallacious, despite the alleged sophistication of the modeling approach.[4] For a more complete discussion of the use of correlational analyses as causative models, see Chapter 11 in "Understanding Educational Research."[5]

PARTIAL CORRELATION

Another correlation problem, almost the reverse of the multiple R, involves establishing an accurate correlation between two variables when both are known to be significantly influenced by a third variable. To solve this problem, a method has been developed that allows the researcher to attempt to rule out the influence of the third variable on the remaining two variables under study. The resulting correlation is called a **partial correlation.**

The equation for the partial correlation is

$$r_{y,1 \cdot 2} = \frac{r_{y,1} - r_{y,2}\, r_{1,2}}{\sqrt{(1 - r_{y,2}^2)(1 - r_{1,2}^2)}}$$

The $r_{y,1 \cdot 2}$ term is read as the correlation between y and 1, with 2 held constant.

Example. It is known that in a certain industry the correlation between corporate profits and workers' salaries (over a 10-year period) is .58. It appears at first that this is an ideal story of free enterprise—the more the workers manufacture and sell, the more they earn. There is, however, a culprit variable in this picture, inflation. That is, we have also discovered that the correlation between worker income and inflation is .80 and that the correlation between corporate profits and inflation is .70. What is the correlation between corporate profits and workers' salaries with inflation partialed out?

Identify the "2" Variable

The first step in this problem is to identify and clearly label the "2" variable, the variable whose influence is being partialed out. Once this is done, it doesn't mat-

---

*For an excellent example of the research value of this technique, see the following study, which was directed by Ellis Page at Duke University: T. Z. Keith and E. B. Page, "Do Catholic High Schools Improve Minority Student Achievement?" *American Educational Research Journal*, vol. 22 (1985), pp. 337–349.

ter to the equation which variable is called "y" and which "1." In the problem above, then, *inflation must be the "2" variable.* We have chosen to call worker income the y variable and corporate profits the "1" variable.

Label the components as follows:

2 = inflation

y = worker income

1 = corporate profits

The second step is to go back and read the problem carefully and then set up the correlations according to your labels.

$r_{y,1}$ = .58 (the correlation between worker income and corporate profits)

$r_{y,2}$ = .80 (the correlation between worker income and inflation)

$r_{1,2}$ = .70 (the correlation between corporate profits and inflation)

We plug the data into the partial correlation equation:

$$r_{y,1 \cdot 2} = \frac{.58 - (.80)(.70)}{\sqrt{(1 - .80^2)(1 - .70^2)}} = \frac{.58 - .56}{\sqrt{(1 - .64)(1 - .49)}} = \frac{.02}{\sqrt{(.36)(.51)}}$$

$$r_{y,1 \cdot 2} = \frac{.02}{\sqrt{.18}} = \frac{.02}{.42} = .05$$

Therefore, what seemed like a fairly strong correlation between worker income and corporate profits literally vanishes when the influence of inflation is partialed out. The remaining correlation of .05 is trivial.

Using the Partial Correlation

Example. A researcher is interested in the relationship between reading speed and reading comprehension. However, since the subjects in the study vary considerably as to IQ, it is decided to run a partial correlation in which the influence of IQ will be nullified.

Since in the example, IQ is being partialed out, it must be designated as the "2" variable. We then arbitrarily decided to call reading speed the y variable and reading comprehension the "1" variable. Thus,

2 = IQ

y = reading speed

1 = reading comprehension

From the problem, then, the correlations are as follows:

$r_{y,1} = .55$ (correlation between reading speed and comprehension)

$r_{y,2} = .70$ (correlation between reading speed and IQ)

$r_{1,2} = .72$ (correlation between comprehension and IQ)

These values are substituted into the equation for partial correlation.

$$r_{y,1\cdot2} = \frac{r_{y,1} - r_{y,2}\, r_{1,2}}{\sqrt{(1 - r_{y,2}^2)(1 - r_{1,2}^2)}}$$

$$r_{y,1\cdot2} = \frac{.55 - (.70)(.72)}{\sqrt{(1 - .70^2)(1 - .72^2)}} = \frac{.55 - .50}{\sqrt{(1 - .49)(1 - .52)}}$$

$$r_{y,1\cdot2} = \frac{.55 - .50}{\sqrt{(.51)(.48)}} = \frac{.05}{\sqrt{.24}} = \frac{.05}{.49} = .10$$

For example, by partialing out IQ in the preceding problem, we have, in effect, created a statistical equivalence within the group being measured. The subjects have, thus, been statistically equated, although after the fact, through the use of this partialing process.

If the study in this example had been done by selecting subjects with identical IQs at the outset (a better procedure), there would have been no need to partial out the IQ variable.

The equation given here is called a first-order partial, that is, one variable is held constant while examining the remaining relationship between the other two variables. Higher-order partials are also calculable, in which two or more variables are simultaneously held constant. For these procedures, consult more advanced texts.

The partial correlation is logically the other side of the multiple R coin. Whereas the multiple R combines variables in order to see what their cumulative influence might be, the partial correlation focuses on peeling away variables in order to see what influences are left over. Although the two examples of partial correlations demonstrated a lowering of the original correlation, the opposite does sometimes occur. For example, the correlation of .58 between IQ and college GPA jumps to .68 when the influence of time spent studying has been partialed out.[6]

Partial Correlation Compared to Covariance

In Chapter 11, it was pointed out that a technique called analysis of covariance is sometimes used in an attempt to create after-the-fact equivalent groups. As we have just seen, the partial correlation technique can also be used for this purpose.

In the covariance technique, groups that are different to begin with are statistically equated by weighting the mean of each group. This weighting process is based on regression predictions. If the actual means differ from the weighted predictions, there is a strong probability that the independent variable caused this difference. Covariance never gives final proof of causation, but it does provide a fairly strong deductive case.

The partial correlation, though not specifically equating group means, also attempts to ferret out a potential causal factor. Analysis of covariance and partial correlation are like a pretrial grand jury; each has the power to indict a causal factor. Only the experimental method, the full trial itself, can pass a final verdict beyond a reasonable doubt.

SUMMARY

Correlation allows the researcher to make better than chance predictions. When the correlation between X and Y is known, the prediction of a value of Y from a given value of X is based on the regression of Y on X. The higher the correlation (the more the correlation deviates from zero, either positively or negatively), the greater is its predictive power.

On a scatter plot, each point represents a pair of scores. The single straight line that comes closest to all the points in the scatter plot is called the regression line. This line is useful for making predictions when three important facts are known: (1) where the line crosses the ordinate, its point of intercept, (2) the slope of the line, and (3) the correlation between the two variables.

Unless the correlation between the variables is perfect (± 1.00), the predictions resulting from the regression equation will contain some degree of error. The predicted Y value can, however, be assessed for accuracy by calculating the standard error of estimate. This value is used for estimating the interval within which the researcher feels *confident* the true score should be. The greater the range of the interval, the higher is the level of confidence.

The general regression equation can also be used for secular trend analysis, where the data are spread over time. The hazard in using such data in order to predict future events is that secular trend analysis assumes that the past and present trends will continue indefinitely.

Whenever more than two variables are to be correlated, the multiple R must be used. If the multiple R value is greater than the separate internal correlations with the criterion, it can be used to further increase the predictive power of the correlations.

Just as the bivariate regression equation is used for predicting Y when X is known, so too can the multiple regression equation be used for predicting Y when several input, or X, variables are known. This prediction is followed by a standard error of multiple estimate for assessing the accuracy of the result.

Finally, when it is assumed that several variables do intercorrelate, a partial correlation can be calculated to establish the degree of correlation that exists between any two variables when the influence of the others is ruled out. This technique is sometimes used in post-facto research in order to create an after-the-fact equivalence of groups of subjects that were originally different.

KEY TERMS AND NAMES

beta coefficient (slope)	partial correlation	secular trend analysis
Galton, Sir Francis	path analysis	standard error of estimate
homoscedasticity	point of intercept	standard error of multiple
multiple R	regression line	estimate
multiple regression		

REFERENCES

1. BLOOM, B. S. (1964). *Stability and change in human characteristics.* New York: Wiley, p. 33.
2. TURNBULL, W. W. (1980). *Test use and validity.* Princeton, NJ: Educational Testing Service, p. 16.
3. YARENKO, R. M., HARARI, H., HARRISON, R. C., & LYNN, E. (1982). *Reference handbook of research and statistical methods in psychology.* New York: Harper & Row, p. 172.
4. GAMES, P. A. (1988). Correlation and causation: An alternative view. *The Score, 11,* pp. 9–11.
5. SPRINTHALL, R. C., SCHMUTTE, G. T., & SIROIS, L. (1990). *Understanding educational research.* Englewood Cliffs, NJ: Prentice-Hall.
6. POPHAM, W. J. & SIROTNIK, K. A. (1973). *Educational statistics.* (2nd ed.). New York: Harper & Row, p. 90.

PROBLEMS

1. The correlation between IQ and the Verbal SAT score among the seniors at West High School is .90; it is significant. The mean IQ is 100 with a standard deviation of 10. The mean SAT score is 500 with a standard deviation of 100. Predict SAT scores for seniors with the following IQs:
   a. 105
   b. 110
   c. 120

2. A random sample of factory workers is selected. Each worker is given an IQ test and also a factual "Current Events" test. The mean IQ for the group is 110 with a standard deviation of 12.85. The mean score on the "Current Events" test is 50 with a standard deviation of 15.
   a. Assume a significant correlation of .95 and predict the IQ of a worker whose "Current Events" score is only 20.
   b. Assume a significant correlation of .25 and again predict the IQ of a worker whose "Current Events" score is only 20.
   c. Why is the IQ prediction in part (b) higher than the prediction in part (a)? (Hint: Note the sizes of the two correlations used and recall the meaning of regression.)

3. A researcher is interested in the possible relationship between two of the sub-tests on the WAIS (Wechsler Adult Intelligence Scale). A random sample of eight Army recruits is selected; they are given both the Vocabulary subtest and the Digit Span (a test of short-term memory) subtest. Their weighted, scaled scores are as follows:

SUBJECT	DIGIT SPAN	VOCABULARY
1	9	11
2	6	8
3	12	13
4	7	6
5	10	10
6	5	6
7	9	11
8	10	9

    a. What is the correlation between the two sets of subtest scores?

    b. If one of the subjects receives a score of 11 on the Digit Span subtest, what is your best estimate of the subtest score that person will receive on Vocabulary?

4. Analyze the following yearly profits per share of a giant conglomerate.

YEAR	PROFITS PER SHARE OF COMMON STOCK (DOLLARS)
1960	.96
1965	1.03
1970	1.23
1975	1.15
1980	1.10
1985	1.40
1990	2.10

How much will the corporation earn in 1995? In 2010?

5. The correlation between height and IQ among the students at a certain Eastern women's college is zero. The mean IQ at the college is 125 with a standard deviation of 8.20. The mean height is 65 inches with a standard deviation of 2.40 inches. What IQ would you predict for a student at this college who is 68 inches tall?

6. Of the men's varsity football players in the mid-America conference, there is a significant correlation between height and weight of .75. The mean height is 73 inches with a standard deviation of 2.10 inches. The mean weight is 210 pounds with a standard deviation of 16.25 pounds. Predict the height of each of the following players.

    a. A 175-pound halfback.

      b.   A 195-pound split end.

      c.   A 240-pound linebacker.

7.   For each of the height predictions in problem 6, give the .95 confidence interval.

8.   Assume that a large sample of adult females was selected and the following Pearson *r* correlations were obtained. The correlation between weight and blood cholesterol levels was .75. The correlation between pulse rate and cholesterol levels was .70, and the correlation between weight and pulse rate was .30. Calculate the multiple *R* for predicting blood cholesterol levels, using both weight and pulse rate as the predictors.

9.   A large sample of adult males is selected, and the following Pearson *r* correlations were obtained. The correlation between height and weight was .60. The correlation between weight and chest size was .70, and the correlation between height and chest size was .50. Find the partial correlation between height and chest size, with the weight variable having been ruled out.

10.   The correlation between the IQs of fathers and their children is a significant .45. The correlation between the IQs of mothers and their children is a significant .55. The correlation between parents' IQs is a significant .40. Find the multiple *R*, where the child's IQ is the criterion variable and the IQs of the parents are the predictor variables.

11.   Using the value for the multiple *R* found in problem 8, predict the IQ of a child whose father's IQ is known to be 115 and whose mother's IQ is known to be 120. For all three IQ distributions (mother, father, and child), assume a mean IQ of 100 and a standard deviation of 15.

12.   Using the predicted value for the child's IQ found in problem 9, find the accuracy of the prediction at

      a.   A confidence level of .95.

      b.   A confidence level of .99.

13.   Among elementary school children, the correlation between height and strength is a significant .65. The correlation between height and age is a significant .82. The correlation between strength and age is a significant .75. What is the resulting correlation between height and strength when the age variable has been partialed out?

14.   The correlation between reading speed and SAT scores is .60. The correlation between reading speed and IQ is .62. The correlation between IQ and SAT scores is .58. Find the partial correlation between reading speed and SAT scores with the influence of IQ ruled out.

Indicate what term or concept is being defined in each of problems 15 through 19.

15.   The amount of increase in *Y* that accompanies a given increase in *X*.

16.   The single straight line that lies closest to all the points on a scatter plot.

17.   The value of *Y* when *X* equals zero.

18.   The resulting correlation between two variables when the effects of a third variable have been ruled out.

19.   The correlation between a criterion variable and several predictor variables.

Fill in the blanks in problems 20 through 25.

20. When the correlation between X and Y is zero, what is the best prediction of a Y score that can be made from a given X score? _____
21. If the regression line has a slope of .50, then each single unit increase in X will be accompanied by how much of an increase in Y? _____
22. Each single point on a scatter plot represents _____.
23. When the regression line slopes from upper left to lower right, then the sign of the correlation must be _____.
24. When all the points on a scatter plot lie directly on the regression line, then the value of the correlation must be _____.
25. In the regression equation, what term denotes the point where the regression line crosses the ordinate? _____

True or False—Indicate either T or F for problems 26 through 32.

26. The higher the correlation, the more a predicted Y value may deviate from the mean of the Y distribution.
27. The more a correlation deviates from zero, the better is its predictive accuracy.
28. For the Pearson r, degrees of freedom are assigned on the basis of the number of *pairs* of scores.
29. A correlation of $+.75$ must be significant, regardless of the degrees of freedom.
30. No correlation is ever greater than $+1$ or less than -1.
31. If X correlates significantly with Y, then X is probably the cause of Y.
32. If X correlates .90 with Y, then Y must also correlate .90 with X.

chapter 15

Before–After and Matched-Subjects Designs with Interval Data

During our earlier discussion of experimental methodology (Chapter 11), three basic designs were presented: the after-only, the before–after (or repeated-measure), and the matched-subjects. Each of these designs is aimed at the creation of equivalent groups of subjects in order that the potential effects of the manipulated independent variable can be ascertained. In the after-only (A/O) design, each subject is randomly and independently selected and independently assigned to either the control or the experimental conditions. That is, the fact that a given subject has been assigned to the experimental group has no influence on who might then be selected for the control group. Because of this principle of independence, analysis of A/O experimental results can be accomplished with some of the tests we have already covered, such as the independent t or F tests for interval data and the chi square for nominal data.

THE PROBLEM OF CORRELATED OR DEPENDENT SAMPLES

The other two experimental designs—before–after (B/A) and the matched-subjects (M/S)—however, pose an inherent statistical problem that must be resolved before analysis of their results can be completed. This problem is that of **correlated samples,** and it arises due to the fact that whenever subjects are paired off, either subjects with themselves as in the B/A design or subjects with their partners in the M/S design, a correlation almost certainly results between the paired scores.

Table 15.1 Results of a before–after weight loss study.

SUBJECT	WEIGHT BEFORE, X_1 (POUNDS)	WEIGHT AFTER, X_2 (POUNDS)
1	220	180
2	160	155
3	140	133
4	112	108
	$\bar{X}_1 = 158$	$\bar{X}_2 = 144$

For example, if a B/A design is used to conduct a weight loss study, each subject is weighed, the independent variable is manipulated, and each subject is weighed again. Even if every single subject does lose weight, however, the relative standing among the subjects might well remain the same in both the before and after measures. (See Table 15.1.)

We see in Table 15.1 that Subject 1, who was the heaviest in the before measure is still heaviest in the after measure, despite the dramatic loss of 40 pounds. Also, Subject 4 is the lightest in both before and after measures. The paired scores, thus, show a high degree of correlation, even though the thrust of our analysis is on the difference between the two weight measures. As with all experimental designs, the appropriate statistical test for B/A data is of the hypothesis of difference, but we must take this correlation into account.

Similarly, in an M/S design, if our matching process has been at all effective, there should be a correlation between the resulting pairs of matched scores. For example, assume that two groups of subjects are matched person for person on the basis of IQ. One group (control) is given a placebo, while the other group (experimental) is given a special math ability–enhancing drug. The two groups then take a math achievement test. (See Table 15.2.)

In this case, we expect the paired math scores to correlate—that is, if in fact matching on IQ is relevant, as it certainly appears to be. Note, too, that although the difference between the means of the two groups is not dramatically

Table 15.2 Results of a matched-subjects math achievement study.

SUBJECT PAIRS	CONTROL MATH SCORES, X_1	EXPERIMENTAL MATH SCORES, X_2
1 (Both with IQs of 130)	90	95
2 (Both with IQs of 75)	55	57
3 (Both with IQs of 110)	85	88
4 (Both with IQs of 100)	70	72
	$\bar{X}_1 = 75$	$\bar{X}_2 = 78$

great, it is rather impressive that all subjects in the experimental group outperformed their counterparts in the control group.

The previous discussion applies only to true repeated-measure and/or matched-subjects designs. The matched-group design, where whole groups are matched on the basis of average scores, does not create the potential for correlation, and the data from such a design should be treated as though the groups were independent.

THE PAIRED t RATIO

When two distributions of interval data are to be compared for possible differences and the data result from either a B/A or M/S design, then the **paired t ratio** must be used. That is, when there is correlation between the pairs of scores, then the independent t is no longer the appropriate test.

The Corrected Equation for the Standard Error of Difference

When the equation for the estimated standard error of difference (the denominator in the t ratio) was first introduced, the equation given was

$$s_{\bar{x}_1 - \bar{x}_2} = \sqrt{s_{\bar{X}_1}^2 + s_{\bar{X}_2}^2}$$

In a way, this equation is not theoretically correct, even though its use in calculating the independent t ratio is certainly justified. We were not lying to you—just protecting you from possible trauma.

The corrected equation for the estimated standard error of difference is

$$s_{\bar{x}_1 - \bar{x}_2} = \sqrt{s_{\bar{X}_1}^2 + s_{\bar{X}_2}^2 - (2r_{1,2}s_{\bar{X}_1}s_{\bar{X}_2})}$$

The element $r_{1,2}$ is read as "r sub one two." It is the correlation value between the first and second sets of measures.

Effects of the Correlation Term. The new element in the equation for the standard error of difference, the correlation term, $2r_{1,2}s_{\bar{X}_1}s_{\bar{X}_2}$, is, in fact, a single product. Therefore, if any value in that product is equal to zero, then the whole term must be equal to zero. (Zero times any value equals zero.) When this equation is applied to independently selected samples, the correlation, $r_{1,2}$, has to be zero. There is no correlation between a pair of independently selected samples. In fact, even if we tried, we could not pair off such scores to calculate a Pearson r. So for the independent t, where no correlation between the measures is possible, the standard error of difference is, in fact,

$$s_{\bar{x}_1 - \bar{x}_2} = \sqrt{s^2_{\bar{x}_1} + s^2_{\bar{x}_2}} - 0$$

or, simply,

$$\sqrt{s^2_{\bar{x}_1} + s^2_{\bar{x}_2}}$$

Also, although the estimated standard error of difference can be calculated for unequal sample sizes, when the correlational term is included there must be the same number of scores in each distribution.

Advantages of the Paired *t* Ratio

For the paired t, then, the full equation for the estimated standard error of difference must be employed. In fact, using the full equation provides a decided statistical advantage. Since the correlation term is subtracted from the first term, the resulting value for the estimated standard error of difference will be *less* than what it would otherwise have been. Also, since t is a ratio of mean differences divided by the estimated standard error of difference, then *reducing the estimated standard error of difference increases the size of t.* For the same numerator, then

$$\text{independent } t = \frac{2}{4 - 0} = \frac{2}{2} = .50$$

$$\text{paired } t = \frac{2}{4 - 3} = \frac{2}{1} = 2.00$$

This is important. The larger the value of the t ratio, the more likely it is that t will equal or exceed the table value and, therefore, be found to be significant.

Using the paired t, then, has three major effects:

1. It reduces the estimated standard error of difference by a factor related to the size of r which in turn
2. increases the size of the t ratio which in turn
3. increases the chances of rejecting the null hypothesis and achieving significance.

These points are valid only when the correlation has a positive value, which it certainly should have with a B/A design. If with an M/S design, we find a negative value for the correlation, then the matching variable is definitely not relevant and should be discarded. Actually, a negative correlation value *increases* the value of the standard error of difference and *decreases* the size of the t ratio. This is a somewhat academic issue, however, since the chances of getting a negative r from an M/S design are about the same as the chances of being hit by an arrow from a crossbow.

Degrees of Freedom for the Paired *t* Ratio

The degrees of freedom for the **paired *t* ratio** are equal to the number of pairs of scores, *N*, minus one:

$$df = N - 1$$

Since the paired *t* makes use of scores that are yoked together, the data do not include as many independent observations as there are with the independent *t*. As someone once said, "Going from the independent to the paired *t* is like getting married—you lose half your degrees of freedom." For a given number of scores, the df for the paired *t* are exactly half what they are if the independent *t* is used. As we know from our previous use of the *t* table, as degrees of freedom decrease, the table value of *t* needed to reject the null hypothesis increases. Thus, the decrease in degrees of freedom makes it more difficult to get a significant value of *t*.

We just discovered in the preceding section that using the correlation term decreases the value of the estimated standard error of difference and increases the *t* ratio. However, it seems now that that advantage has been thrown away into the degrees-of-freedom scrap heap. The situation is rosier than it appears. When that correlation term is large, the increase in the resulting *t* ratio *more than compensates for the loss of degrees of freedom.*

It's all Greek to me, sir!

Example. A researcher wishes to discover whether or not the intake of orange juice affects the potassium level in the bloodstream. A group of 12 elderly patients are selected from those in a nursing home, where previous diet has been controlled. Potassium blood levels are measured for each subject. Then each subject is given a quart of orange juice, and, two hours later, potassium levels are again measured. The data are as follows (the scaled scores represent potassium blood levels):

SUBJECT	BEFORE X_1	X_1^2	AFTER, X_2	X_2^2	$X_1 X_2$
1	26	676	25	625	650
2	25	625	28	784	700
3	24	576	27	729	648
4	23	529	26	676	598
5	23	529	25	625	575
6	21	441	23	529	483
7	19	361	21	441	399
8	17	289	19	361	323
9	17	289	16	256	272
10	16	256	19	361	304
11	15	225	18	324	270
12	14	196	17	289	238
	$\Sigma X_1 = 240$	$\Sigma X_1^2 = 4992$	$\Sigma X_2 = 264$	$\Sigma X_2^2 = 6000$	$\Sigma X_1 X_2 = 5460$

To calculate the paired t ratio (from the estimated population standard deviation), use the following steps:

1. Calculate the means for each distribution.

$$\bar{X}_1 = \frac{\Sigma X_1}{N} = \frac{240}{12} = 20.00$$

$$\bar{X}_2 = \frac{\Sigma X_2}{N} = \frac{264}{12} = 22.00$$

2. Calculate the estimated population standard deviation.

$$s_1 = \sqrt{\frac{\Sigma X_1^2 - (\Sigma X_1)^2/N}{N-1}} = \sqrt{\frac{4992 - 240^2/12}{11}} = \sqrt{\frac{4992 - 57,600/12}{11}}$$

$$= \sqrt{\frac{4992 - 4800}{11}} = \sqrt{\frac{192}{11}} = \sqrt{17.45} = 4.18$$

$$s_2 = \sqrt{\frac{\Sigma X_2^2 - (\Sigma X_2)^2/N}{N-1}} = \sqrt{\frac{6000 - 264^2/12}{11}} = \sqrt{\frac{6000 - 69,696/12}{11}}$$

$$= \sqrt{\frac{6000 - 5808}{11}} = \sqrt{\frac{192}{11}} = \sqrt{17.45} = 4.18$$

The two distributions, strictly by a chance coincidence, happen to have identical variability.

3. Calculate the Pearson r. This is the same equation shown earlier, except that X_2 is substituted for Y.

$$r = \frac{\dfrac{\Sigma X_1 X_2 - [(\Sigma X_1)(\Sigma X_2)]/N}{N - 1}}{s_{X_1} s_{X_2}} = \frac{(5460 - 5280)/11}{(4.18)(4.18)}$$

$$= \frac{16.36}{17.47} = .94$$

4. Calculate the estimated standard error of each mean.

$$s_{\bar{X}_1} = \frac{s_1}{\sqrt{N}} = \frac{4.18}{\sqrt{12}} = \frac{4.18}{3.46} = 1.21$$

$$s_{\bar{X}_2} = \frac{s_2}{\sqrt{N}} = \frac{4.18}{\sqrt{12}} = \frac{4.18}{3.46} = 1.21$$

5. Calculate the estimated standard error of difference. (Chain multiply the correlation term, that is, 2 times .94 times 1.21 times 1.21 = 2.75.)

$$s_{\bar{X}_1 - \bar{X}_2} = \sqrt{s_{\bar{X}_1}^2 + s_{\bar{X}_2}^2 - (2r_{1,2} s_{\bar{X}_1} s_{\bar{X}_2})}$$
$$= \sqrt{1.21^2 + 1.21^2 - 2(.94)(1.21)(1.21)}$$
$$= \sqrt{1.46 + 1.46 - 2.75} = \sqrt{2.92 - 2.75} = \sqrt{.17} = .41$$

6. Calculate the paired t ratio.

$$t = \frac{\bar{X}_1 - \bar{X}_2}{s_{\bar{X}_1 - \bar{X}_2}} = \frac{20.00 - 22.00}{.41} = \frac{-2.00}{.41} = -4.88$$

7. Check our calculated value for significance, using 11, or $N - 1$, as the degrees of freedom

$$t_{.01(11)} = \pm 3.11$$

$$t = -4.88 \quad \text{Reject } H_0; \text{ significant at } P < .01.$$

Thus, the subjects do differ significantly in their before and after potassium blood levels. The independent variable is effective, since the group, originally selected from one population, now represents a different population. If the direction of this difference had been predicted, that is, if the alternative hypothesis had been stated as $\mu_1 < \mu_2$ (or that potassium levels will increase as a result of the independent variable), then the one-tail t table could have been used for the critical value of t, as follows:

$$t_{.01(11)} = -2.72$$

$$t = -4.88 \quad \text{Reject } H_0; \text{ significant at } P < .01.$$

Thus, the null hypothesis would still have been rejected, in fact, with even more room to spare.

If the use of the actual sample standard deviation is preferred, the steps are as follows:

1. $\bar{X}_1 = \dfrac{\Sigma X_1}{N} = \dfrac{240}{12} = 20.00$

$\bar{X}_2 = \dfrac{\Sigma X_2}{N} = \dfrac{264}{12} = 22.00$

2. $SD_1 = \sqrt{\dfrac{\Sigma X_1^2}{N} - \bar{X}_1^2} = \sqrt{\dfrac{4992}{12} - 20.00^2}$

$SD_1 = \sqrt{416 - 400} = \sqrt{16} = 4.00$

$SD_2 = \sqrt{\dfrac{\Sigma X_2^2}{N} - \bar{X}_2^2} = \sqrt{\dfrac{6000}{12} - 22.00^2}$

$SD_2 = \sqrt{500 - 484} = \sqrt{16} = 4.00$

3. $r = \dfrac{\Sigma X_1 X_2 / N - \bar{X}_1 \bar{X}_2}{SD_1 SD_2}$

$r = \dfrac{5460/12 - (20.00)(22.00)}{(4.00)(4.00)} = \dfrac{455 - 440}{16.00} = \dfrac{15}{16} = .94$

4. $s_{\bar{X}_1} = \dfrac{SD_1}{\sqrt{N-1}} = \dfrac{4.00}{\sqrt{11}} = \dfrac{4.00}{3.32} = 1.20$

$s_{\bar{X}_2} = \dfrac{SD_2}{\sqrt{N-1}} = \dfrac{4.00}{\sqrt{11}} = \dfrac{4.00}{3.32} = 1.20$

5. $s_{\bar{X}_1 - \bar{X}_2} = \sqrt{s_{\bar{X}_1}^2 + s_{\bar{X}_2}^2 - 2r_{1,2} s_{\bar{X}_1} s_{\bar{X}_2}} = \sqrt{1.20^2 + 1.20^2 - 2(.94)(1.20)(1.20)}$

$s_{\bar{X}_1 - \bar{X}_2} = \sqrt{1.44 + 1.44 - 2.71} = \sqrt{2.88 - 2.71} = \sqrt{.17}$

$s_{\bar{X}_1 - \bar{X}_2} = .41$

6. $t = \dfrac{\bar{X}_1 - \bar{X}_2}{s_{\bar{X}_1 - \bar{X}_2}} = \dfrac{20.00 - 22.00}{.41} = \dfrac{-2.00}{.41} = -4.88$

The Direct-Difference Method

A special equation has been derived for the paired t ratio that considerably shortens its computation.

Example. Using the data from the preceding example, we set the problem up as follows:

SUBJECT	BEFORE, X_1	AFTER, X_2	DIFFERENCE, $D = X_1 - X_2$	D^2
1	26	25	+1	1
2	25	28	−3	9
3	24	27	−3	9
4	23	26	−3	9
5	23	25	−2	4
6	21	23	−2	4
7	19	21	−2	4
8	17	19	−2	4
9	17	16	+1	1
10	16	19	−3	9
11	15	18	−3	9
12	14	17	−3	9
	$\Sigma X_1 = 240$	$\Sigma X_2 = 264$	$\Sigma(+D) = +2$ $\Sigma(-D) = -26$	$\Sigma D^2 = 72$

The direct-difference method for calculating the paired t ratio involves the following steps:

1. In setting up the data table, establish a difference column of values of $X_1 - X_2$, including the sign of each value. Add all the positive values to obtain $\Sigma(+D) = +2$. Add all the negative values to obtain $\Sigma(-D) = -26$. Add these two sums together, being careful to include their signs, to obtain $\Sigma D = -24$.

2. Calculate the mean difference.

$$\bar{X}_D = \Sigma D / N = -24/12 = -2.00$$

As a check of this step, we can calculate the two means and subtract.

$$\bar{X}_1 = \frac{\Sigma X_1}{N} = \frac{240}{12} = 20.00$$

$$\bar{X}_2 = \frac{\Sigma X_2}{N} = \frac{264}{12} = 22.00$$

$$\bar{X}_D = \bar{X}_1 - \bar{X}_2 = 20.00 - 22.00 = -2.00$$

3. For our data table, we squared all the difference values and placed the squared differences in a column. The total of these values is $\Sigma D^2 = 72$. Using this, calculate the standard deviation of the differences, treating the difference values as though they were any distribution of raw scores.

$$SD_D = \sqrt{\frac{\Sigma D^2}{N} - \bar{X}_D^2} = \sqrt{\frac{72}{12} - (-2.00)^2}$$

$$SD_D = \sqrt{6.00 - 4.00} = \sqrt{2.00} = 1.41$$

4. Calculate the standard error of the mean difference. This equals the value obtained for the estimated standard error of difference obtained in the method shown previously.

$$s_{\bar{x}_D} = \frac{SD_D}{\sqrt{N - 1}}$$

$$s_{\bar{x}_D} = \frac{1.41}{\sqrt{12 - 1}} = \frac{1.41}{\sqrt{11}} = \frac{1.41}{3.32} = .42$$

(In the long method, we obtained $s_{\bar{x}_1 - \bar{x}_2} = .41$. The difference in the second place after the decimal is due to rounding.)

5. Calculate the paired t ratio.

$$t = \frac{\bar{X}_D}{s_{\bar{x}_D}}$$

$$t = \frac{-2.00}{.42} = -4.76$$

6. Check the calculated value for significance.

$$t_{.01(11)} = \pm 3.11$$

$$t = -4.76 \quad \text{Reject } H_0; \text{ significant at } P < .01.$$

The difference between the values obtained with the two methods is not a cause for concern.

Within rounding errors, the answers are identical. The *PH-STAT* computer program yields a paired t ratio of -4.69.

The Paired *t* and Power

Previously, in Chapter 9, we discussed the concept of power, which was defined as the test's sensitivity in detecting significant results. It was mentioned then that a variety of methods could be employed for increasing power, such as increasing the sample size, increasing the alpha error, using all the information the data provide, and *fitting the statistical test to the research design*. A discussion of the inner workings of the paired t can be especially illustrative of this final point—fitting the test to the design.

Let's compare the workings of the independent and paired (or dependent) t tests. On the surface they seem to be very similar—both use interval-ratio data, both test for the possibility of a difference between two sample means, both require a normal distribution of scores in the underlying population, and so on. However, they differ dramatically in both the procedures involved and the kind of results each may be expected to give. The independent t demands that the two sample measures be absolutely independent of each other or that the selection of one sample in no way influences the selection of the second sample.

The paired, or dependent t, on the other hand, demands that the two samples be somehow related to each other, where the selection of a subject for one sample group determines who will be selected for the second group. An example of an experimental design where this holds true would be the matched-subjects design.

What if, in the preceding example, the independent t had been used? That is, what if the researcher had not taken advantage of the correlation between the pairs of scores—could the decision to reject the null hypothesis still have been made? We can easily find out. The standard error of difference, without the correlation term included, would have been as follows:

$$s_{\bar{X}_1 - \bar{X}_2} = \sqrt{s_{\bar{X}_1}^2 + s_{\bar{X}_2}^2} = \sqrt{1.20^2 + 1.20^2} = \sqrt{1.44 + 1.44} = \sqrt{2.88} = 1.70$$

Then t would have been as follows:

$$t = \frac{\bar{X}_1 - \bar{X}_2}{s_{\bar{X}_1 - \bar{X}_2}} = \frac{-2.00}{1.70} = -1.18$$

Now, no matter how many degrees of freedom, no matter whether a one- or two-tail test is used, this is an acceptance of H_0. The important point is that it should not be. Therefore, by not taking advantage of the correlation term, the decision would have been to accept *when it should have been to reject.* Using the wrong t test caused the beta error to occur, that is, accepting the null hypothesis when we should be rejecting it.

On the other hand, suppose that a researcher decides that the paired t works so beautifully that it should be used in all research situations, even those where the observations are independent. Again, the beta error will be increased; since the observations are independent, there is no correlation with which to reduce the estimated standard error of difference.

In short, using the wrong statistical test increases the beta error and, therefore, reduces the power of the test. A paired t is more powerful than is an independent t when the subjects are truly dependent on each other; the independent t is more powerful when the subjects are independently selected and assigned.

Some Cautions Regarding the Paired *t*

Although the paired *t*, as we have seen, is an appropriate test for analyzing results of matched-subjects and before–after designs, it is not the best test to apply to a before-after design with an independent control group. In this experimental design, as we know, two groups of subjects are randomly selected, and both groups are measured in the before and after conditions. The groups differ only in the extent to which they experience different levels of the independent variable. This design is becoming increasingly popular, and justifiably so.

As we learned previously, the use of the before–after design with separate control reduces the possibility of the occurrence of one of the major research errors, confounding the independent variable. We saw that with a one-group before–after design, factors other than the independent variable may inadvertently act on the subjects to change them. One example of such confounding is the Hawthorne effect, where the subjects change as a result, not of the independent variable, but as a result of the flattery and attention they receive from the researcher(s). With a separate control group, any possible confounding variables should influence both groups equally, and, thus, the pure effects of the independent variable can be more accurately factored out.

A Significant Change in the Control Group Before and After

With two groups, each measured twice, the paired *t* test may not always work. For example, although it is possible, despite the increase in alpha error, to do *t* tests between the experimental group's pre- and postmeasures and the control group's pre- and postmeasures, what happens when both *t*'s are significant? That is, the control group may also change significantly from the before to the after condition, even though the change is much less than that observed for the experimental group. After all, the possibility that there might be a difference in the magnitude of the change scores between the two groups is the very reason for using a separate control group in the first place.

For example, suppose that we wish to test a new method for training learning disabled children to read. Two groups are randomly selected and are given a reading test. One group (experimental) then undergoes six weeks of special training, while the other group (control) does not. At the end of the six-week period, both groups are given the reading test again. It is possible that the control group will improve, because of maturation, outside influences, or whatever. However, the experimental group's improvement may be far greater—a clear indication that the training program works. The comparison, therefore, should be between each group's *change scores*. And since the two groups have been randomly and independently assigned, then the change scores themselves are independent of each other. We must remember that the paired *t* can only be used when scores can be paired off with each other. In this study, because the groups are independently selected, there is no way to match up the change scores. The appropriate technique in this case is to compare the change scores using the independent t, assessing the degrees of freedom on the basis of the number of separate change scores.

    Example. Two groups of 10-year-old children are randomly selected from a population of learning disabled children and then assigned randomly to experimental and control conditions. Both groups are given a reading test. The experimental group then undergoes six weeks of special training. Both groups are again given the reading test. The data are as follows:

Experimental Group			Control Group		
BEFORE, X_1	AFTER, X_2	CHANGE, $X_2 - X_1$	BEFORE, X_1	AFTER, X_2	CHANGE, $X_2 - X_1$
23	31	+8	20	23	+3
22	29	+7	19	21	+2
24	30	+6	20	22	+2
22	28	+6	22	24	+2
20	25	+5	25	26	+1
22	27	+5	24	24	0
25	23	-2	23	18	-5

Because it makes no difference for the t test whether the scores are plus or minus, we can establish the change-score distribution with values of $X_2 - X_1$, giving us mostly positive change values and making the calculation easier.

    Now we compare only the change score distributions, using the independent t:

Experimental Group		Control Group	
CHANGE, X_{EG}	SCORE, X_{EG}^2	CHANGE, X_{CG}	SCORE, X_{CG}^2
+8	64	+3	9
+7	49	+2	4
+6	36	+2	4
+6	36	+2	4
+5	25	+1	1
+5	25	0	0
-2	4	-5	25
$\Sigma X_{EG} = 35$	$\Sigma X_{EG}^2 = 239$	$\Sigma X_{CG} = 5$	$\Sigma X_{CG}^2 = 47$

We calculate the mean for each group.

$$\bar{X}_{EG} = \frac{\Sigma X_{EG}}{N} = \frac{35}{7} = 5.00$$

$$\bar{X}_{CG} = \frac{\Sigma X_{CG}}{N} = \frac{5}{7} = .71$$

Using the steps shown in Chapter 9, calculate the independent t ratio.

$$t = \frac{\bar{X}_{EG} - \bar{X}_{CG}}{s_{\bar{X}_{EG}-\bar{X}_{CG}}} = \frac{5.00 - .71}{1.60} = \frac{4.29}{1.60} = 2.68$$

We check our calculated value for significance. First, we compare it with the critical value from the two-tail t table, obtaining the df by adding the number of change scores in the first group to the number of change scores in the second group, minus two $(7 + 7 - 2 = 12)$:

$$t_{.05(12)} = \pm 2.18$$

$t = 2.68$ Reject H_0; significant at $P < .05$.

or against the one-tail t table value (since most change studies do predict the direction of the difference):

$$t_{.01(12)} = 2.68$$

$t = 2.68$ Reject H_0; significant at $P < .01$.

Thus, although both groups changed, the experimental group changed significantly more. The training method did cause an increase in reading ability.

Using the Paired t on a Before–After Design with Matched Control Group

Another variation on the before–after design theme is to create the separate control group by matching the subjects person by person with those in the experimental group. In this way it is assumed that the two groups are equivalent at the outset. The data from this type of design are handled in a fashion similar to that used for the B/A design with control group, in that again the focus is on the change scores rather than the raw scores. However, since the change scores can be paired between the matched subjects in this variation, analysis of the data can be accomplished using the paired t. In the following example, the direct-difference method is used.

Example. Suppose that in the preceding example, the two groups of learning disabled students matched on IQ. Then we would have the following data.

	Experimental Group			Control Group		
PAIR	BEFORE, X_1	AFTER, X_2	CHANGE, $X_2 - X_1$	BEFORE, X_1	AFTER, X_2	CHANGE, $X_2 - X_1$
1	23	31	+8	20	23	+3
2	22	29	+7	19	21	+2
3	24	30	+6	20	22	+2
4	22	28	+6	22	24	+2
5	20	25	+5	25	26	+1
6	22	27	+5	24	24	0
7	25	23	-2	23	18	-5

The change scores are set up in a separate table, as before.

PAIR	EXPERIMENTAL CHANGE SCORE	CONTROL CHANGE SCORE	DIFFERENCE, D	D²
1	+8	+3	+5	25
2	+7	+2	+5	25
3	+6	+2	+4	16
4	+6	+2	+4	16
5	+5	+1	+4	16
6	+5	0	+5	25
7	-2	-5	+3	9
	$\Sigma X_{EG} = 35$	$\Sigma X_{EG} = 5$	$\Sigma(+D) = 30$	$\Sigma D^2 = 132$
	$\bar{X}_{EG} = 5.00$	$\bar{X}_{EG} = .71$	$\Sigma(-D) = 0$	
			$\Sigma D = +30$	

1. $\bar{X}_D = \dfrac{\Sigma D}{N} = \dfrac{30}{7} = 4.29$

2. $SD_D = \sqrt{\dfrac{\Sigma D^2}{N} - \bar{X}_D^2} = \sqrt{\dfrac{132}{7} - 4.29^2} = \sqrt{18.86 - 18.40}$

   $= \sqrt{.46} = .68$

3. $s_{\bar{X}D} = \dfrac{SD_D}{\sqrt{N-1}} = \dfrac{.68}{\sqrt{7-1}} = \dfrac{.68}{\sqrt{6}} = \dfrac{.68}{2.45} = .28$

4. $t = \dfrac{\bar{X}_D}{s_{\bar{X}D}} = \dfrac{4.29}{.28} = 15.32$

For this analysis the df would be the number of paired change scores, minus one $(7 - 1 = 6)$. The value for the paired t shows a tremendous increase over the value of the independent t done on the same data. This is more evidence that when there is a substantial correlation between pairs of scores, the paired t is a very powerful test.

THE WITHIN-SUBJECTS *F* RATIO

When a before-after or matched-subjects design is being used and results in more than two sets of scores, the t test is no longer the appropriate test. Actually, the term before–after is a little misleading when we are discussing three or four distributions of scores, unless we are careful to say "before–after–after–after." Typically, when more than two distributions result, the before–after design is referred to as the repeated-measure design, or the within-subjects design. Also, although a matched-subjects design can theoretically be set up involving virtually any number of groups, matching does become increasingly difficult as the number of groups increases. It can be difficult enough to obtain matched pairs of subjects, let alone matched trios or quartets. Some statisticians[1] demand proof that the matching is, indeed, relevant to the dependent variable, pointing out that unless the matching process is proven, ANOVA procedures for repeated-measure and matched-subjects designs are different.

Calculating the Within-Subjects *F* Ratio

Once it has been proven that the matching process has, in fact, produced equivalent groups of subjects, the basic computational method for the **within-subjects *F* ratio** is identical for data from both repeated-measure and matched-subjects designs. Therefore, the following example with a repeated-measure design serves also to illustrate the procedure if the design had utilized matched subjects. However, in this example, notice that each subject is serving in every treatment condition.

In a repeated-measure design, each subject, therefore, receives all levels of the independent variable, so there are always going to be more scores than there are subjects. This method is also referred to as the "treatments by subjects" design.

Example. A researcher wishes to find out if buying TV time for the repeated showing of a political propaganda film is worth the cost. Do attitudes change more the more times a person sees the film, or does buying the repeated TV time simply give different people the opportunity to see it? An experiment is designed in which a group of registered voters is randomly selected and asked to complete an "Attitude Toward the Candidate" questionnaire (high scores indicate a pro attitude). Then they see the 20-minute film and fill out the question-

naire again. They see the film yet another time and again fill out the questionnaire. Thus, they fill out the questionnaire three times. The data are as follows:

SUBJECT	First Time X_1	First Time X_1^2	Second Time X_2	Second Time X_2^2	Third Time X_3	Third Time X_3^2	ΣX_r
1	1	1	2	4	3	9	6
2	2	4	4	16	6	36	12
3	3	9	3	9	3	9	9
4	4	16	5	25	6	36	15
5	5	25	6	36	7	49	18
$\Sigma = $	15	55	20	90	25	139	60

1. Calculate the total sum of squares (as shown in Chapter 12). Again circle the correction factor, C.

$$SS_t = \Sigma X^2 - \frac{(\Sigma X)^2}{N} = \Sigma X^2 - C$$

$$SS_t = 55 + 90 + 139 - \frac{(15 + 20 + 25)^2}{15} = 284 - \frac{60^2}{15}$$

$$SS_t = 284 - \frac{3600}{15} = 284 - \boxed{240} = 44$$

2. Calculate the sum of squares between columns (shown in Chapter 12 as the SS_b).

$$SS_{bc} = \frac{(\Sigma X_1)^2}{n_1} + \frac{(\Sigma X_2)^2}{n_2} + \frac{(\Sigma X_3)^2}{n_3} - C$$

$$SS_{bc} = \frac{15^2}{5} + \frac{20^2}{5} + \frac{25^2}{5} - 240 = \frac{225}{5} + \frac{400}{5} + \frac{625}{5} - \boxed{240}$$

$$SS_{bc} = 45 + 80 + 125 - \boxed{240} = 250 - \boxed{240} = 10$$

3. Calculate the sum of squares between rows. Add the scores across each row, so
that $\Sigma X_{r1} = 6$, $\Sigma X_{r2} = 12$, and so on. Square each row total and divide by the
number of scores in each row.

$$SS_{br} = \frac{(\Sigma X_{r1})^2}{n_{r1}} + \frac{(\Sigma X_{r2})^2}{n_{r2}} + \frac{(\Sigma X_{r3})^2}{n_{r3}} + \frac{(\Sigma X_{r4})^2}{n_{r4}} + \frac{(\Sigma X_{r5})^2}{n_{r5}} - C$$

$$SS_{br} = \frac{6^2}{3} + \frac{12^2}{3} + \frac{9^2}{3} + \frac{15^2}{3} + \frac{18^2}{3} - \boxed{240}$$

$$SS_{br} = \frac{36}{3} + \frac{144}{3} + \frac{81}{3} + \frac{225}{3} + \frac{324}{3} - \boxed{240}$$

$$SS_{br} = 12 + 48 + 27 + 75 + 108 - \boxed{240}$$
$$SS_{br} = 270 - \boxed{240} = 30$$

4. Calculate the residual (r × c) sum of squares (this is the variability that is left over, after the column and row variabilities have been taken out).

$$SS_{r \times c} = SS_t - SS_{bc} - SS_{br}$$
$$SS_{r \times c} = 44 - 10 - 30 = 4$$

5. Set up the ANOVA table, with df assigned as follows:

column df = number of columns minus one = 2

row df = number of rows minus one = 4

residual df = column df times row df = 8

We check the degrees of freedom by adding them together, $2 + 4 + 8 = 14$, to be sure they equal $N - 1$, or $15 - 1 = 14$.

SOURCE	SS	df	MS	F
Between columns	10	2	$\dfrac{SS}{df} = \dfrac{10}{2} = 5$	$\dfrac{MS_{bc}}{MS_{r \times c}} = \dfrac{5}{.50} = 10$
Residual (r × c)	4	8	$\dfrac{SS}{df} = \dfrac{4}{8} = .50$	

Thus, the F ratio is equal to 10.00, calculated from the mean square between columns divided by the residual mean square.

Interpreting the Within-Subjects F Ratio

Since the numerator (mean square between columns) of the F ratio in the preceding example had 2 df and the denominator (residual mean square) had 8 df, we use these values to look up the critical value of F in Table G.

Across the top row of the table are the degrees of freedom associated with the numerator, which in this case equal 2, and down the far left column, the df associated with the denominator, which in this case equal 8. At the intersection of the two, we find for an alpha level of .01, a critical F value of 8.65. Again, the statistical decision is written as follows:

$$F_{.01(2,8)} = 8.65$$

$F = 10.00$ Reject H_0; significant at $P < .01$.

Therefore, for the preceding example, subjects who viewed the propaganda film more than once have attitudes increasingly favorable to the candidate.

The Importance of Correlation Within Subjects

Like the paired t, the within-subjects F is an extremely powerful statistical tool, *when there is, in fact, correlation within subjects across the rows*. If there is no correlation, the F ratio value may plummet into nonsignificance.

Example. Using the data from the preceding example, we keep the column totals the same, but rearrange the rows to eliminate the correlation.

X_1	X_1^2	X_2	X_2^2	X_3	X_3^2	ΣX_r
1	1	5	25	3	9	9
2	4	4	16	6	36	12
3	9	3	9	7	49	13
4	16	6	36	6	36	16
5	25	2	4	3	9	10
15	55	20	90	25	139	60

$$SS_t = 284 - \frac{60^2}{15} = 284 - 240 = 44$$

$$SS_{bc} = \frac{15^2}{5} + \frac{20^2}{5} + \frac{25^2}{5} - 240 = 10$$

$$SS_{br} = \frac{9^2}{3} + \frac{12^2}{3} + \frac{13^2}{3} + \frac{16^2}{3} + \frac{10^2}{3} - 240 = 250 - 240 = 10$$

$$SS_{r \times c} = SS_t - SS_{bc} - SS_{br} = 44 - 10 - 10 = 24$$

SOURCE	SS	df	MS	F
Between columns	10	2	5	$\dfrac{MS_{bc}}{MS_{r \times c}} = \dfrac{5}{3} = 1.67$
Residual (r × c)	24	8	3	

$$F_{.05(2,8)} = 4.46$$

$$F = 1.67 \quad \text{Accept } H_0; \text{ not significant.}$$

Therefore, the same data with the rows rearranged to destroy the correlation results now in an acceptance of the null hypothesis.

Comparing the Within-Subjects and the Factorial ANOVAS

The within-subjects F is in some ways similar to the factorial (two-way) ANOVA presented in Chapter 12. The factorial ANOVA uses the row variability to assess possible treatment effects, and the within-subjects F uses the same variability to get at possible differences among the subjects. Also, whereas the factorial ANOVA uses the mean square within groups as the error term, the F ratio's denominator, the within-subjects F uses the residual or interaction mean square as its error term (which reflects the variability left over after any systematic variability has been removed). For other experimental designs, such as the repeated-measure with separate control group (either independent or matched), other ANOVA analyses are possible. These are presented in many advanced texts.

TESTING CORRELATED EXPERIMENTAL DATA

The experimental designs reviewed in this chapter have in common the fact that the data to be analyzed are, in some way, associated. The statistical tests used are all aimed at testing the hypothesis of difference (as they must be in experimental research), but by taking the score correlations into account, these tests become extremely sensitive. With the paired t or within-subjects F, very small differences may be found to be significant differences.

As we have seen, applying the t or F test for independent measures to data that are really correlated, inevitably lowers the chance of rejecting the null hypothesis. Thus, when there is correlation, the researcher can take advantage of it, thereby reducing the beta error and increasing the power of the test. The moral for the researcher is that to use the wrong test is to bet against yourself.

SUMMARY

Whenever subjects are paired, either with themselves in a B/A design or with equated partners in an M/S design, a correlation almost certainly results between the paired scores. For this reason, different statistical tests are used for analyses of these data than are used for independently selected samples. This chapter covers some of the statistical procedures for testing the hypothesis of difference when the data are in at least interval form and the samples are correlated.

When two distributions of interval scores are being compared and the data result from either a B/A or M/S design, then the paired t must be used. For a given-sized difference between two sample means, the paired t will produce a higher numerical value than will the independent t. This is due to the fact that the estimated standard error of difference (the t ratio's denominator) is reduced in value by the subtraction of the correlation term. However, the use of the paired t also reduces the degrees of freedom by a factor of one-half. To some extent, then, the benefits from increasing the t value may be offset by the reduction in the degrees of freedom. Typically, however, the increased t value more than compensates for the loss in degrees of freedom. This means that when the data are correlated, the paired t is a more powerful test than is the independent t. (The more powerful the test, the less is the likelihood of accepting H_0 when it should have been rejected.) The independent t, however, is more powerful than the paired t when the samples are independent. This is because, although independent samples should produce no correlation (and, therefore, no reduction in the estimated standard error of difference), the independent t does have more degrees of freedom. (The more degrees of freedom, the smaller is the t value needed to reject H_0.)

When the experiment involves a B/A design with a separate control group, then another approach to the analysis of the data must be used. Rather than comparing the individual raw scores, the "change" scores should be evaluated. If the separate control group is independent of the experimental group, the comparison of the change scores can be assessed via the independent t. If the subjects in the separate control group are matched to those in the experimental group, then the analysis of the data is accomplished by using the paired t ratio. When more than two sets of scores are involved, either from a repeated-measure or matched-subjects design, then the within-subjects F ratio (treatments by subjects) should be used, again assuming at least interval data. The within-subjects F ratio is a more powerful test than is the independent F when the experimental design is either repeated measure or matched subjects.

KEY TERMS

correlated samples   within-subjects F ratio
paired t ratio

REFERENCE

1. EDWARDS, A. E.   (1972). *Experimental design in psychological research.* (4th ed.). New York: Holt, Rinehart and Winston. (See especially Ch. 14).

PROBLEMS

1.   Calculate a paired t ratio between the following sets of pre- and postscores. Use the long method, where the Pearson r is calculated separately.

SUBJECT	PRE	POST
1	11	8
2	15	15
3	16	13
4	14	11
5	10	11
6	15	10
7	6	6
8	11	8

2. Calculate a paired t ratio on the data from problem 1 using the direct-difference method.
3. Calculate a within-subjects F ratio on the following scores from a repeated-measure design.

SUBJECT	A	B	C	D
1	15	12	11	9
2	17	15	13	10
3	18	16	15	11
4	17	17	14	10
5	20	18	16	11
6	15	13	12	8

4. A researcher wishes to find out if typing speed is affected by the kind of typewriter (electric versus manual) used. A group of student typists, equally experienced on both types of machines, are randomly selected and are matched on the basis of typing speed (error-free words per minute). One group is then tested on an electric machine and the other group on a manual machine. The data are as follows:

PAIR	ELECTRIC	MANUAL
1	50	42
2	65	60
3	72	65
4	90	85
5	48	50
6	62	60
7	75	60
8	50	51
9	68	59

Test the hypothesis using the direct-difference method.
5. A psychologist wishes to establish whether or not room color has any effect on anxiety level. Three groups of college students are randomly selected, and the

subjects are matched on the basis of their Taylor Manifest Anxiety Test scores. All students had their flu shots the previous day, but are told to come back to the infirmary because there was "something wrong" with the vaccine. The subjects are then brought into one of three rooms: one decorated in an off-blue (cool color), one in gray (neutral color), and one in bright red (warm color). Each subject is connected to a galvanometer (which measures the skin's electrical conductivity—presumably a measure of situational anxiety). A nurse then pretends to give each subject another shot. At the moment when the rubber-tipped hypodermic needle (the tip hidden from the subject's view) touches the subject's arm, the galvanic skin response (G.S.R.) is recorded, with high scores indicating greater anxiety. G.S.R. scores are recorded in decivolts. After the G.S.R. scores are recorded, the subjects are told the real purpose of the study.

TRIAD	BLUE ROOM	GRAY ROOM	RED ROOM
1	2	5	6
2	3	6	7
3	5	7	9
4	6	8	10
5	3	5	8
6	2	4	6
7	1	3	5

Test the hypothesis.

6. Hypothesis: Frustration affects attitudes. All the subjects in problem 5 are given an "Attitudes Toward Psychologists" test both before and after participating in the room-color study, with high scores indicating a positive attitude. Test the hypothesis using the direct-difference method.

SUBJECT	BEFORE	AFTER
1	44	20
2	20	10
3	35	30
4	42	26
5	35	30
6	30	20
7	34	30
8	30	22
9	19	21
10	17	20
11	25	17
12	30	15
13	32	25
14	31	26
15	34	30
16	20	25

SUBJECT	BEFORE	AFTER
17	31	24
18	37	19
19	32	30
20	33	28
21	16	15

7. A researcher is interested in discovering whether a role-reversal procedure will influence men's attitudes toward women. Two random samples of married men were independently selected and both groups were given a test measuring their attitudes toward women's roles (high scores indicating a traditional attitude). The men in Group A were then instructed to reverse roles with their wives for the next two weekends (he doing her work and she doing his). The men in Group B were told nothing. Two weeks later both groups took the test again. The data are as follows:

Group A			Group B		
SUBJECT	PRETEST	POSTTEST	SUBJECT	PRETEST	POSTTEST
1	163	150	1	160	152
2	142	130	2	150	142
3	152	140	3	165	160
4	149	140	4	143	140
5	150	142	5	150	147
6	160	158	6	158	156

Was there a significant difference between the two groups regarding how much their attitudes changed?

8. For a given difference between the means and equal numbers of scores, which test, the paired t or the independent t, has more statistical power? (For this question, assume that for both tests all their assumptions are met.)

9. Of the three basic experimental designs, A/O, B/A, and M/S, on which type or types can the paired t ratio be used?

10. When calculating the paired t ratio, what effect does a substantial correlation have on the size of the estimated standard error of difference?

11. State how the degrees of freedom compare between the independent and paired t ratios.

12. When calculating the paired t ratio, what effect does a substantial correlation have on the size of the resulting t ratio?

13. A within-subjects F ratio performed on data from a matched-subjects design results in a (higher or lower) F ratio than would be obtained by an independent, one-way ANOVA performed on the same data.

14. The fact that the paired t has fewer degrees of freedom than does its independent counterpart, *and that fact alone,* has what effect on the probability of achieving significance?

Fill in the blanks in problems 15 through 17.

15. In matched-subjects designs, the subjects should be equated on some variable(s) which are related to the (dependent or independent) _____ variable.
16. When all its assumptions are met, the paired t is (more able or less able) _____ than the independent t to reject null when only a small difference exists between the sample means.
17. The paired t and the within-subjects F should only be used when the data are in the form of at least (interval, ordinal, or nominal) _____ measures.

True or False—Indicate either T or F for problems 18 through 27.

18. For equal numbers of scores, the paired t has more degrees of freedom than does the independent t.
19. The paired t has as its ultimate goal the detection of differences between two sets of interval scores when the data sets are dependent.
20. The paired t may only test the hypothesis of association, whereas the independent t may test the hypothesis of difference as well.
21. The more degrees of freedom a given t ratio has, the higher the likelihood of rejecting the null hypothesis.
22. For both t and F, whether from correlated or independent designs, the more subjects being tested, the greater the number of degrees of freedom.
23. The higher a test's power, the less likely one is of committing the beta error.
24. The more powerful a test is, the greater the possibility of detecting differences when there really are differences in the population.
25. The paired t ratio may never be used for making population inferences.
26. Statistical tests which use interval data are inherently more powerful than are those which utilize ordinal or nominal data.
27. Beta error occurs when the null hypothesis is rejected when it should have been accepted.

chapter 16

Nonparametrics Revisited: The Ordinal Case

As we learned in Chapter 1, the ordinal scale gives information regarding greater than or less than, *but not how much greater or how much less.* That is, a score with the rank of 1 is known to be greater than a score with the rank of 2, but whether 1 is inches or miles in front of 2 is not known. In short, then, for ordinal data, the distance between successive scale points is unknown.

In Chapter 10, we learned a technique for testing the hypothesis of association with ordinal data, the Spearman r_S. In this chapter, tests of the hypothesis of difference for ordinal data are presented—one test for each of the basic research situations. As we know, all data gathered by any of the experimental methods must be analyzed by testing the hypothesis of difference. Data from the post-facto method, however, may be analyzed by testing either the hypothesis of association or the hypothesis of difference. When we test for association, we can use the r_S, but when we test for difference, one of the tests to follow may be used. Thus, each and every test in this chapter tests the hypothesis of difference. (For each of the research situations presented, a counterpart test suitable for interval data is mentioned. For instance, whereas the Spearman r_S is used to test for correlation with ordinal data, the Pearson r is used for the same purpose when the data are in interval form.)

Ordinal Data Require Nonparametric Tests

All the statistical tests used on ordinal data are considered to be **nonparametric,** or distribution free. Whereas the major interval data tests, t, F, and r, all make careful assumptions regarding the characteristics of the population to which the results are to be generalized, the tests for ordinal (and as we saw in Chapter 13, nominal) data make no such assumptions. These tests do not make *any assumptions* regarding μ, the mean of the population, nor do they assume a normal distribution in the population. Therefore, if we obtain interval data from a population known not to be normal, these interval scores may be converted into ranks, and a test for ordinal data can then be performed. Although, as has been stated many times, the nonparametric tests are not as powerful as their parametric cousins, they are much safer when the population characteristics are at all suspect.

When to Use the Nonparametric Tests. As with the r_S, the nonparametric tests for ordinal data should be used whenever the distributions to be compared are such that

1. Both, or all, are originally presented in ordinal form.
2. One distribution is presented in ordinal form and the other(s) in interval form. In this situation, we convert the interval data to ordinal and perform a test for ordinal data. (We must never attempt, as was pointed out in Chapter 10, to convert ordinal to interval data.)
3. Both, or all, distributions are in at least interval form, but the populations from which the samples were selected are known to lack normality, or the samples show significant differences in variability.

In these situations, convert both (all) sets of interval scores into ordinal ranks, and use one of the nonparametric ordinal tests that are to be covered in this chapter.

When converting interval scores into ordinal ranks, handle all tied scores as shown previously in Chapter 10.

THE MANN–WHITNEY *U* TEST FOR TWO ORDINAL DISTRIBUTIONS WITH INDEPENDENT SELECTION

With interval data, when the hypothesis of difference is to be tested between two independently selected samples, the obvious statistical test is the independent t. The ordinal answer to the independent t is the **Mann–Whitney *U* test.** Therefore, any time the research situation dictates the use of the independent t, but the data are in ordinal form, the Mann-Whitney U can be used.

The Mann–Whitney U test assesses whether two sets of ranked scores are representative of the same population. If they are, the two distributions should be random and H_0 is accepted. If, however, the value of U detects a nonrandom pattern, then H_0 is rejected.

Calculating the Mann-Whitney *U*

As is true of the r_S, all the tests of ordinal data utilize a number of constants—values that do not change, regardless of the data.

Example.   A political analyst wishes to establish whether or not a differ-ence in income exists between registered Republicans and registered Democrats. Random samples are selected of 10 Republicans and 11 Democrats, and the an-nual income for each subject is obtained. Because the income distribution in the population is known to be skewed, the interval scores are converted to ordinal ranks. In this process, the ranks are assigned to both sample distributions *combined* rather than ranking each distribution separately, as is done for the Spearman r_S. The reason we rank the combined distributions is to find out if one set of ranks is significantly lower than the other.

The income scores and the resulting ranks are as follows:

Republicans		Democrats	
X_1	R_1	X_2	R_2
40,000	8	16,000	21
41,000	7	17,000	20
43,000	5	20,000	19
42,000	6	21,000	18
190,000	1	39,000	9
44,000	4	38,000	10
55,000	3	36,000	11
60,000	2	35,000	12
31,000	14	34,000	13
30,000	15	29,000	16
	$\Sigma R_1 = \overline{65}$	28,000	17
$n_1 = 10$		$n_2 = 11$	

To calculate the Mann–Whitney U, the only data values needed are those for R_1, n_1, and n_2. We carry out the following steps.

1. Add the ranks for the first distribution ($\Sigma R_1 = 65$). We use this value and the two sample sizes, n_1 and n_2, to solve for U.

$$U = n_1 n_2 + \frac{n_1(n_1 + 1)}{2} - \Sigma R_1$$

$$U = (10)(11) + \frac{10(11)}{2} - 65$$

$$U = 110 + 55 - 65 = 100$$

2. Using the value of U and the sample sizes again, solve for z_U.

$$z_U = \frac{U - n_1 n_2/2}{\sqrt{[n_1 n_2 (n_1 + n_2 + 1)]/12}}$$

$$z_U = \frac{100 - [(10)(11)]/2}{\sqrt{[(10)(11)(10 + 11 + 1)]/12}} = \frac{100 - 55}{\sqrt{2420/12}}$$

$$z_U = \frac{45}{\sqrt{201.67}} = \frac{45}{14.20} = 3.17$$

3. Compare the obtained value of z_U with the z score value that excludes the extreme 1% of the distribution, $z_{.01} = \pm 2.58$. If z_U is equal to or greater than this value, reject the null hypothesis. (For an alpha error level of .05, compare our value with $z_{.05} = \pm 1.96$.)

$z_{.01} = \pm 2.58$

$z_U = 3.17$ Reject H_0; significant at $P < .01$.

Our conclusion, therefore, is to reject the null hypothesis (H_0: $R_1 = R_2$) that the two sets of ranks represent a single population. The data from our samples show a significant difference in income rank between Republicans and Democrats. We accept the alternative hypothesis, that $R_1 \neq R_2$. Thus, the two sets of ranked scores represent different income populations.

***Sample Size and* U.**   Note that in making the comparison between z_U and z, there are no degrees of freedom involved. This is because the distribution of U values, when n_1 and n_2 are each at least 9, is assumed to be close enough to normality to be directly compared to the z score distribution. Remember though that this is only true when there are *at least 9* ranked scores in each sample. (For very small samples, that is, 8 or less, separate tables of critical U values are available.)[1]

Interpreting the *U* Test Results

In the preceding example, because H_0 was rejected, we concluded that Republicans do earn more than Democrats. Great care, however, must be taken when interpreting the results. The example was a post-facto study, with the independent variable (party affiliation) an assigned subject variable rather than manipulated. Therefore, although it turned out that Republicans do earn more than Democrats, the causal basis of this difference was not established. Are persons who announce themselves as Republicans more apt to get promotions and raises? Or are persons who do get raises more apt to then become Republicans? Or could a third factor, perhaps parents' socioeconomic level, cause a person to be both a Republican and an earner of a higher income? (After all, if your father is president of General Motors, you may start out on a pay scale somewhat above minimum wage.) Finally, age was not controlled, so that it is possible that older persons, who earn more because of seniority, tend to be Republicans. Post-facto research can answer none of these questions with certainty.

Of course, since it tests the hypothesis of difference, the Mann–Whitney U can be used for the analysis of experimental data. However, as was the case with the independent t test, the experimental design must be completely randomized.

The after-only design, for example, satisfies the independent samples restriction for the use of U.

THE KRUSKAL–WALLIS *H* TEST FOR THREE OR MORE ORDINAL DISTRIBUTIONS WITH INDEPENDENT SELECTION

With more than two sets of interval scores, and where the samples are independently selected, the hypothesis of difference is tested using the one-way ANOVA. When the data are ordinal, these same research situations are analyzed by the **Kruskal–Wallis *H* test.** The Kruskal–Wallis *H*, then, *is a one-way ANOVA for ordinal data.* Thus, whenever the independent variable has at least three levels, and the data are ordinal, *H* is the proper statistic to use. Since *H* demands that the several sample groups be *independently* selected, then, of the various experimental designs, *H* can only be applied to studies in which the subjects have been independently assigned to the groups.

Also, as with the Mann–Whitney *U*, if the data represent skewed distributions, the interval scores from the combined distributions must be rank ordered together. Whether the scores are ranked high to low or low to high does not affect the value of the Kruskal–Wallis *H*. The choice made here was to rank high to low as was done previously with the Spearman r_S.

Calculating the Kruskal–Wallis *H*

Example. Suppose that the Federal Aviation Administration is interested in discovering if differences in flying ability are a function of pilot age. A dispute arises. One hypothesis states that younger persons, due to their better reflexes and general physical conditioning, are better pilots. Another hypothesis insists that older persons, due to their longer flying experience, are better pilots. A random sample of 24 licensed pilots is selected, with 6 pilots from each of 4 age categories. An FAA examiner takes each pilot on a test flight and rank orders all 24 on their flight skills. Since the ranking is high to low in this study, the rank of 1 identifies the best pilot, whereas the rank of 24 designates the worse pilot.

Group 1 (21–30 years old), R_1	Group 2 (31–40 years old), R_2	Group 3 (41–50 years old), R_3	Group 4 (51–60 years old), R_4
2	20	23	24
4	6	11	17
18	8	15	22
1	5	13	21
3	9	10	19
7	12	14	16
$\Sigma R_1 = 35$	$\Sigma R_2 = 60$	$\Sigma R_3 = 86$	$\Sigma R_4 = 119$
$n_1 = 6$	$n_2 = 6$	$n_3 = 6$	$n_4 = 6$

The only data values needed for this analysis are the sum of the ranks for each group and the group, or sample, sizes. We perform the H test in the following steps.

1. Add the ranks in each column (group) to obtain $\Sigma R_1 = 35$, $\Sigma R_2 = 60$, $\Sigma R_3 = 86$, and $\Sigma R_4 = 119$.

2. Substitute the values for ΣR, N, and n into the H equation and solve.

$$H = \frac{12}{N(N+1)} \left(\frac{\Sigma R_1^2}{n_1} + \frac{\Sigma R_2^2}{n_2} + \frac{\Sigma R_3^2}{n_3} + \frac{\Sigma R_4^2}{n_4} \right) - 3(N+1)$$

$$H = \frac{12}{24(24+1)} \left(\frac{35^2}{6} + \frac{60^2}{6} + \frac{86^2}{6} + \frac{119^2}{6} \right) - 3(24+1)$$

$$H = \frac{12}{24(25)} \left(\frac{1225}{6} + \frac{3600}{6} + \frac{7396}{6} + \frac{14{,}161}{6} \right) - 3(25)$$

$$H = \frac{12}{600} (204.17 + 600 + 1232.67 + 2360.17) - 75$$

$$H = .02(4397.01) - 75 = 87.94 - 75 = 12.94$$

3. Compare the calculated value of H with the critical values in the chi square table (Table I). The degrees of freedom are equal to k, the number of columns (groups), minus one, that is, df = 3.

$$\chi^2_{.01(3)} = 11.34$$

$$H = 12.94 \quad \text{Reject } H_0; \text{ significant at } P < .01.$$

We thus reject the null hypothesis (H_0; $R_1 = R_2 = R_3 = R_4$) and conclude that the four sample groups do indeed represent different population levels of pilot skill. It appears that younger pilots perform better than older pilots, but the key factor may not be age. We must be careful of our interpretation of these results, since this is again an example of post-facto research. The independent variable, age, was a subject variable, and, therefore, other factors may be involved. One may presume that the older (yet according to this study, worse) pilots have had more flying hours, but this is only conjecture. Perhaps some of the older pilots only recently earned their licenses and have few flying hours. More precise studies might be done where the pilots are matched according to age and then assigned to categories on the basis of flying time, or matched on flying time and then categorized on age. Studies of this nature, however, cannot be analyzed by H, since the use of H requires independent sample groups.

***Sample Size and* H.** With at least three sample groups, and a minimum of 6 subjects per sample, H may be assessed for significance with the chi square table. With smaller samples, a special table of H values is available,[2] but attaining significance with these tiny samples becomes quite difficult. With three-group designs, the researcher should select a total sample of at least 18 subjects, 6 per group.

THE WILCOXON *T* TEST FOR TWO ORDINAL DISTRIBUTIONS WITH CORRELATED SELECTION

With interval data, when two correlated groups are to be compared for possible differences, the appropriate test is the paired *t* test. With ordinal data, the same situation can be handled by the **Wilcoxon *T* test**. This means that the analysis of ordinal data from either the before–after or matched-subjects designs can be appropriately tested with the Wilcoxon *T*.

Procedure for the Wilcoxon *T* Test

The procedure for the Wilcoxon *T* is demonstrated in the following example.

Example. Suppose a golf pro creates a new method, including videotape replays, of teaching golf. A random sample of golfers at a certain country club is selected, and their average golf scores are ascertained. The subjects are then placed into matched groups on the basis of their average scores. That is, one golfer who averages 85 is placed in the experimental group, and another golfer who also averages 85 goes into the control group. The subjects in the experimental group are then given a week's instruction using the new teaching method; those in the control group are taught in the traditional way. At the end of the week's training, both groups play a round of golf and their scores are compared. In looking over the two sets of golf scores, it is discovered that the distributions are badly skewed, since in each group there were a few new members with extremely high scores. So despite the fact that the data are originally in interval form (golf scores), because of the skew, an ordinal test, Wilcoxon *T* test is chosen for the analysis.

The data for 10 matched pairs of subjects are as follows:

Pair	EXPERIMENTAL GROUP (NEW METHOD), X_1	CONTROL GROUP (OLD METHOD), X_2	DIFFERENCE, $X_1 - X_2$	RANK OF DIFFERENCE	SIGNED RANK	RANKS WITH LESS FREQUENT SIGN
1	85	86	−1	1	−1	—
2	90	95	−5	6	−6	—
3	92	96	−4	5	−5	—
4	93	93	0 ←	(Dropped)		—
5	93	95	−2	2.5	−2.5	—
6	94	96	−2	2.5	−2.5	—
7	95	98	−3	4	−4	—

Pair	EXPERIMENTAL GROUP (NEW METHOD), X_1	CONTROL GROUP (OLD METHOD), X_2	DIFFERENCE, $X_1 - X_2$	RANK OF DIFFERENCE	SIGNED RANK	RANKS WITH LESS FREQUENT SIGN
8	95	101	−6	7	−7	—
9	140	133	+7	8	+8	8
10	150	135	+15	9	+9	9
						$T = 17$

$T_{.05(9)} = 6$

$T = 17$ Accept H_0; not significant.

Important Note: With the Wilcoxon T, the null hypothesis is rejected *only* when the calculated T is equal to or *less than* the tabled value of T.

1. *Obtain the differences.* We set up the difference column, $X_1 - X_2$, being careful to retain the correct sign.
2. *Rank the differences.* We rank order the absolute values of the differences. In this step, the sign of the difference is irrelevant. The difference value of −1 receives a rank of 1, or first place, not because it was negative, but because it was the *smallest* difference. Note also that two of the differences (−2) are tied for second and third place. As for all conversions to ordinal ranks, we add the tied ranks (2 + 3), divide by the number of ranks tied in that position [(2 + 3)/2 = 2.5], and assign each the resulting average rank. Finally, whenever there is a zero difference between a pair of scores, as in the case of Pair 4 (where each subject scored a 93), the scores for these subjects are *dropped* from the analysis.
3. *Sign the ranks.* In this step, we simply affix the sign of the difference to the rank for that difference. Thus, the ranked differences appear in a separate column, but now have whichever sign appears in the preceding difference column. Thus the difference of −1 (for Pair 1), ranked first, and the rank now gets a negative sign, because the value in the difference column is negative. Similarly, the largest difference, which ranked ninth, is obtained from Pair 10 where the difference value is positive.
4. *Add the less frequent signed ranks.* Finally, we determine which sign, plus or minus, occurs less frequently among the ranks. The *plus* sign occurs less often (only twice, compared to seven minus signs). Then, to obtain the value of T, we merely add the ranks having the less frequent sign; $T = 17$.
5. *Check for significance.* We compare the calculated value of T with the critical table value of T with $N = 9$. (See Table J.)

$T_{.05(9)} = 6$

$T = 17$ Accept H_0; not significant.

(Note that the null would be accepted even if this had been a one-tail, unidirectional test of significance. N stands for the number of signed ranks; although we started with 10 pairs of scores, we lost pair 1 because of its zero difference.)

Now, *unlike any other test* in this book, with the Wilcoxon T test the null hypothesis is rejected only when the calculated value of T is *equal to or less than the table value.* For T, smaller means more significant.

We thus conclude that there is no significant difference between our two golfing groups. The independent variable (teaching method) had no effect on golfing performance. The two groups, originally selected from a single population, still represent that same population.

Sample Size and the Wilcoxon **T.** The procedure just outlined for the Wilcoxon T is appropriate for use in many common research situations, where the sample sizes range from 6 to 25 pairs of scores. When more than 25 paired ranks are available, however, the distribution of Wilcoxon T values approaches normality. Then the following equation must be used:

$$z_T = \frac{T - [N(N + 1)]/4}{\sqrt{N(N + 1)(2N + 1)]/24}}$$

The resulting z_T value is compared with a critical z of ± 1.96 for the .05 alpha error level or ± 2.58 for the .01 alpha error level. The null hypothesis is rejected if the obtained value of z_T equals or exceeds one of these z values. This is exactly the same as the procedure described earlier for the Mann–Whitney U test with z_U.

THE FRIEDMAN ANOVA BY RANKS FOR THREE OR MORE ORDINAL DISTRIBUTIONS WITH CORRELATED SELECTION

When either matched-subjects or repeated-measure designs are used, and the hypothesis of differences is to be tested on three or more ordinal distributions, the appropriate test is the **Friedman ANOVA by ranks.** This is analogous to the within-subjects F when nonskewed distributions of interval data are involved.

In the example demonstrating the Mann–Whitney U test, the results informed us that Republicans do earn more money than Democrats. The interpretation of the results was unclear, however, because so many other variables were left uncontrolled, not the least of which was age. Now any age distribution is certainly composed of at least interval data, but like income distributions, it has to be badly skewed in the population. (There are far fewer 90-year-olds than there are 2-year-olds.) Therefore, whenever age or income is a dependent variable, we must consider converting to ordinal ranks before testing for significance.

Calculating the Friedman ANOVA by Ranks

Example. A researcher is interested in whether or not there is an age difference among Independent, Democratic, or Republican voters. However, because older persons are apt to be earning more money, perhaps the key to the age-party affiliation relationship is economic. Perhaps richer persons are more apt to be Republicans, regardless of their age. To test this, random samples of 10 Independents, 10 Democrats, and 10 Republicans were selected and *matched on income.* That is, trios made up of 1 Independent, 1 Democrat, and 1 Republican, all earning roughly the same yearly income, are put together. They are then checked for age.

The data are as follows:

TRIAD	INDEPENDENT, X_1	DEMOCRAT, X_2	REPUBLICAN, X_3	INDEPENDENT, R_1	DEMOCRAT, R_2	REPUBLICAN, R_3
1	26	30	22	2	1	3
2	29	31	28	2	1	3
3	55	60	54	2	1	3
4	27	26	24	1	2	3
5	70	69	74	2	3	1
6	21	23	32	3	2	1
7	33	35	34	3	1	2
8	40	39	38	1	2	3
9	41	42	43	3	2	1
10	45	44	46	2	3	1
				$\Sigma R_1 = 21$	$\Sigma R_2 = 18$	$\Sigma R_3 = 21$

To solve the equation for the Friedman ANOVA, we need the values for ΣR, N (the number of rows, that is, of matched groups of subjects), and k (the number of columns of ranked scores). We go through the following steps.

1. In each row of interval scores, that is, each triad, rank order the scores from high to low. Thus, in the first row (Triad 1), the age score of 30 is ranked 1, the age score of 26 is ranked 2, and the age score of 22 is ranked 3. (There are only 3 scores per row here, so the ranks must run 1, 2, 3 in each and every row.)

2. Sum the columns of ranked scores to obtain

$$\Sigma R_1 = 21 \qquad \Sigma R_2 = 18 \qquad \Sigma R_3 = 21$$

3. Plug in the values for $\Sigma R_1, \Sigma R_2,$ and ΣR_3, along with the values for N (the number of rows), that is, 10, and for k (the number of rank columns), that is, 3, into the equation and solve for χ_r^2.

$$\chi_r^2 = \frac{12}{Nk(k+1)} (\Sigma R_1^2 + \Sigma R_2^2 + \Sigma R_3^2) - 3N(k+1)$$

$$\chi_r^2 = \frac{12}{(10)(3)(3+1)} [(21)^2 + (18)^2 + (21)^2] - 3(10)(3+1)$$

$$\chi_r^2 = \frac{12}{120} (441 + 324 + 441) - 120$$

$$\chi_r^2 = (.10)(1206) - 120 = 120.60 - 120 = .60$$

4. We check the calculated value of χ_r^2 against the table value (Table I) for df $= k -$ 1 $= 2$. If the obtained value is equal to or greater than the critical value, we reject the null hypothesis.

$$\chi_{.05(2)}^2 = 5.99$$

$\quad \chi_r^2 = .60$ Accept H_0; not significant.

Here, the null hypothesis (H_0: $R_1 = R_2 = R_3$) is accepted. The three distributions of age scores represent a single population. Since the subjects were matched on income, this means that when income is not a factor, there are no age differences among persons of different political affiliations. This is another example of post-facto research, so even if the age differences had been found to be significant, only rather tentative conclusions could be drawn.

Sample Size and the Friedman ANOVA.   To use the chi square table for assessing the significance of χ_r^2, we must have a minimum of 10 scores per column when there are 3 columns of ranked scores. With 4 columns of ranked scores, only 5 scores per column are necessary. For smaller sample sizes, such as for an N of from 2 to 9 with a k of 3, or an N of from 2 to 4 with a k of 4, special tables are available.[3]

ADVANTAGES AND DISADVANTAGES OF NONPARAMETRIC TESTS

None of the tests of ordinal data makes any assumptions regarding the parameters of the population. For this reason, they are called nonparametric tests. Neither do these tests make any assumptions regarding the shape of the underlying population distribution. The population distributions can be skewed right, skewed left, or even be bimodal, and these tests may still be used. For this reason, they are also called distribution free. Since they seem to be so safe (it is hard to violate assumptions if there are none), why not always use them instead of t and F? The answer is because they are less powerful.

Power is equal to $1 - \beta$ (one minus the beta error). Whenever beta error is increased, power is reduced. (Beta error is the probability of being wrong when accepting the null hypothesis, that is, accepting H_0 when it should have been rejected.) Nonparametric tests all tend to increase the beta error. That is, the nonparametrics are less sensitive to smaller differences, less able to detect that these differences might be significant. For equal sample sizes, a given difference between groups that proves to be significant with a t test might not be significant with a Mann–Whitney U test.

Thus, whenever the assumptions of the parametric tests can be reasonably met, we should use them! When the data are in ordinal form, or when the distribution of interval scores is obviously skewed, we should use the nonparametric tests. As someone has said, good alternative hypotheses are sometimes hard to come by and should not be needlessly thrown away. The nonparametrics may be quick and easy to calculate, but the time saved does not equal the lost significance.

SUMMARY

Tests of significance for ordinal data are those which analyze measurements that are in rank-order form, that is, measures that provide information regarding greater than or less than status, but *not* how much greater or how much less. Since tests of ordinal data need not predict the population parameter, μ, they are collectively known as nonparametric tests. Although ordinal data may be tested for correlation (see Chapter 10), the focus of this chapter was on testing ordinal data to assess differences.

1. When two independently selected sets of ordinal scores are to be tested for differences, use the Mann–Whitney U test. The U test is the ordinal equivalent of the independent t test for interval data.
2. When three or more independently selected sets of ordinal scores are to be tested for differences, use the Kruskal-Wallis H test. The H test is the ordinal equivalent of the one-way ANOVA for interval data.
3. When two sets of ordinal scores are to be tested for differences, and the samples are correlated, use the Wilcoxon T test. This T is the ordinal analog of the paired t test for interval data.
4. When three or more sets of ordinal scores are to be tested for differences, and the samples are correlated, use the Friedman ANOVA by ranks. The Friedman ANOVA by ranks is the ordinal equivalent of the within-subjects F ratio for interval data.

These tests, all basically designed for ordinal data, can also be used on interval data when the underlying distributions are known to deviate significantly from normality, or there are large differences in variability among the sample groups. To do this, the interval scores must first be converted to ordinal ranks.

KEY TERMS

Friedman ANOVA by ranks   nonparametric
Kruskal–Wallis H test   Wilcoxon T test
Mann–Whitney U test

REFERENCES

1. SIEGEL, S. (1988). *Nonparametric statistics.* (2nd ed.).   2. Ibid.
   New York: McGraw-Hill.   3. Ibid.

PROBLEMS

1. Calculate a Mann–Whitney U for the following data obtained from two independent sets of ordinal measures.

R_1	R_2
1	2
3	5
4	8
6	11
7	14
9	16
10	18
12	17
13	15

2. Calculate a Kruskal–Wallis H for the following data obtained from three independent sets of ordinal measures.

R_1	R_2	R_3
1	2	4
3	6	8
5	10	11
7	13	17
9	15	21
12	16	20
14	18	19
22	23	24

3. Calculate a Wilcoxon T for the following data obtained from two correlated and skewed distributions of interval measures.

X_1	X_2
50	52
40	48
55	60
48	52
42	40
45	51
51	82
70	55

4. Calculate a Friedman ANOVA by ranks for the following data obtained from three correlated and skewed distributions of interval scores.

X_1	X_2	X_3
15	17	21
14	16	25
16	18	15
15	17	19
15	18	20
16	18	19
12	14	20
14	21	20
16	35	21
31	11	57

5. A researcher is interested in establishing whether or not the type of background music causes a difference in how ice-skating performances are judged. A random sample of 24 skaters is selected from among the students of a large skating club. The skaters are randomly assigned to one of three conditions: A, background music—jazz; B, background music—classical; C, no background music. Each skater performs for 10 minutes. All are rank ordered by judges as to their performance. (Each skater receives the median rank assigned by three judges.)

CONDITION A, R_1	CONDITION B, R_2	CONDITION C, R_3
3	1	8
4	2	15
6	5	18
9	7	21
11	10	20
13	12	19
16	14	17
22	23	24

Test the hypothesis.

6. Hypothesis: There is a significant difference in the Breathalyzer readings from suspected drunken drivers as a function of the time between apprehension and the onset of the testing procedure. A random sample of 10 suspected drunken drivers are tested on the Breathalyzer under three conditions: first, 20 minutes after their arrival at the police station (suspects must be observed for 20 minutes to verify that they do not place anything, solid or liquid, in their mouths before testing begins); second, after 1 hour and 20 minutes; third, after 2 hours and 20 minutes. (Scores on the Breathalyzer range from 0 to .40 and indicate the percentage of alcohol in the bloodstream. Thus, these scores are

in interval form. Since a score of .35 indicates a comatose condition, few subjects score that high. Thus, the distribution is typically skewed to the right.)

SUBJECT	FIRST CONDITION	SECOND CONDITION	THIRD CONDITION
A	.12	.11	.10
B	.17	.15	.14
C	.12	.11	.10
D	.11	.10	.09
E	.15	.14	.13
F	.12	.11	.10
G	.13	.12	.11
H	.20	.19	.16
I	.30	.28	.25
J	.15	.14	.13

Test the hypothesis.

7. A researcher is interested in discovering if competency test scores for high school teachers can be increased by the introduction of a three-day inservice workshop in the teacher's own specialty area. A random sample of 10 high school teachers is selected. They are given the competency test, sent to the workshop, and then given the competency test again. A nonparametric statistical analysis of the data is undertaken as it became obvious that the distribution of test scores is severely skewed to the left.

TEACHER	BEFORE SCORES	AFTER SCORES
1	84	87
2	88	89
3	85	83
4	92	96
5	90	92
6	88	90
7	78	75
8	41	40
9	87	90
10	31	29

Test the hypothesis.

8. Hypothesis: Students are more likely to show verbal aggressiveness (assertively challenging the professor, etc.) in small classes than they are in large classes. Random samples of 10 large classes (50 students or more) and 10 small classes (less than 25 students) are selected. A three-judge panel visits each classroom on five separate occasions and then rank orders all classes as to the extent of student assertiveness.

SMALL CLASS	LARGE CLASS
2	6
7	13
11	5
3	9
10	18
1	17
4	20
8	19
12	16
14	15

Test the hypothesis.

For the research situations described in problems 9 through 12, indicate which statistical test would be most appropriate.

9. Data: ordinal
   Design: two groups, after-only
10. Data: skewed, interval
   Design: one group, before-after
11. Data: skewed interval
   Design: four groups of matched subjects
12. Data: ordinal
   Design: three groups, after-only

True or False—Indicate either T or F for problems 13 through 17.

13. The Mann–Whitney U test is less powerful than the independent t test.
14. The ordinal equivalent of the paired t test is the Wilcoxon T test.
15. With equal sample sizes, nonparametric tests are just as powerful as parametric tests.
16. The larger the numerical value of the Wilcoxon T, the higher is the likelihood of achieving significance.
17. Nonparametric tests can never be used on interval data, no matter what the shape of the underlying distribution.

Fill in the blanks in problems 18 through 20.

18. With ordinal data, the nonparametric counterpart of the independent t test is the _____.
19. A less powerful statistical test is one in which it is less likely that the _____ hypothesis will be rejected.
20. With ordinal data, the nonparametric counterpart of the within-subjects F ratio is the _____.

chapter 17

Computers and Statistical Analysis

Now, during the last decade of the twentieth century, the computer revolution (at least as important as the Industrial Revolution) is fully upon us. Computers and their "magic chips" now do everything from making out the payroll and scanning income tax returns to running cars, watches, and cameras. Are computers affecting our lives? Absolutely! Every time we use a credit card, make an airplane or hotel reservation, go to a bank—the list is endless—a computer is hiding somewhere in the wings, crunching out the data *and keeping track of you.* Sometimes the computer can even become an intractable big brother, who won't take no for an answer. In one celebrated case, a man kept getting a computerized bill for $0.00. Phone calls, visits to the department store, letters, whatever, failed to stop this electronic harassment. Only when the man capitulated, and sent in a check for $0.00 did the computer stop its cunning dunning. And just as there has been a tremendous increase in the use of computers, there has also been a dramatic drop in the cost of computers over the past few years. In fact, it has been estimated that had automobile technology kept pace with computer technology over the past 35 years, a Rolls-Royce would now cost less than three dollars, and it would get about 2 million miles per gallon of gas. It's little wonder that the demands of society now virtually insist that every educated person become to some degree computer literate, and perhaps nowhere is this message more compelling than in the field of statistical analysis. In order to organize and understand large groups of data, or even to run sophisticated statistical tests of significance on small-sample data bases, the computer has fast become the statistician's best friend.

If you, like so many others, are approaching the computer for the very first time, fear not! You don't have to be a rocket scientist to get the computer up and running, and if it is your first time you'll probably be amazed at how little effort it will take to get the computer to give you an instant standard deviation or t ratio. Think of it more as a calculator with a typewriter-like keyboard than as an imposing, science-fiction creation, laden with cryptic dials, blinking lights, and arcane operations.

In the late nineteenth century Sir Francis Galton argued plaintively that the field of statistics is one where scientists have to spend enormous amounts of time creating their own subject matter before beginning the arduous task of analysis. According to Galton, "the work of statisticians is that of the Israelites in Egypt—they must not only make the bricks but find the materials."[1] One can only guess as to what analytic strides Galton and his friend, Karl Pearson, would have made had they been able to avail themselves of even the most inexpensive of today's microcomputers. As you will see, you don't have to be a computer expert to get results from a computer. You should, however, have at least a modicum of "computer literacy."

COMPUTER LITERACY

No doubt about it, "computer literacy" has become a catchword phrase of the 1990s. Even some first-grade children are being introduced to the computer and are being taught elementary programming skills. To argue against computers today is viewed as akin to seeing the first automobiles roll down the road and saying "I'd rather have a horse," or "the only accelerator I'll ever need is a buggy whip."

But what about this phrase "computer literacy"? How literate must one become to take advantage of today's technology? A few short years ago computer literacy was thought by many as being virtually synonymous with achieving an advanced degree in computer science. Students, it was thought, should all have courses in everything from machine language to FORTRAN, from computer architecture to advanced Pascal, and much, much more. Mercifully, that view is now held only by a minority, and computer literacy has been downsized to far more modest proportions. Programming skills are no longer seen as a necessary component. Computer literacy is now accepted as being more like automobile literacy: you need to learn only how to drive it, not all the electronics and physics involved in its operation. Also, today there are even programs, such as QUICK-PRO, which have been designed to write programs for you.[2] Instead of learning the programming language directly, you simply answer a series of easy questions that appear on your screen, and the program then writes the new program.

Taking a course or two in BASIC may be like taking a couple of French courses. Although it might be fun to go to a French restaurant and order your meal directly from the menu, if you really need a highly accurate translation of a technical journal article, you'd better hire an expert. Besides, in programming, a little learning could be dangerous. If you start trying to break into a prepackaged program to change a line or two, you might just ruin part of the program. One enterprising young student broke into a "canned" statistical package in order to "make it better." Inadvertently, however, the student inserted an endless loop, and the program could no longer get past this spot. (A simple example of an endless loop is where the program at A says GOTO B, and at B the program says GOTO A.) Imagine how expensive this might be if you were using this program on a time-sharing mainframe. Even learning binary math is no longer seen as essential. Spending a lot of time polishing up your ability to add, subtract, multiply, and divide in base two may become a hip dinnertime conversation piece, but may not be that much help in conquering a new word-processing, spreadsheet, or statistical program. The machine may have to "think" in on-off terms, but you don't. What, then, should you have to learn in order to become computer literate in the 1990s? Minimal computer literacy should probably include the following:

1. Knowing a little about the history of computers, the sequence from the giant, vacuum-tubed monsters of yesterday to the miniaturized microcomputers of today.
2. Knowing the basic parts of the computer: from input, to storage, to output.
3. For the microcomputer, knowing how to handle, format, and copy disks as well as how to protect them from damage.
4. Knowing at least one word-processing program (you'll wonder how you lived without it), a spreadsheet program such as LOTUS, and a math or statistical package. (As a bonus, research strongly suggests that the use of these programs will enhance your writing and math skills.[3])
5. Knowing how to handle the computer safely and intelligently, and as much as possible, learning the basic damage-prevention techniques.

THE STATISTICAL PROGRAMS

There are a myriad of prepackaged statistical programs available today, especially for the microcomputer. Some micro programs, can be purchased for as little as $25, such as Epistat,[4] while others, such as the complete SPSS/PC+ (Statistical Analysis Package for the Social Scientist) micro version, cost considerably more.[5] Still others, such as our own PH-STAT, are given away as part of the price of the textbook.

The big, traditional programs, such as SPSS, SAS (Statistical Analysis System), BMDP (Bio Medical Computer Program), and MINITAB, are also available in mainframe (or minicomputer) form, and one of these programs may already be available to you in your college's computer lab.

Who's in Charge?

The most powerful statistical programs, SPSS, SAS, and BMDP, can handle virtually any job the professional statistician will ever be called on to perform. Hundreds of variables and many thousands of scores can be analyzed in what seem (from the manuals) to be almost endless ways. These programs also, however, can be used to generate the simplest of calculations, means, medians, and modes. Mid-sized programs, such as MINITAB, although not as professionally complete as SPSS, are, to be sure, easier to learn. In fact, MINITAB uses a straightforward, spreadsheet format and is fully interactive with the user.[6] A small program, such as our own PH-STAT, is one that is completely menu driven and can be learned by the average student in less than an hour. PH-STAT may be small, compared with SPSS, but it will do most of the jobs the student statistician will ever be asked to do, and includes the descriptive techniques, means, medians, modes, and z scores, as well as the most popular inferential techniques, such as t tests (independent and paired), correlation and regression, ANOVAS (including factorial and within-subjects), and a host of nonparametrics, including chi square.

Despite the power of these statistical packages, you must constantly keep in mind that you, not the computer, are the one who's in full charge of the analysis. The choice of which statistical test to perform is in your hands. You're the one who must remember that chi square demands nominal data, that the Freidman ANOVA asks for ordinal data, and that t, F, and r must be fed at least interval data. You're the one, also, who must remember how these values may be interpreted and which statistical tests are appropriate to which research designs. In short, the computer will be your lightning-fast slave who will do (only) exactly what it is told to do. And if you tell it to do dumb things, it will surely provide you with quick, but incredibly dumb, solutions. You simply can't blindly trust any result the "genius computer" spits out. Here is a sampling of the kinds of thorny situations created by the user, not the computer—for in each case the computer did exactly what it was told to do.

1. A student decided to do a correlational study using a large number of variables and a large sample size. Among the variables chosen was GPA, and among the subject descriptions was each subject's social security number. Later, the printout revealed a Pearson r (or as SPSS called it, a PEARSON CORR) of $-.016$ between GPA and social security number, a difficult value to interpret. The GPA may be an interval measure, but the social security number surely is not (and SPSS didn't care). The computer was simply following orders. Most of the statistical programs are totally insensitive to errors in the scale of measurement, and will quickly crunch the data and spew out uninterpretable values if you feed in uninterpretable data. It would be like asking for a measure of skewness on ordinal measures. The computer will oblige, but whatever does it mean?

2. A fledgling researcher decided to assign subjects to one of two groups according to gender. Each subject was coded on gender, a "1" representing females and a "2" representing males. The program was then asked to T-TEST, and, not surprisingly, the two groups were found to differ significantly on gender—another embarrassingly difficult result around which to build a research paper.

3. In another example, a student familiar with SPSS and the T-TEST, but not with statistical logic, fed the computer two sets of scores for each of over 50 subjects. The first distribution listed heights in inches, and the second, weights in pounds. Since the t ratio was obviously significant, the only possible conclusion was that the subjects were heavier than they were tall. The program, of course, assumed that both sets of measures were of the same trait.

4. Finally, it is absolutely crucial that you read the manual carefully before using any of these popular statistical packages. Even for those familiar with statistical assumptions it is important to find out whether a given program's version of a particular analysis conforms to the user's expectations. For example, in the SAS version of ANOVA, the procedure does not handle repeated measure designs, or samples with unequal n's. SAS does include this type of analysis, but only under "GLM," which, incidentally, can also handle covariance, multiple regression, and partial correlations.[7]

Computer Overkill

Sometimes the novice researcher, flushed with the excitement of having so many computer-generated techniques to choose from, simply enters the data and tries them all. In one case an SPSS cross tab was created which included the following information on each subject: age, class, sex, parental income, student's summer earnings, major, and several more. Now, CROSSTABS can certainly produce the table and, in this case, subtables within tables, but getting a table of age by class by sex by major by GPA and so on would not only be enormous in size, but, even worse, literally impossible to understand.

Also, just because a given program can theoretically handle data from extremely complicated designs does not necessarily mean that the results are always readily open to interpretation. For example, BMDP can easily analyze ANOVA designs with 10 independent variables, but making sense out of the many interactions is at best harrowing to contemplate.

In designing experiments, it is well to be reminded of the acronym KISS: keep it simple, stupid. The research design should always be simple enough to generate results that have the potential for answering the research question. The computer, alas, can perform far more calculations than are always logically possible to interpret. Therefore, try to avoid that dreaded disease called "conspicuous computing," which is an obsessive and irrational lust for overusing the computer in order to generate an aura of scientific sophistication.

Bugs, Glitches, and Things That Go Bump in the Night

Despite their elegant sophistication, any of the statistical packages have the potential for producing bizarre and even scary results. In fact, under some conditions, you may get no results at all. Until your data have been saved, they simply reside in the volatile world of RAM (the computer's internal memory, called random access memory). If, for example, there were suddenly a power failure, all the RAM data would disappear and be gone forever. This is an especially traumatic event if you have already spent hours on the data entry. A prudent rule to follow is to be disciplined enough to stop about every 15 minutes and save the files. Remember, when the power goes out, the lights will come back, but your data won't. As you will quickly learn, things can and do go wrong. Entry errors are probably the number one culprit, but other dangers are definitely lurking in the background. A speck of dust, especially on a hard drive, can even cause a total crash of the system and leave you looking around for a hand calculator or perhaps even an abacus. Particles of dust or smoke can also cause extremely subtle problems either to the stored program or to your data files. Even the computer itself may have a hardware problem that affects only part of the program being run. These are the most insidious problems, because most of the program will still run flawlessly, but one or two sections have been damaged and are suddenly producing wildly inaccurate results. The worst part of this tale is that since most of the program still yields results that check out perfectly, you, of course, assume that the entire program is still in working order.

Even with hardware in perfect condition, bugs can remain in programs for months or even years. Hidden programming errors often don't announce themselves, until you, the user, do something that the program doesn't expect. A fairly simple case would be when you enter a letter where the program has to have a number, or you mistakenly put a question mark before the value. Rather than aborting, or telling you that you've made an error, the program may go into convulsions and provide you with some rather weird answers. As Covvey and McAlister have said:

> Most completed software, then, harbors error. In programs of any complexity, it is literally impossible to ever eliminate all bugs. An operating system, for example, will always contain several serious errors when first released. Efforts to correct bugs in systems programming may introduce new bugs. Only as users subject the system to an enormous variety of uses will some of the bugs emerge from the woodwork.[8]

In fact, one of the authors of the quote above created his own statistical package, a program that seemed to work so well that he passed it on to his colleagues in the department. About a year later, one of his colleagues detected a serious error in the independent t test, and they all then realized that for over a year they had been relying on a faulty piece of software. Some of the t ratios, alas, may have even been published. Journal editors, after all, don't have time to check calculations, even when they have the raw data. Therefore, the prudent user of any program should scan the results and do at least some spotchecking before rushing the results to some journal.

None of this is intended to make you overly cynical, just cautious. The major statistical programs have been thoroughly tested. In fact, in assessing the percentage of time spent on program development, it has been estimated that about 11% is spent in the design phase, 7% in converting the design into a computer language, and 67% in the testing of the program.[9] Nevertheless, although you shouldn't be overly cynical, you should adopt a healthy skepticism, and at least run your results through some logical checkpoints.

Logic Checkpoints

There are a series of logic checks that you can implement when reading your printouts that should at least alert you to the possibility of some of the most flagrant types of errors. Following is a list of 15 computer printouts, all of which have in some way been beset with flawed results. In each case, this powerful, beautiful, technologically sophisticated machine has acted like a fast idiot, and your job as the smart leader is to ferret out its mistakes. At the end of this section there will be an analysis of each error and a citation as to where in the text you may find the appropriate logic check.

1. The minimum value is 2.000
   The maximum value is 10.000
   The range is 8.000
   The mean is 6.000
   The true standard deviation of this distribution is 12.828
2. Please enter lower z score: .... -1.45
   Please enter upper z score: .... -1.05
   The area is 78.1913% or a probability of 0.7819
3. Two-Sample Independent t Test
   The standard deviation of the first set is 2.0548
   and its standard error of the mean is 3.9189
   The standard deviation of the second set is 2.4095
   and its standard error of the mean is 1.0775
   The value of the t statistic is .... -3.884
   and the standard error of difference is 1.4162
   Type return to continue
4. Pearson Correlation
   Enter two numbers per line
      1: 9   10
      2: 2   5
      3: 12 15

      .
      .
      .

    10: 4   6
   The Pearson r correlation is 2.456
5. One-Way Analysis of Variance

SOURCE OF VARIATION	DEGREES OF FREEDOM	SUM OF SQUARES	MEAN SQUARES	F
Between (treatment)	3	-147.000	49.000	42.000
Within (error)	12	14.000	1.167	
Total	15	161.000		

6. Two-Way Analysis of Variance

SOURCE OF VARIATION	DEGREES OF FREEDOM	SUM OF SQUARES	MEAN SQUARES	F
Rows	1	344.450	344.450	120.860
Columns	1	54.450	54.450	19.105
Interaction	5	2.450	2.450	0.860
Within	16	45.600	2.850	

7. Chi Square Tests
   Enter the observed frequency first, then the expected frequency
   1: 40 25
   2: 30 25
   3: 20 25
   4: 10 25
   The value of chi square is −20.000

8. The minimum value is   200.00
   The maximum value is   800.000
   The range is   600.000
   The mean is   500.000
   The true standard deviation of this distribution is   100.000
   The distribution is severely leptokurtic

9. The minimum value is   55.000
   The maximum value is   145.000
   The range is   90.000
   The mean is   100.000
   The median is   115.000
   The true standard deviation of this distribution is   15.000
   The distribution is normal

10. Two-Sample Independent t Test
    The standard deviation of the first set is   14.970
    and its standard error of the mean is   3.469
    The standard deviation of the second set is   16.7245
    and its standard error of the mean is   4.001
    the value of the t statistic is   2.967
    and the standard error of difference is   −1.4162
    Please type return to continue

11. Calculating a Regression Line and the Pearson r
    The regression equation is

$$Y = 0.921X + 1.421$$

    The correlation is −0.861

12. Within-Subjects F

SOURCE OF VARIATION	DEGREES OF FREEDOM	SUM OF SQUARES	MEAN SQUARES	F
Between columns	0	10.00	5.000	10.000
Residual	8	4.00	.500	

13. Two-Way Analysis of Variance

SOURCE OF VARIATION	DEGREES OF FREEDOM	SUM OF SQUARES	MEAN SQUARES	F
Rows	2	0.620	0.310	0.139
Columns	2	229.672	114.836	25.706
Interaction	2	2.602	1.301	0.291
Within	54	241.235	4.467	

14. Chi Square Test
---entering the first row---
Enter the two values : 7   10
---entering the second row---
Enter the two values : 15   8
The value of chi square is   2.2827
The value of chi square modified with the Yates correction factor is   4.0667
15. One-Sample t Test
Mean and standard deviation calculated
from data
Please enter the hypothesized mean   : 80.000
The sample mean is   : 80.000
The estimated population standard deviation is   :   8.489
The t statistic is   : 18.602

ANSWERS

1. The true standard deviation of 12.828 is too high. The true standard deviation of any set of sample scores may not be greater than half the range, or in this case 4.000. (See page 51.)
2. The area of 78.1913% is too large. Since both of the z scores fall on the same side of the mean, the area between them cannot be greater than 50% and the probability may not be greater than .50. (See page 67.)
3. The standard error of the mean for the first set of scores is too large. The value of the standard error of the mean cannot be greater than the standard deviation of the sample scores. (See page 143.)
4. The Pearson r value of 2.456 exceeds its limits. The Pearson r may never exceed 1.00. (See page 197.)

5. The one-way analysis of variance is producing a negative sum of squares of − 147.000. Since the sum of squares is a measure of variability, it may never be negative. (See page 262.)

6. The two-way analysis of variance is indicating an error among the degrees of freedom. If the degrees of freedom for the rows and columns are correct, then the interaction must equal 1, not 5. (See page 280.)

7. Without running a calculation check on the value of chi square, you still can tell that the answer of − 20.00 cannot be correct. No chi square value may ever be negative. (See page 290.)

8. If the standard deviation of this distribution is correct, then the distribution cannot be severely leptokurtic. If, however, the distribution is leptokurtic, the computer-generated standard deviation is too high. (See page 53.)

9. If the distribution is normal, the computer values for the mean and/or median are incorrect. If, however, the mean and median values are accurate, the distribution is not normal. (See page 58.)

10. The standard error of difference, since it estimates variability, can never be negative. The computer solution of − 1.4162 must be incorrect. (See pages 167, 168.)

11. Since the regression equation is indicating a positive slope of .921, the correlation of − .861 cannot be correct. When the slope is positive, the correlation must also be positive. If the correlation in this printout is correct, the slope value must be wrong. (See page 315.)

12. The within-subjects F is showing 0 degrees of freedom between columns. This must be an incorrect value, since an ANOVA of this type is based on repeated measures, and the measures should be set up in the columns. (See page 361.)

13. For this two-way analysis of variance, the degrees of freedom are being incorrectly allocated. The interaction df of 2 must be in error if the row and column df's are correct. Or, perhaps the df's are incorrect for either the rows, the columns, or both. (see page 280.)

14. For any 2 × 2 chi square, the value for the Yates correction may never be greater than the value of the chi square itself. Thus, either the chi square value is too high, or the Yates correction is too high. (See page 296.)

15. If the sample and hypothesized means have been correctly entered, then the t ratio is too high. Since there is no difference between the means, the t ratio itself must be equal to zero. Or, if the t ratio is correct, one or both of the means must be in error. (See page 172.)

REFERENCES

1. KEVLES, D. J. Annals of eugenics (part 1). *The New Yorker.* Oct. 8, 1984, p. 60.

2. Quick Pro+II (1987). *Automatic Program Writer.* Orange Park, FL: ICR Future Soft.

3. BRACEY, G. Computers in education: What the research shows. *Electronic Learning.* November/December, 1982, pp. 51–54.

4. Epistat (1984). *Statistical Program.* Round Rock, TX: Epistat Services.

5. NORUSIS, M. J. (1988). *SPSS/PC+ Studentware.* Chicago: SPSS Inc.

6. RYAN, T. A., JOINER, B. L., & RYAN, B. F. (1976). *Minitab Student Handbook.* North Scituate, MA: Duxbury Press.

7. SAS Institute, Inc. (1982). *SAS Introductory Guide.* Cary, NC: SAS Institute Inc.

8. COVVEY, H. D. & McALISTER, N. H. (1982). *Computer choices.* Reading, MA: Addison Wesley, p. 90.

9. Ibid.

chapter 18

Research Simulations: Choosing the Correct Statistical Test

In this final chapter, all the various statistical tests, both parametric and nonparametric, for all the measurement scales, nominal, ordinal, interval, will be matched with a series of simulated research situations. Since all the interval data statistical tests—t, F, r, and so on—can also handle ratio data, we will not make the interval-ratio distinction throughout this chapter. For the purist, the phrase "at least interval data" may be inserted wherever interval data are mentioned. The ability to calculate the tests is important, but the more meaningful competency consists of knowing *when* to calculate which one. This decision making can be greatly simplified by using the checklist and charts included in this chapter. With these aids, the decision as to which test to use becomes totally automatic. When confronted with any problem, the only way we can get the right answer is to ask the right question. The specific, checklist questions, when asked and answered in sequence, serve to narrow the range of appropriate statistical tests until there is one—one best test for the particular research situation. A flow chart is presented for each form of data, and as you answer the questions and follow the various routes offered on the chart, the eventual solution should, it is hoped, become clear.

METHODOLOGY: RESEARCH'S BOTTOM LINE

Before discussing the three specific questions, we must consider the one overriding factor that literally transcends all the others. That one is the bottom-line question as to what type of research is involved—experimental or post-facto. Although the answer to this question does not affect which statistical test to apply, it does dictate the type of conclusions that can validly be drawn from the results of the study.

We should not be like the researcher who, so the story goes, trained groups of grasshoppers to respond on command. When the researcher yelled "Jump," they all obediently jumped. Then, each day the researcher pulled out one leg from each of the grasshoppers, again cried "Jump," and measured how high they leaped. After six sequential trials, when the grasshoppers had all been totally delegged, the researcher found that, despite the strident command, the grasshoppers would no longer jump. A series of fancy, repeated-measure F ratios was calculated, and all were found to be highly significant. The researcher's conclusion? When one pulls out the legs of a grasshopper, it becomes deaf. Selecting the appropriate statistical test and then drawing erroneous conclusions is like winning the daily double at the races and then having your pocket picked on the way home—an empty victory, indeed.

Drawing logically derived conclusions does not depend on which statistical test has been used, but rather on the type of research methodology—experimental or post-facto. And this categorization depends solely on the independent variable. When the independent variable is manipulated, the research is experimental. When the independent variable is an assigned subject characteristic, the research is post-facto.

There are two types of post-facto research. One type tests the hypothesis of difference—groups already assumed to differ on some important variable(s) are compared in order to discover if they also differ systematically on other variables. Another type of post-facto research tests the hypothesis of association by attempting to find out if a correlation exists between certain variables. As we have learned, correlation does not prove causation, but *neither does it rule it out.* A discovered correlation may later be shown, through an experimental method, to be involved in causation. For example, there is obviously a high, positive correlation between ambient temperature and the amount a person sweats. However, other important factors are also involved, such as the amount of humidity, the person's body size, how active the person is, and so on. An experiment can be set up in which these other factors (humidity, amount of activity) are held constant (not allowed to be variables), and heat and *only heat* is manipulated as the independent variable. If the subjects now exhibit significantly different sweat levels as a function of changes in the heat level (*manipulated* independent variable), then it is probable that a cause-and-effect relationship has been uncovered. Thus, post-facto research, although designed to lead to better than chance predictions, may provide important clues that might later be used to home in on causation.

THE CHECKLIST QUESTIONS

Question 1: What Scale of Measurement Has Been Used? The choice of the correct statistical test depends heavily on which scale of measurement the data represent. When the data are in nominal form, that is, presented as frequencies of occurrence within mutually exclusive categories, then chi square or one of its many variations should be used. Chi square, as is true with all tests for nominal and ordinal data, is a nonparametric test. As such, it is able to sidestep any questions concerning the shape, manner, or form of the population distribution. (All the nonparametric tests are also called distribution-free tests.)

When the scores are in ordinal form, that is, the distances between successive scale points are not known or are known not to be equal, then the ordinal tests, U, H, T, χ_r^2, and r_s must be used. These tests of ordinal data are called nonparametrics, because no assumptions regarding the population characteristics are made. As a result, these tests are also appropriate for interval data when the underlying distribution of interval scores is not normal. In this situation, the interval scores are first rank ordered and then the test analysis is performed.

Finally, when the scores are in interval form, that is, the distances between successive scale points are assumed to be equal, then, in most instances, t, F, and r should be used. We limit this rule to most instances because there is one dramatic exception, and that is when the interval scores do not distribute normally in the population. Fortunately, most measures of human attributes *do*, in fact, distribute normally, but the exceptions, such as age and income, must be constantly watched for. The tests of interval data are called parametric tests, because they assume some knowledge of the population, for example, that its underlying distribution is normal. Also, they are used to estimate certain population parameters, such as the mean and the standard deviation.

Question 2: Which Hypothesis Has Been Tested? There are only two hypotheses that can be statistically tested—the hypothesis of difference and the hypothesis of association. Whenever the research is experimental, then the hypothesis of difference is the one that must be tested. This hypothesis states that the populations from which the sample groups have been selected are in some way different from each other. If, however, the research is post-facto, then the hypothesis under scrutiny might be one of either difference or association. The hypothesis of association states that a correlation exists in the population from which the sample has been selected. This correlation may exist between different measures taken on the same group of subjects (for example, a single group of subjects being measured on both height and weight) or between the same measure taken on different subjects (for example, obtaining IQ scores from pairs of identical twins). Testing the hypothesis of association requires different statistical tests than does testing the hypothesis of difference.

Question 3: If the Hypothesis of Difference Has Been Tested, Are the Samples Independent or Correlated?* Whenever the hypothesis of difference is tested, whether in experimental or post-facto research, it must be clearly determined whether the sample groups are independent or correlated. If the selection of one sample is in no way influenced by the selection of another, then the samples are

---

*Pertains *only* when the hypothesis of difference is being tested.

independent. This occurs when each sample is randomly selected. If, on the other hand, the subjects to be measured are in any way paired off, either by using the same subject more than once or by equating subjects on the basis of some relevant variable, then the groups are correlated. Attempting to isolate differences between correlated sample measures requires different statistical tests than when analyzing differences between independent sample measures.

Question 4: How Many Sets of Measures Are Involved?   When only two measures are being compared, one set of statistical tests is available, but when more than two sets of measures have been taken, other choices can be made. For example, when two groups of interval scores are being tested for difference, the t test is available, whereas when three or more groups of interval scores are being compared, the F ratio must be used. In experimental research the number of measurement sets always refers to the measures taken on the dependent variable.

CRITICAL DECISION POINTS

Once the checklist questions are answered, we turn to the three flow charts provided here—one for each of the three scales of measurement. Each flow chart presents a series of critical-decision points going from top to bottom and based on the checklist questions just outlined. At each critical-decision point, the answer to the next question dictates which route should be followed to the next point. The routes, branching at each decision point, lead inevitably to an appropriate statistical test.

RESEARCH SIMULATIONS: FROM A TO Z

This section presents a series of research simulations, a total of 26, each designed to be used with the checklist questions and the accompanying flow charts to aid you in choosing an appropriate statistical test for each situation.

Simulation A

It has been traditional for the man rather than the woman to receive the check when a couple dines out. A researcher wondered if this would still be true if the woman was clearly in charge, asking for the wine list, and so on. A large random sample of restaurants was selected. One couple was used in all restaurants, but in half the man assumed the traditional in-charge role, and in the other half the woman was in charge. The data were in the form of the number of times the check was presented to each member of the couple.

Analyzing the Methodology. This is experimental research; the independent variable (man or woman in charge) was clearly manipulated. The dependent variable was whether the man or the woman received the check. Thus, if differences in the dependent variable occur here, they can be attributed to the action of the independent variable. Of the experimental designs, this was an after-only (no pretest was given, and no matching took place).

Answering the Checklist Questions: The Critical Decisions

1. *Scale of measurement?* The measurements in this study are in the form of nominal data—the responses being categorized on the basis of which member of the couple received the check. Use Fig. 18.1.
2. *Hypothesis?* The hypothesis being tested is that of difference—whether or not a difference as to who received the check would occur as a function of who appeared to be in charge. (Since this was the experimental method, it had to test the hypothesis of difference.)
3. *If the hypothesis of difference is tested, are the groups independent or correlated samples?* The subjects (waiters or waitresses) were independently selected, since the restaurants were randomly chosen.
4. *How many sets of measurements?* There are two sets of nominal measurements based on whether the man or the woman gets the check.

Solution. Use chi square, in this instance a 2 × 2 chi square, with the independent variable (who is in charge) in the rows and the dependent variable (who gets the check) in the columns.

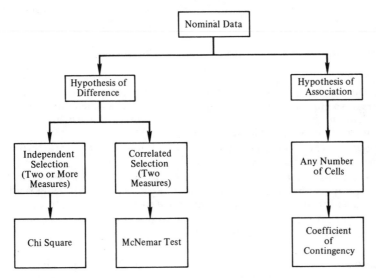

Figure 18.1

	WOMAN GETS CHECK	MAN GETS CHECK
MAN IN CHARGE	a	b
WOMAN IN CHARGE	c	d

N

Simulation B

A researcher, noting the positive correlation between socioeconomic status and amount of education, assumes there will be a difference in the amount of TV viewing done by the principal wage earners in high and low socioeconomically placed families. A random sample of families is selected and categorized as having either high or low socioeconomic status on the basis of a number of measures (income of principal wage earner, occupation, and location of residence). The principal wage earners of all families were then contacted and asked for their hours per week of TV viewing. The hours per week for each socioeconomic class were compared.

Analyzing the Methodology. This is an example of post-facto research; the independent variable (socioeconomic status) is a subject variable, not manipulated. Regardless of whether or not the statistical test is significant, no definitive cause-and-effect statement can be made. Even if the upper-class group is shown to watch significantly less, the reason for this difference can only be suspected. (Perhaps upper-class persons read more, or work more hours, or have expensive time-consuming hobbies, or . . . the list of possibilities is endless.)

Answering the Checklist Questions: The Critical Decisions

1. *Scale of measurement?* The measurements here are in at least interval form. If one person watches TV for 25 hours and another for 24, we know not only that one person spends more time watching TV, but also how much more time. With no evidence to the contrary, the distribution of number of hours spent watching TV is assumed to be normal. Use Fig. 18.2.
2. *Hypothesis?* The hypothesis tested is that of difference—that the two groups would *differ* as to the extent that they watched TV.
3. *If the hypothesis of difference is tested, are the groups independent or correlated samples?* The two groups in this case are independent of each other, each being randomly selected and then assigned to the high or low socioeconomic status.
4. *How many sets of measurements?* There are two sets of measurements, one for upper status and one for lower status.

Solution. Use the independent *t* test. Had it been predicted that one group would watch more TV than the other, then test as a one-tail *t*.

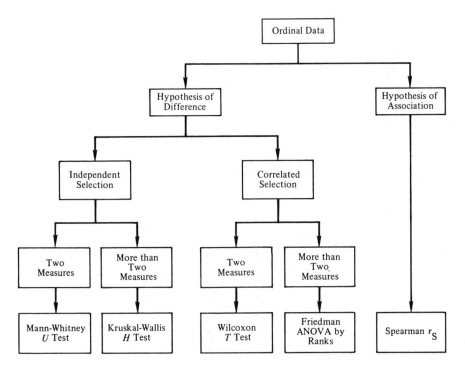

Figure 18.2

Simulation C

A researcher wished to test the hypothesis that taller men are more likely than shorter men to be judged as leaders. A random sample of 30-year-old men was selected and measured for height. The men were then brought before a panel of personnel managers and rank ordered on the basis of perceived leadership qualities. Each subject was assigned the median rank of the panel's decisions.

Analyze the Methodology. This is post-facto research. The independent variable (height) was not manipulated (although it could have been by using elevator shoes, or even hidden stilts). Therefore, the issue of causation is not relevant here, although the probability of making accurate leadership predictions is very much at issue.

Answering the Checklist Questions: The Critical Decisions

1. *Scale of measurement?* Despite using the panel's consensus on rankings, the data are still in ordinal form. Although we may know that Subject 1 is perceived as being imbued with more of the leadership image than is Subject 2, we do not know how much more. Also, although height is clearly an interval measurement, it must be

Figure 18.3

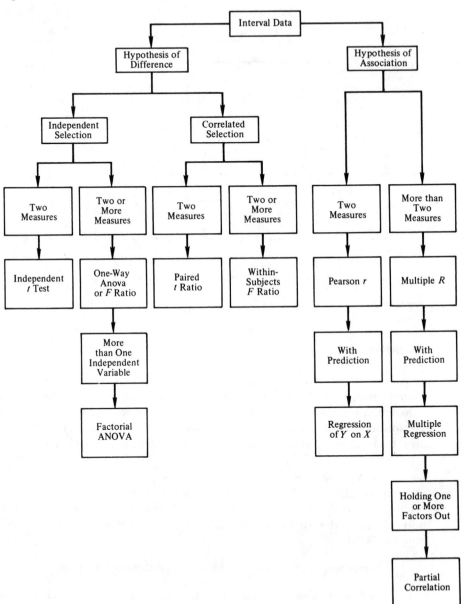

converted to ordinal rank before the statistical analysis can be completed (the men are simply rank ordered as to height). Use Fig. 18.3.

2. *Hypothesis?* The hypothesis being tested here is that of association. It is gratuitous to try and establish a difference between such already disparate measures as height and perceived leadership qualities.

3. *If the hypothesis of difference is tested, are the groups independent or correlated samples?* Not applicable; the hypothesis of association is being tested.

4. *How many sets of measurements?* There are two sets of measurements, height and leadership.

Solution.   Use the Spearman r_s. Note that in this problem, both age and sex were ruled out as variables, since only men of the same age were selected.

Simulation D

A market researcher, working for a manufacturer of hair coloring, wished to establish whether or not blondes do indeed have more fun. A large random sample of female college sophomores was selected and categorized as to hair color—blonde, brunette, or red. Each subject was then asked to answer "yes" or "no" to the question, "On balance, would you say you've been having fun this semester?"

Analyzing the Methodology.   This is post-facto research. The independent variable (hair color) was a subject variable, not manipulated. (Here again, this could have been designed as an experiment, by perhaps randomly dividing the brunettes into two groups and then giving all members of one group a blonde rinse and allowing the other group to remain as they were.) As a post-facto study, however, no cause-and-effect statement can be made. The groups may differ on a variety of other variables related to happiness, number of dates, grade-point average, whatever. Also, it may be that some of the women are already using the blonde rinse, and this itself may be a function of the woman's personality. Perhaps socially oriented women are more apt to use a rinse, and to be more optimistic.

Answering the Checklist Questions: The Critical Decisions

1. *Scale of measurement?* Since the women are measured on the basis of answering the question either "yes" or "no," and then the frequencies of these answers are tallied, the data are nominal. Use Fig. 18.1

2. *Hypothesis?* This is one of those studies in which the hypothesis could conceivably be classified either way—difference or association. It is likely that the researcher meant this as a difference study since the hypothesis was stated as, "Blondes have more fun." (Had it been the hypothesis of association, it probably should have been stated as, "There is a correlation between hair color and having fun.")

3. *If the hypothesis of difference is tested, are the groups independent or correlated samples?* The samples (blondes, brunettes, and redheads) are independent of one another. The fact that one woman was placed in the blonde group neither caused another woman to be placed in, nor precluded her placement in, a different group.

4. *How many sets of measurements?* There are two sets of scores based on whether the subjects report having or not having fun.

Solution.   Use chi square, in this case a 3 × 2 chi square.

	HAVING FUN	NOT HAVING FUN
BLONDE	a	b
BRUNETTE	c	d
REDHEAD	e	f

N

Since this is a 3 × 2 chi square, the special 2 × 2 computational method cannot be used. This means that all of the values for f_e (frequencies expected) must be separately calculated. Had this been a test of association, the chi square could have been followed up with the coefficient of contingency.

Simulation E

A researcher wished to test the hypothesis that older men sleep less than younger men do. Random samples of 30-year-old men, 50-year-old men, and 70-year-old men were selected. Each subject was brought to a sleep laboratory and measured as to how many hours of sleep per night occurred.

Analyzing the Methodology.   This is a post-facto study; the independent variable (age) was a subject variable, not manipulated. (Age can, of course, never be a manipulated independent variable.) Even if the results prove to be significant, great care must be taken in interpreting them. If it is found that the older men sleep less, it may be they did so as young men too. As youngsters, these men may have been more apt to rise early as a result of the differing cultural patterns that typified their younger days. (This is actually cross-sectional research, the hazards of which have been pointed out.)

Answering the Checklist Questions: The Critical Decisions

1. *Scale of measurement?* The dependent variable (hours of sleep) provides at least interval data, with an underlying distribution that is probably normal. Use Fig. 18.2
2. *Hypothesis?* The hypothesis being tested is one of difference—that different age groups have *different* sleep habits.
3. *If the hypothesis of difference is tested, are the groups independent or correlated samples?* These sample groups are independent of each other. The fact that a given man is

selected for the 30-year-old group has no bearing on who is being selected in the 50- or 70-year-old groups.

4. *How many sets of measurements?* There are three sets, one for each age group.

Solution. Use the one-way ANOVA, the *F* ratio. If *F* proves to be significant, proceed with Tukey's HSD.

Simulation F

A researcher wanted to test the hypothesis that racial prejudice is a function of personal authoritarianism. A random sample of college students was selected and measured on the F scale, an index of personal rigidity and authoritarianism. All subjects were then given the A-S (for Anti-Semitic) scale, a measure of prejudice toward Jews.

Analyzing the Methodology. This is a post-facto study; the independent variable (authoritarianism) is a subject variable, not manipulated. Even if the hypothesis is validated, there will be no way to tell whether authoritarianism affects prejudice or prejudice affects authoritarianism. It is even possible that a third variable, such as a family's child-rearing practices, produces both authoritarianism and prejudice.

Answering the Checklist Questions: The Critical Decisions

1. *Scale of measurement?* Both the F and A-S scales are considered to be interval measures distributed normally in the population. Use Fig. 18.2
2. *Hypothesis?* The hypothesis in this case is one of association. (One group is being measured on two different response dimensions.) We can never test for differences between completely unrelated measures.
3. *If the hypothesis of difference is tested, are the groups independent or correlated samples?* Not applicable; the hypothesis of association is being tested.
4. *How many sets of measurements?* There are two sets, one for F scale and one for A-S scores.

Solution. Use the Pearson *r*. If found to be significant, the *r* could be followed by a regression equation, with which specific A-S scores could be predicted from given F scores.

Simulation G

A researcher working for a large corporation wished to test the hypothesis that the company's toothpaste, containing fluoride, reduces dental caries. A random sample of 18-year-olds was selected, and all subjects checked for caries. A dentist then filled the cavities for all subjects having them. For the next three years, all subjects received free monthly supplies of the fluoride toothpaste. Finally, at age

21, the subjects are again checked for whether or not they have dental caries. The researcher then compared the number of persons with caries found in the first dental checkup with the number of persons with caries found at age 21.

Analyzing the Methodology.   This is experimental research, before-after design. The independent variable (toothpaste) is manipulated (rather than being assigned on the basis of whether or not the subjects, on their own, were using it). This is, however, a shaky design, because maturation is a variable and may confound the independent variable. Perhaps 18-year-olds, as a group, are more apt to have caries (the cavity-prone years) than are 21-year-olds. Perhaps, therefore, there might have been a significant result even without the introduction of the independent variable. It would have been better to have had a separate control group, checked for caries at both ages, but given, instead of the fluoride brand, toothpaste that looked the same but did not contain fluoride.

Answering the Checklist Questions: The Critical Decisions

1. *Scale of measurement?* These are nominal data. The subjects were categorized, as a group, at each age as to whether or not caries were detected. We do not know if any subject had more or fewer caries than another, only whether any caries were or were not present. Frequencies of occurrence for each age were compared. Had the subjects been rank ordered in terms of amount and/or severity of caries, the data would have been ordinal. Perhaps even interval measures could have been designed, where the subjects received scaled scores based on the number and severity of caries found. Use Fig. 18.1

2. *Hypothesis?* This is the hypothesis of difference—the assumption being that after using the fluoride toothpaste the frequency of caries within the group will diminish.

3. *If the hypothesis of difference is tested, are the groups independent or correlated samples?* The groups are as correlated as possible, since the same group was used as its own control. The samples are always correlated in a before-after design.

4. *How many sets of measurements?* There are two sets, one taken at age 18 and the other at age 21.

Solution.   Use the McNemar test, which is a chi square based on the change scores.

Simulation H

A researcher wished to find out if the perception of a person's height depends on that person's perceived status. A random sample of army inductees was selected. The subjects were then equally divided into four groups. An actor gave a short address to each group separately, extolling the joys of army life. For the first group, the actor was dressed as a private; for the second, as a sergeant; for the third, as a captain; and, finally, for the fourth group, as a colonel. The inductees were asked to fill out a questionnaire evaluating the speech. Among the questions was one asking for an estimate of the lecturer's height.

Analyzing the Methodology. This is experimental research, after-only design. The independent variable (perceived status) was manipulated by having the same actor wear different uniforms. Using the same actor was a good idea, since otherwise, differences in personal characteristics might have confounded the independent variable.

Answering the Checklist Questions: The Critical Decisions

1. *Scale of measurement?* The dependent variable (estimated height) provided at least interval data in this study. Use Fig. 18.2.
2. *Hypothesis?* The researcher is looking for *differences* in the height estimates made by the four groups.
3. *If the hypothesis of difference is tested, are the groups independent or correlated samples?* The four groups are independent of each other. The selection of one soldier had no bearing on whether another was or was not selected. This is an after-only experimental design, in which sample groups must be independent of each other.
4. *How many sets of measurements?* There are four sets, one for each treatment condition.

Solution. Use the one-way ANOVA, the *F* ratio. If *F* is significant, follow it up with Tukey's HSD.

Simulation I

A researcher in Detroit wished to demonstrate differences in automotive fuel efficiency as a function of both a new carburetion system and the octane level of the gasoline used. A sample of 40 new cars was randomly selected from the assembly line and sent to the company's proving grounds for testing. The same test driver was used for all cars, and the cars were kept at the same speed. Half of the cars were fitted with the new carburetor. Of those 20, 10 were fueled with high-octane gasoline, and the others, low-octane gasoline. Of the cars with regular carburetors, again, 10 were fueled with high-octane gasoline and the other 10 with low-octane gasoline. Each car was evaluated for fuel efficiency on the basis of its miles-per-gallon rate.

Analyzing the Methodology. This is experimental research involving *two* manipulated independent variables, carburetion system and octane level of fuel. The experimenter determined which cars received which combination of treatments. Ideally, the driver was kept unaware of which treatment conditions a given car had been subjected to. Using the same driver helped to prevent confounding due to different driving techniques.

Answering the Checklist Questions: The Critical Decisions

1. *Scale of measurement?* The dependent variable (miles-per-gallon rate) provides at least interval data, which is assumed to be normally distributed. Use. Fig. 18.2.

2. *Hypothesis?* The researcher attempted to show differences in fuel efficiency as a result of the independent variables.

3. *If the hypothesis of difference is tested, are the groups independent or correlated samples?* The subjects (cars) were independently selected for the four combinations of treatment conditions. This was an after-only experimental design.

4. *How many sets of measurements?* There are four sets, one for each of the four treatment conditions.

Solution. Use the factorial ANOVA, with the data set up in four cells.

	Factor A	
	HIGH OCTANE	LOW OCTANE
NEW CARBURETOR	a	b
OLD CARBURETOR	c	d

Factor B

N

Test for the main effects, columns and rows, separately, and then test for the interaction. Because each independent variable had only two levels in this problem, Tukey's HSD is not needed. The between-columns value of F will indicate if octane made a difference, and the between-rows value of F if carburetion made a difference. The value of $F_{r \times c}$ will determine if an interaction occurred.

Simulation J

A researcher for an electronics corporation wished to establish whether or not, other things being equal, the tonal quality of a hi-fi set is judged to be better as the size of the speaker enclosure is increased. A random sample of subjects was selected and asked to listen to the same record played on "different sound systems." Actually, the amplifier, the size and quality of the speaker and baffle, the tone arm, the pick-up, and so on remained the same. Only the size of the speaker enclosure was allowed to vary. Three enclosure sizes were used—small, medium, and large. The subjects were asked to rank order their preferences, from 1 (best) to 3 (worst). The order in which the subjects were presented with the various speaker sizes was counterbalanced, so that some subjects had the large enclosure first, others the small enclosure first, and so on.

Analyzing the Methodology. This is experimental research, repeated-measure design (before–after–after). The independent variable (enclosure size) was manipulated by the experimenter. If significant results are obtained, cause-and-effect inferences can be made.

Answering the Checklist Questions: The Critical Decisions

1. *Scale of measurement?* The dependent variable (judgment of tonal quality) is in ordinal form, that is, the subjects' rank ordering of the three listening conditions. Use Fig. 18.3.
2. *Hypothesis?* This study tests the hypothesis of difference—that different judgments of sound quality will occur as enclosure size is changed.
3. *If the hypothesis of difference is tested, are the groups independent or correlated samples?* The groups in this study are definitely correlated, as the same group is used in all three treatment conditions. Groups are always correlated in repeated-measure designs.
4. *How many sets of measurements?* There are three sets, one for each treatment condition.

Solution. Use the Friedman ANOVA by ranks. Compare the ranks (1, 2, and 3) of each subject under the three listening conditions.

Simulation K

A researcher was interested in establishing whether attendance in a preschool program affects the social maturity level of children. A random sample of 30 kindergarten children was selected and watched closely by trained observers for one full week. The children were then rank ordered on the basis of perceived social maturity. The children were then divided into two groups on the basis of whether or not they had previously attended a day-care center.

Analyzing the Methodology. This is post-facto research; each child's parents, not the researcher, decided whether or not the child was to attend a day-care center. The independent variable (whether or not the child attended the day-care center) was thus a subject variable, not manipulated. Perhaps parents are more apt to send a less mature (or, more mature, who knows?) child to the day-care center in the first place.

Answering the Checklist Questions: The Critical Decisions

1. *Scale of measurement?* This is an example of ordinal data, since the dependent variable is the ranking of the child's social maturity level. Use Fig. 18.3.
2. *Hypothesis?* This is the hypothesis of difference—that maturity levels should differ as a function of day-care attendance.
3. *If the hypothesis of difference is tested, are the groups independent or correlated samples?* The groups are independent of each other. The fact that one child was placed in a given category had no influence on where another child was placed.
4. *How many sets of measurements?* There are two sets, those of children who had attended the day-care center versus those who had not.

Solution. Use the Mann–Whitney U test, which compares the ranks of two independent groups.

Simulation L

A university researcher wished to find out whether a dependable relationship exists between the amount of financial aid granted to a student and the amount that that student earned during the summer. Furthermore, the researcher wished to exclude grade-point average as a variable. A large random sample of financial aid recipients is selected, and each student is checked for amount of aid granted, amount of summer earnings, and grade-point average.

Analyzing the Methodology.   This is post-facto research; the independent variables are subject variables, not manipulated. If the results prove to be significant, predictions, not cause-and effect inferences, are possible.

Answering the Checklist Questions: The Critical Decisions

1. *Scale of measurement?* The three sets of measurements (financial aid, amount earned, and grade-point average) are all at least interval measures. Furthermore, all three measures distribute in a fairly normal fashion. (The summer earnings of students, unlike family incomes, are not apt to be badly skewed. Few, if any, students earn so much during the three-month summer period so as to cause a dramatic shift from normality.) Use Fig. 18.2.
2. *Hypothesis?* This is the hypothesis of association—that a correlation exists between the disparate measures of financial aid and summer earnings.
3. *If the hypothesis of difference is tested, are the groups independent or correlated samples?* Not applicable; the hypothesis of association is being tested.
4. *How many sets of measurements?* There are three sets—financial aid, summer earnings, and GPA.

Solution.   Use the Pearson r. In fact, calculate three values of the Pearson r

1. Between financial aid and summer earnings.
2. Between financial aid and grade-point average.
3. Between summer earnings and grade-point average.

Calculate a partial correlation between financial aid and summer earnings, thereby nullifying (partialing out) the effects of grade-point average.

Simulation M

A study was designed to test whether presenting one side or both sides of an argument is more effective in changing attitudes. Perhaps presenting just the pro side would be more effective because an audience might not be fully aware of the anti side. Or, perhaps, to appear more fair and, also, to prevent the members of the audience from silently going over to the anti side and, therefore, tuning out the pro message, it would be more effective to at least present some of the anti arguments.

A large random sample was selected, and the subjects were assigned to one of two conditions. Group A heard only the pro side of the issue, whereas Group B heard the entire pro side plus a few minutes of anti arguments. Both presentations were made by the same person. A questionnaire, tapping attitudes toward the issue, was then filled out by each subject.

Analyzing the Methodology. This is experimental research. The independent variable (one-sided versus two-sided presentations) was manipulated by the researcher. As no matching occurred and no attitude testing was done prior to the presentation, this was an after-only design. If the results prove significant, causal inferences can be drawn.

Answering the Checklist Questions: The Critical Decisions

1. *Scale of measurement?* The questionnaire was scored as interval data, and the assumption of a normal distribution was made. Use Fig. 18.2.
2. *Hypothesis?* As in all experimental research, the hypothesis of difference was tested. Presumably, *differences* in attitudes between the two groups can be attributed to the independent variable.
3. *If the hypothesis of difference is tested, are the groups independent or correlated samples?* As is true of all after-only experimental designs, the groups were independently selected.
4. *How many sets of measurements?* There are two sets of scores to be compared.

Solution. Use the independent *t*. As no prediction regarding direction was even suggested, check the *t* as a two-tail test.

Simulation N

A researcher wishes to test the hypothesis that male business majors earn more in later life than do either male liberal arts or education majors. A random sample of alumni was selected from the university files from each of the three subject-major categories. To attempt to control for length of experience on the job, all subjects were selected from the same graduating class—the class that graduated 10 years ago. All the selected alumni were contacted and asked to indicate their yearly incomes. The men were promised that the information would be held in strict confidence and would not be given to the chairman of the upcoming alumni fund drive. Because a few of the subjects reported enormously high incomes, the resulting distribution became so skewed that it was decided to rank order the incomes.

Analyzing the Methodology. This is post-facto research. The independent variable, college major, was a subject variable, not manipulated.

Answering the Checklist Questions: The Critical Decisions

1. *Scale of measurement?* Although income is an interval measurement, the skewed distribution forced a rank ordering of the data, thus, creating a series of ordinal measures. Use Fig. 18.3.

2. *Hypothesis?* The researcher was testing for differences among the income ranks of the three groups.

3. *If the hypothesis of difference is tested, are the groups independent or correlated samples?* The groups are independent. The assignment of alumni into subject major categories is strictly independent. The selection of one person from the "education" category did not demand or preclude another subject being selected from the "liberal arts" category.

4. *How many sets of measurements?* There are three sets, one for each of the subject major categories.

Solution. Use the Kruskal–Wallis *H* test for three or more independent groups and ordinal data.

Simulation O

A researcher wishes to increase the predictability of student pilot scores on the FAA's written general aviation exam. Dependable relationships were found to exist between number of hours of ground school and FAA exam scores and also between IQ and exam scores. Finally, a small, but significant, relationship was found to exist between IQ and number of hours of ground school. (Note that for the private pilot's license, the number of hours of ground school is not fixed by the FAA. A few student pilots put in many hours, and a few study on their own and never attend at all.)

Analyzing the Methodology. This is post-facto research. Although the pilots did experience different conditions (attending ground school or not), this was their choice, not the choice of the experimenter. Also, IQ can never be a manipulated variable. Thus, the two independent variables (ground school and IQ) were subject variables, not manipulated.

Answering the Checklist Questions: The Critical Decisions

1. *Scale of measurement?* The three measures (hours of ground school, IQ, and scores on the FAA exam) all yielded at least interval scores. The distributions all appear to be normal. Use Fig. 18.2.

2. *Hypothesis?* The researcher has attempted to test for associations among the three measures.

3. *If the hypothesis of difference is tested, are the groups independent or correlated samples?* Not applicable; the hypothesis of association is being tested.

4. *How many sets of measurements?* There are three sets—hours of ground school, FAA examination scores, and IQ.

Solution.   Use the multiple R. The three separate values of the Pearson r (between ground school and the exam scores, between IQ and the exam scores, and between IQ and ground school), should all be used together to determine if their combinations increase the predictive efficiency. If the value of the multiple R is larger than the separate correlations with the exam scores, solve the multiple regression equation.

Simulation P

A firearms manufacturer hired a researcher to establish whether or not a new handgun increases accuracy. A group of law enforcement agents was randomly selected and brought to the firing range. First, all subjects used the same traditional service revolver, and their error scores (in inches from the bull's eye) were determined. Then, they all fired again, using the new weapon, and their error scores were again determined.

Analyzing the Methodology.   This is experimental research; the independent variable (type of weapon) was manipulated by the experimenter. Since the same subjects are used in both treatment conditions, the design is before-after. This is not the best design for this study, because it is possible that scores might improve the second time as a result of practice. This could act to confound the independent variable. It would have been better to set up a separate group that used the old weapon twice, another group that used the new weapon first and then the old weapon, and another group who had used the new weapon twice.

Answering the Checklist Questions: The Critical Decisions

1. *Scale of measurement?* The dependent variable (error measured on the basis of inches from the bull's eye) provides at least interval data. The researcher claimed a normal distribution for these error scores. Use Fig. 18.2. Had the distribution not been normal, a different statistical test should have been used. (See solution).
2. *Hypothesis?* As in all experimental research, the hypothesis of difference (between error scores) was tested.
3. *If the hypothesis of difference is tested, are the groups independent or correlated samples?* The groups are correlated; the same group is used as its own control.
4. *How many sets of measurements?* There are two sets, one for each treatment condition.

Solution.   Use the paired t ratio, probably as a one-tail test since the manufacturer undoubtedly has some reason for believing in the superiority of the new weapon. (If the results had gone the other way, they possibly would have been filed away in the back of a drawer.) Had the distribution of error scores been skewed (many officers hitting the bull's-eye, but a few missing the target altogether), then the scores should have been rank ordered and the Wilcoxon T test performed.

Simulation Q

An investigator wished to establish whether or not a dependable relationship exists between height at age 3 and height at age 21. A random sample of 3-year-olds was selected, and height measures were taken on each. The researcher then patiently waited 18 years and measured the subjects again.

Analyzing the Methodology. This is post-facto research, the independent variable (height at age 3) is a subject variable, not manipulated. This is also called longitudinal research, since the same subjects are followed through the years and are used again. (A less patient researcher could have obtained adult heights and then checked personal records for the infant heights.)

Answering the Checklist Questions: The Critical Decisions

1. *Scale of measurement?* The data are in at least interval form, and the distributions for each age level are probably normal. Use Fig. 18.2.
2. *Hypothesis?* This is strictly the hypothesis of association. (Testing the hypothesis of difference in this situation, that is, that 21-year-olds are significantly taller than 3-year-olds, would hardly add much to the book of knowledge.)
3. *If the hypothesis of difference is tested, are the groups independent or correlated samples?* Not applicable; the hypothesis of association is being tested.
4. *How many sets of measurements?* There are two sets of measurements, one taken at age 3 and the other at age 21.

Solution. Use the Pearson r. If it is significant, set up the regression equation of Y on X. Thus predictions of adult height can be made from height at age 3.

Simulation R

A Wall Street analyst discovered a dependable relationship between the price of a share of General Motors common stock and the Dow Jones average. Also, a relationship was noted between M1 (a measure of the money supply in the United States) and the Dow Jones average. The analyst wished to use the price of General Motors common stock and M1 to predict the Dow Jones average. (All values are in at least interval form).

Analyzing the Methodology. This is post-facto research. The analyst has bet that the trends discovered from past events will continue in the future, although there is no guarantee that this will be the case.

Answering the Checklist Questions: The Critical Decisions

1. *Scale of measurement?* All data are in at least interval form. Use Fig. 18.2.
2. *Hypothesis?* The researcher is looking for *associations* among the three measures.

3. *If the hypothesis of difference is tested, are the groups independent or correlated samples?* Not applicable; the hypothesis of association is being tested.
4. *How many sets of measurements?* There are three sets.

Solution. Use the Pearson r calculated between General Motors stock and the Dow Jones average, between M1 and the Dow Jones average, and between General Motors stock and M1. The correlations can then be plugged into the multiple R equation. If the multiple R value is larger than either of the values of the Pearson r generated on the Dow Jones average, it can be used for increasing the predictive accuracy. Set up a multiple regression, using both the price of General Motors stock and M1 for predicting the Dow Jones average.

Simulation S

A cultural anthropologist became interested in discovering whether or not differences in the age of menarche (the age when young women have their first menstrual cycle) are a function of climate. Two groups of young women were selected—one from a northern climate (Norway) and one from a southern climate (Italy). The subjects were matched according to both height and weight, and their ages at menarche were compared. The age distribution was found to be skewed.

Analyzing the Methodology. This is post-facto research; the independent variable (climate) was a subject variable, not manipulated. Thus, even if significance is established, no positive causal statement can be made. The subjects obviously differ on a host of variables (diet, genetic background, medical care, etc.) other than climate.

Answering the Checklist Questions: The Critical Decisions

1. *Scale of measurement?* Although age is at least an interval measure, the lack of normality in the underlying distribution forces a conversion of the age scores into ordinal data. Use Fig. 18.3.
2. *Hypothesis?* The researcher is testing the hypothesis of difference—that age at menarche *differs* as a function of climate.
3. *If the hypothesis of difference is tested, are the groups independent or correlated samples?* The groups are correlated, having been matched on both height and weight.
4. *How many sets of measurements?* There are two sets of measurements, one taken in Norway and the other in Italy.

Solution. Use the Wilcoxon T test.

Simulation T

A researcher wanted to find out whether or not IQ is a function of family size. The speculation was that among families with fewer children, each child receives more parental attention and intellectual stimulation and should, therefore, have a higher IQ than would a child reared in a larger family. A large random sample of two-child families was selected as well as a similar sample of six-child families. The IQs of all children were measured, and the two sample groups were compared.

Analyzing the Methodology. This is post-facto research. The independent variable (family size) was a subject variable, not manipulated. (Natural forces or their own decision, not the decision of the experimenter, determined which families had small or large numbers of children.) Thus, even if significance is established, the causal factor remains in the realm of speculation. Could it be, instead, that lower-IQ parents have more children?

Answering the Checklist Questions: The Critical Decisions

1. *Scale of measurement?* IQ scores are considered to be interval measures, and the underlying distribution to be fairly normal. Use Fig. 18.2.
2. *Hypothesis?* The researcher is looking for IQ *differences* among children from small and large families.
3. *If the hypothesis of difference is tested, are the groups independent or correlated samples?* The sample groups are independent. The selection of a given family depended on its size, not on whether or not some other family had been selected.
4. *How many sets of measurements?* There are two sets of IQ scores, one taken from large families and one taken from small families.

Solution. Use the independent t. If the score for each child is to be used separately, use the equation for unequal values of N (there are three times as many IQ scores in the six-child families). If the children's IQ scores are to be averaged within each family, then equal values of N can be maintained.

Simulation U

A sports analyst wanted to know whether or not team performance is influenced by crowd encouragement at home games. A random sample of 10 university football teams was selected and rank ordered on the basis of at-home wins over a four-year period. Observers attended each team's home games and rank ordered the fans from 1, most enthusiastic, to 10, most apathetic.

Analyzing the Methodology. This is post-facto research. The independent variable (crowd enthusiasm) was not orchestrated by the experimenter. The independent variable, therefore, was a subject variable. Even if significance is attained, we will not know if crowd encouragement produces victory or if victory produces rabid fans (or, even, whether some third variable, say, general campus atmosphere, produces both). There is also another difficulty to be confronted here—since all observers could not have visited all the home games, do they all evaluate crowd enthusiasm in the same way?

Answering the Checklist Questions: The Critical Decisions

1. *Scale of measurement?* Rank ordering always produces ordinal data. Use Fig. 18.3.
2. *Hypothesis?* By comparing disparate measures (team performance and crowd enthusiasm) the researcher is looking for an association.
3. *If the hypothesis of difference is tested, are the groups independent or correlated samples?* Not applicable; the hypothesis of association is being tested.
4. *How many sets of measurements?* There are two sets.

Solution. Use the Spearman r_s, and use great caution in interpreting results.

Simulation V

A political analyst attempted to find out if the political slant of a newspaper affects the voting preference of the readers. In a large eastern city, a random sample of homes was selected where Newspaper L (Liberal) was delivered. Also a random sample of homes receiving Newspaper C (Conservative) was selected. The voting preference of each head of household was obtained and categorized as Republican, Democrat, or Other.

Analyzing the Methodology. This is post-facto research. The subjects themselves chose which newspaper (independent variable) to have delivered. If significant results are obtained, does it mean the newspaper affected voting preference, or was voting preference the key to which newspaper was ordered?

Answering the Checklist Questions: The Critical Decisions

1. *Scale of measurement?* This is the nominal case. The measures are in the form of *how many* persons subscribe to which newspaper and *how many* persons vote in which category. Frequency of occurrence within mutually exclusive categories defines the nominal case. Use Fig. 18.1.
2. *Hypothesis?* The researcher is interested in differences in voting preference.
3. *If the hypothesis of difference is tested, are the groups independent or correlated samples?* The groups are independent of each other. The choice of which group a subject was placed in was determined by the newspaper being delivered, not by the group another subject was placed in.
4. *How many sets of measurements?* There are three sets based on voting preference.

Solution. Use chi square set up as a 2 × 3.

| | Voting Preference | | |
	REP.	DEM.	OTHER
LIBERAL	a	b	c
CONSERVATIVE	d	e	f

Newspaper

N

Simulation W

An investigator tried to shed light on the hypothesis that perception shapes attitudes. A large random sample was selected, and each subject was then randomly assigned to one of three groups. Each group then heard an identical speech, given by the same speaker, on the topic of land reform in Cuba. To Group A, the speaker was introduced as a prominent political scientist; to Group B, as a member of the U.S. State Department; and to Group C, as a member of the Soviet delegation to the UN. After the speech, the subjects all took an "Attitude Toward Cuba" test. The distribution of test scores was normal.

Analyzing the Methodology. This is experimental research. The independent variable (perception of the speaker) was manipulated by the experimenter. Other factors, such as content of speech, personality of speaker, and so on, were not allowed to vary. If significance is obtained, a causal factor may be isolated.

Answering the Checklist Questions: The Critical Decisions

1. *Scale of measurement?* Attitude test scores provide interval data. As has been stated, the scores distribute normally. Use Fig. 18.2.
2. *Hypothesis?* The researcher is testing the hypothesis of difference—that different attitudes will result from different perceptions.
3. *If the hypothesis of difference is tested, are the groups independent or correlated samples?* The three sample groups are independent of each other.
4. *How many sets of measurements?* There are three sets, one from each treatment condition.

Solution. Use the one-way ANOVA, the F ratio. If F is significant, follow it up with Tukey's HSD.

Simulation X

A researcher for the Registry of Motor Vehicles became interested in whether recidivism among convicted drunken drivers is affected by the judicial outcome. A random sample of convicted drunken drivers was selected. The drivers were then randomly divided into two groups. The members of Group A received heavy fines and temporarily lost their licenses. The members of Group B were not fined, but were placed in a six-month rehabilitation program and temporarily lost their licenses. Two years later, all subjects were checked for repeat convictions. The groups were compared on the basis of the number of subjects in each group found to have and not have repeat convictions.

Analyzing the Methodology. This is experimental research, after-only design. The independent variable (whether or not subjects go into the rehabilitation program) is manipulated by the researcher. The potential for uncovering a causal relationship is, thus, available.

Answering the Checklist Questions: The Critical Decisions

1. *Scale of measurement?* The dependent variable (repeat convictions versus no repeat convictions) is measured in nominal form. Use Fig. 18.1.
2. *Hypothesis?* The researcher is testing the hypothesis of difference—that new conviction ratios differ as a function of which group the subjects were in.
3. *If the hypothesis of difference is tested, are the groups independent or correlated samples?* The groups are independent, the subjects having been randomly assigned to the two groups.
4. *How many sets of measurements?* There are two sets based on whether or not the subjects participated in rehabilitation programs.

   Solution.   Use chi square. This is a 2 × 2 chi square, so the computational method can be used.

Simulation Y

A sports physiologist was interested in whether or not extended periods of jogging reduce the resting pulse rate. A random sample of accounting majors (none of whom had ever been involved in jogging) was selected from the senior class at a large university. Resting pulse rates were taken on every subject at the following intervals: first, before starting the jogging program; second, after 4 weeks of the program; third, after 8 weeks; and, finally, after 12 weeks.

   Analyzing the Methodology.   This is experimental research; the independent variable (amount of time spent jogging) was manipulated by the experimenter. Significant results of this study would pave the way for a cause-and-effect conclusion.

Answering the Checklist Questions: The Critical Decisions

1. *Scale of measurement?* The dependent variable (resting pulse rate) provides normally distributed data, which are at least interval. Use Fig. 18.2.
2. *Hypothesis?* The researcher is testing the hypothesis of difference, as is always the case in experimental research.
3. *If the hypothesis of difference is tested, are the groups independent or correlated samples?* The groups are correlated; the same subjects were used in each treatment condition.
4. *How many sets of measurements?* There are four sets of measurements of pulse rate.

   Solution.   Use the within-subjects F ratio for repeated measures.

Simulation Z

An investigator wished to test the hypothesis that reading speed is a function of how extensively a student reads. A random sample of high school seniors was selected in September, and the subjects were asked how many books they had read during the summer. The subjects were then categorized in the following groups: Group 1, no books read; Group 2, 1–3 books; Group 3, 4–6 books; and Group 4, more than 6 books. Reading speed tests were then administered, the scores being in the form of words per minute.

Analyzing the Methodology.   This is post-facto research; the independent variable (number of books read) was not manipulated. The subjects were *assigned* to groups on the basis of how many books they themselves had chosen to read. A unidirectional interpretation of this study will, therefore, be impossible. Does extensive reading increase reading speed, or do fast readers prefer to read more? Or is it a combination of the two? One can never know from this study.

Answering the Checklist Questions: The Critical Decisions

1. *Scale of measurement?* The dependent variable (reading speed in words per minute) provides at least interval data. The distribution is close enough to normality to use interval tests. Use Fig. 18.2.
2. *Hypothesis?* The researcher is looking for *differences* in reading speed among the subjects.
3. *If the hypothesis of difference is tested, are the groups independent or correlated samples?* The groups are independent of each other. The placement of a subject in a given group depends on the number of books read, and not on the placement of some other subject.
4. *How many sets of measurements?* There are three sets of measures of reading speed.

Solution.   Use the one-way ANOVA, the F ratio. If F is significant, proceed with Tukey's HSD.

THE RESEARCH ENTERPRISE

Now that we have met the research simulations from A to Z, choosing the correct test for analysis of the data from 26 studies, we need to take the time for a little reflection on the whole idea of statistical research. While confronting each research situation, we were involved in the case study learning approach; that is, the specific examples should provide the fodder for some more global generalizations. There are three especially important general characteristics that must be highlighted here.

Statistical Research Is Empirical.   In each of the research problems, measurements were taken, and the data from these measures were analyzed for possible significance. Measurements always imply *observation,* and observation is the key to the empirical approach. Whenever knowledge comes to us through the direct observation of the world around us (as opposed to knowledge obtained only through the powers of reason—without direct, sensory experience), we are using empirical methods. Statistical research must, therefore, always be empirical. The research enterprise demands good, solid, specific, empirical observations. No statistical analysis, no matter how elegant and sophisticated, can compensate for bad data. Remember, G.I.G.O. (garbage in, garbage out).

Statistical Research Is Inductive.   The logical technique of *induction,* that is, arguing from the specific to the general, is absolutely crucial to statistical research. The empirical measures mentioned must be *specific.* Each separate measure must be individually observed. The statistical analysis then determines whether these specific sample measures can be generalized to the population. Philosophers call this the *inductive leap,* and although, as we have seen, statistical generalizations may be erroneous, at least their probability of error can be assessed. This is really the main thrust of statistical analysis—determining the probability of error of the induced generalizations.

Statistical Research Should Be Interpretable.   Once the data are gathered and the appropriate statistical test used for the data analysis, the results should be open to clear, unequivocal interpretation. It is of little use to do an elaborate study involving enormous amounts of empirical data only to find that the resulting generalizations are laden with ambiguities. This is like saying "the operation was a success, but the doctor died." Ambiguities are almost certain to arise unless great care is taken in structuring the logic and design of the research. No clear statistical analysis is possible, for example, if the measures taken on some of the subjects are correlated and the measures taken on other subjects are independent. Neither can unequivocal interpretations of experimental data be obtained if the independent variable has been confounded.

A FINAL THOUGHT: THE BURDEN OF PROOF

In the world of science, certain ground rules have been long established. One of the most important of these is that the burden of proof is on the innovator. Whenever a new hypothesis is introduced, it is up to the person making the introduction to give accompanying proof. It is not left to the scientific community to disprove it. If an "innovator" were to come along and loudly proclaim that there are indeed unicorns living in some remote region of Nepal, it would be the duty of that individual to produce the evidence. It is not enough for the proponent of the theory simply to offer the challenge: "Prove me wrong." Other scientists are not expected to drop their calculators and test tubes and hurry off to Nepal to disprove a new unicorn theory. Even if this were done, and many years later the now wizened scientists returned empty-handed, the innovator could smugly reply, "You didn't look behind the right rock." Just as in a court of law the burden of proof is on the prosecutor, so, too, in the "court" of scientific inquiry the person making the charge must come up with the proof.

It has been a long road from means and medians to factorial ANOVAs and multiple regressions. However, the road has been well lit and logically straight. You have come a long way—you can read research studies, and you can do some research. Of course, there are more advanced topics to be covered before you can consider doing a senior research project, but, just as certainly, these topics will not be as forbidding as they once may have seemed. You may not yet have all the answers, but you are now in a position to know what many of the questions are.

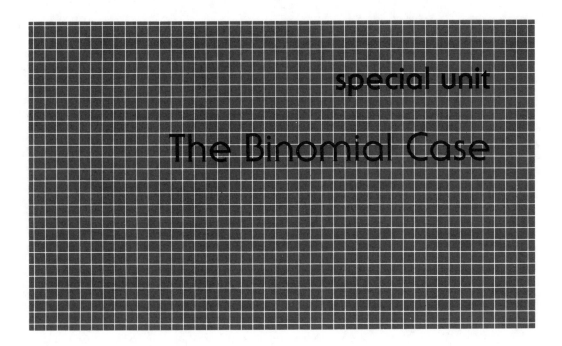

special unit

The Binomial Case

This special unit will cover the basic elements of the binomial distribution and how it may be used in the various research situations. The focus will still be on the *practical applications* of the binomial case, and will not suddenly involve you in a heavy course in higher mathematical theory. The math will be kept as absolutely simple as possible. You and your trusty calculator will find that solving binomial problems will be no more painful than doing z scores or a Pearson r.

This unit is divided into three sections, with problems presented at the end of each of these sections. (Answers to all the problems, not just the odd items, are given at the end of the unit). In Section 1 we cover the ABCs of binomial probability and the relationship between the binomial distribution and the standard normal curve. Section 2 involves hypothesis testing, using the z test to evaluate differences between sample proportions and known population proportions. In Section 3 we again look at the testing of hypotheses but this time the spotlight will be on the difference between one sample proportion when it is being pitted against the proportion occurring in a second sample. The unit concludes with a brief look at the relationship between the binomial test and the chi square test.

The unit may be undertaken in its entirety if you have already read Chapter 6 on probability, Chapter 7 on the z test, and Chapter 13 on chi square. If not, the sections may be treated separately and read as each of the appropriate text chapters have been completed: that is, read Section 1 after Chapter 6, Section 2 after Chapter 7, and Section 3 after Chapter 13.

In any case, rest assured that some understanding of the binomial case will definitely increase your overall appreciation of probability theory, hypothesis testing, and perhaps even, if you're so inclined, general skill in the "games of chance."

SECTION ONE

Binomial Probability

When an event has only two possible outcomes on each of a number of independent occasions, it is called *binomial* (literally, "two names" or "two terms"). The coin-flipping problems discussed in Chapter 6 are examples of binomial problems. You may find that a more detailed look at these kinds of problems, however, is not only worthwhile in its own right but may also make the normal probability curve more understandable. For this section, the book's normal rounding rule of two decimal places will be suspended, since many of the upcoming examples will yield probability values of less than .01.

When tossing a coin there are obviously only two possible outcomes, a head (H) or a tail (T) and as we have seen, each of these outcomes has a probability of .50. Also, as was shown, if we want the probability of tossing two coins and getting both to turn up heads, we use the MULT-AND rule and multiply the separate probabilities.

> With two coins, $P = .50 \times .50 = .50^2 = .25$
>
> With three coins, $.50 \times .50 \times .50 = .50^3 = .125$

Now, although this procedure works perfectly well when asking for the probability of *all heads* or *all tails,* it won't directly handle the situation of finding the probability of getting, say, 2 heads out of 5 tosses. In this type of problem we can't simply multiply the separate probabilities, because now we have *more* tosses than hits. (A "hit" is a success, or when you call the outcome, say a head, and then actually get that outcome, that is, the head turns up.) For problems like these we use the following formula, where n is equal to the number of coins, r is equal to the number of hits, and ! stands for "factorial." Factorial means successively multiplying a given number by one less than the preceding number until you get to 1. Hence, $5! = 5 \times 4 \times 3 \times 2 \times 1 = 120$. Also, by convention, $0! = 1$.

$$P = \frac{n!}{(r!)(n-r)!} \, p^r q^{n-r}$$

where p is the probability of the event occurring or, as we have called it, the probability of a "hit," and q is the probability of the event not occurring, or the probability of a "miss." Therefore, q is always equal to $p - 1$.

In the coin-tossing example, p must always equal .50, and q, which is $1 - p$, must also be equal to .50. Thus, $p = q = .50$ and the equation can be written as

$$P = \frac{n!}{(r!))(n - r)!} (.50)^r (.50)^{n-r}$$

This formula works just as well when assessing the probability of *all* heads out of n tosses, as it does in the situation where we have more coins than hits. For example, to get the probability of *all* heads with 3 coins, which as we saw was .125, this formula gives us,

$$P = \frac{3!}{(3!)(0!)} (.50)^3 (.50)^0$$

$$P = \frac{6}{(6)(1)} (.125)(1) = .125$$
$\qquad\qquad\qquad\qquad$ ↑ (any value taken to the 0 power = 1)

Or, to get the probability of getting exactly 2 heads out of 3 coins,

$$P = \frac{3!}{(2!)(1!)} (.50)^2 (.50)^1$$

$$P = \frac{6}{(2)(1)} (.25)(.50) = .375$$

The same analysis can be made on data obtained from a true-false exam, in which the number of "true" items is equal to the number of "false" items. Assume that there are 8 questions; then, on the basis of chance, the probability of selecting 5 "true" items would be

$$P = \frac{8!}{(5!)(3!)} (.50)^5 (.50)^3$$

$$P = \frac{40320}{(120)(6)} (.03125)(.1250) = .2188$$

To appreciate how helpful the formula above really is, we will now apply it to the seemingly simple situation of including only 4 items, rather than 8. Even here, however, the formula above is a blessing, since working out all those separate possibilities can get rather complicated. Assume that you are taking a 4-item true-false test, where the possible outcomes total 2^4, or 16. The following shows these possible outcomes:

All possible Outcomes (total number of events) for a 4-item true-false test.

T	T	T	T	F	T	T	T	F	F	F	T	F	F	F	F
T	F	T	T	T	T	F	F	T	T	F	F	F	F	T	F
T	T	F	T	T	F	T	F	T	F	T	F	F	T	F	F
T	T	T	F	T	F	F	T	F	T	T	F	T	F	F	F

Number of true "hits"   4 3 3 3 3 2 2 2 2 2 2 1 1 1 1 0

NUMBER OF "HITS"	SPECIFIC OUTCOMES	P-SPECIFIC OUTCOMES/TOTAL POSSIBLE OUTCOMES
4	1	1/16 = .0625
3	4	4/16 = .2500
2	6	6/16 = .3750
1	4	4/16 = .2500
0	1	1/16 = .0625

Or, to find the probability of the 2 hits, which is shown above as .3750, the use of the above formula simplifies the problem to

$$P = \frac{n!}{(r!)(n-r)!}\, p^r q^{n-r}$$

$$P = \frac{4!}{(2!)(2!)} (.50)^2(.50)^2 = \frac{24}{4} (.25)(.25)$$

$$P = 6(.25)(.25) = .3750$$

Discrete and Continuous Distributions

The binomial distribution is a discrete distribution. That is, the steps between adjacent values are totally separated. If you flip 3 coins, you can get 0 heads, 1 head, 2 heads, or 3 heads, but you can't get 2 ½ heads or 1.395 heads. A continuous distribution, on the other hand, yields values that may fall at any point along an unseparated scale of points. Therefore, when fractional values are permissible and have meaning, we are dealing with a continuous measure. Height, for example, is measured on a continuous scale, so that a person's height can take on a value of 65 inches, or 65.25 inches, or 65.2514 inches (or you can go out as many decimal places as the accuracy of the measurement technique will allow). Since so many measures are continuous, we tend to round them at some convenient level, like the nearest inch, or nearest degree (for temperature), or nearest dollar (for income). Because the number of decimal places chosen for any continuous measure is, in the final analysis, arbitrary, mathematicians describe each continuous value as falling within an interval of values. If length is measured to the nearest inch, for example, a measure of 72 inches is assumed to be somewhere in an interval ranging from 71.5 inches to 72.5 inches. If, however, length were being measured only to the nearest foot, a value of 6 feet is assumed to lie between 5.5 feet and 6.5 feet. The z distribution is on a continuous scale, whereas the binomial, as shown above, is discrete.

A Binomial Distribution

Let's look at the resulting distribution when 8 coins are tossed, and, as is always the case with fair coins, where $p = q = .50$. The probability of getting a head (in this case a "hit") is

				Exact Number of Hits				
0	*1*	*2*	*3*	*4*	*5*	*6*	*7*	*8*
$p = .0039$	.0313	.1094	.2188	.2734	.2188	.1094	.0313	.0039

Notice that the probabilities at the end points, 0 heads and 8 heads, are extremely small, whereas the probability in the middle, 4 heads, is much higher. Thus, the probabilities get increasingly less as you go away from the middle of the distribution in either direction. Using our previous formula, let's prove the probability value for 4 heads (which is listed above as .2734).

$$P = \frac{n!}{(r!)(n - r)!} (.50)^r (.50)^{n-r}$$

$$P = \frac{8!}{(4!)(4!)} (.50)^4 (.50)^4 = \frac{40320}{(24)(24)} (.0625)(.0625)$$

$$P = .70 (.0625)(.0625) = .2734$$

If we were to display this distribution as a histogram, it would look like this (with the normal distribution superimposed).

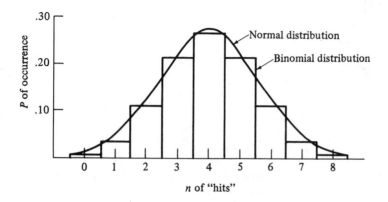

The Binomial and Normal Distributions: A Comparison

As you can see, this distribution has the same general shape as the normal distribution. In fact, as n in the binomial increases and the steps between the adjacent bars in the histogram become ever smaller, the binomial distribution, which is discrete, tends to resemble ever more closely the z distribution (which, as you know, is continuous). In fact, the approximation of the binomial to the z is fairly close, even with an n as small as 8, *as long as the* p *value is .50* (or close to it). As P departs from .50, the approximation is still possible, but it then must take high n values.

The mean and standard deviation of a binomial distribution for a sample of n trials can be calculated on the basis of the following:

$$\bar{X} = np \text{ (n equals the number of events or "trials" and p the probability of a hit)}$$

Thus, with 8 coins and a p of .50, $\bar{X} = 4$:

$$\bar{X} = 8(.50) = 4$$
$$SD = \sqrt{npq} \quad \text{(q equals $1 - p$, or the probability of a "miss")}$$

With the 8 coins, then, we get a standard deviation of 1.41:

$$SD = \sqrt{(8)(.50)(.50)} = \sqrt{2} = 1.41$$

With the mean and standard deviation known, we can now calculate z scores for the various numbers of "hits," and from these z scores we can use the normal distribution to approximate the exact probabilities given under the binomial distribution. It must again be pointed out, however, that the z distribution is continuous. A continuous value, such as the mean value of 4, is really some value that lies between 3.5 and 4.5, which are its literal limits. Using a value of 4, then, is in effect like saying that 4 is the midpoint of continuous interval whose real limits are 3.50 and 4.50.

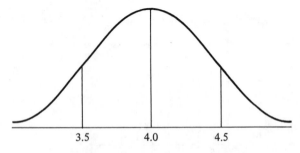

To find the normal distribution's probability of an event falling between the scores 3.50 and 4.50, we first calculate the two z scores.

$$z = \frac{X - \bar{X}}{SD}$$

$$z = \frac{4.50 - 4.00}{1.41} = .35$$

$$z = \frac{3.50 - 4.00}{1.41} = -.35$$

Under the normal curve, as shown on Table A, the percentage values are 13.68% for each of these z scores,

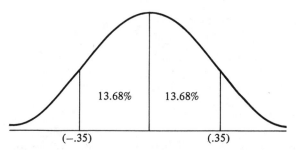

and the area between the z scores is 13.68% + 13.68% = 27.36%. Converting to a probability value, we get .2736. As can be seen, this is an extremely close approximation of the probability value of .2734 for the 4 hits shown under the binomial distribution on page 427.

SECTION ONE PROBLEMS

1. Determine the probability of tossing all heads with 5 coins.
   Answer: .03125
2. Find the probability of getting 3 heads with 5 coins.
   Answer: .3125
3. Using z scores, find the normal curve's approximation of the probability of obtaining 7 hits using a total of 8 coins. Compare this answer with the exact probability shown under the binomial distribution.
   Answer: .03 (a very close approximation)
4. On a 10-item true-false test, where the number of true items equals the number of false items, what is the exact probability of getting
   a. 4 true items?
   b. 5 true items?
   (For this problem, use the binomial formula shown on page 427.)
   Answers:
   a. .2047
   b. .2460

5. Using the data from problem 4, use z approximations for finding
   a. The mean of the distribution.
   b. The standard deviation of the distribution.
   c. The probability of getting 4 true items.
   d. The probability of getting 5 true items.
   e. Compare these normal curve approximations with the exact probabilities found in problem 4.
   Answers:
   a. Mean equals 5.
   b. Standard deviation equals 1.58.
   c. .2035
   d. .2510
   e. Approximations are extremely close.

SECTION TWO

The z Test and Binomial Proportions

In some studies the researcher is confronted with the problem of comparing the observed proportions that occur in a sample with those true proportions which are known to occur in the population. For example, it is known that in an infinite population of tosses using an unbiased coin that the true proportion of heads is 1/2, or .50. Suppose, however, that a friend slipped in a biased coin (loaded to produce more than its share of heads), and your job as the statistician is to show the likelihood of such a coin conforming to a chance explanation.

As shown on page 428, the standard deviation of the binomial distribution for n events can be calculated as

$$SD = \sqrt{npq}$$

where n is the number of events and p is the probability of a "hit,"

$q = p - 1$, or the probability of not getting a "hit"

However, in the case of proportions, where the total population of events can be infinite, the *population* standard deviation is defined as

$$\sigma = \sqrt{pq}$$

We can now use this population standard deviation to calculate the standard error of the entire sampling distribution of proportions, just as we did when calculating the standard error of the mean, by dividing through by the square root of n. This standard error of the proportion will be symbolized as σ_p.

$$\sigma_p = \frac{\sigma}{\sqrt{n}} = \frac{\sqrt{pq}}{\sqrt{n}} = \sqrt{\frac{pq}{n}}$$

Getting back to our friend with the loaded coin, assume that on a run of 20 tosses this coin has produced heads 16 times, for an observed proportion of $16/20 = .80$. Although it is known that an unbiased coin produces a population proportion of .50 heads, your friend still insists that this was just a streak of luck and that the coin is perfectly fair.

To test this allegation, first calculate the standard error of the proportion for the true population parameter.

$$\sigma_p = \sqrt{\frac{pq}{n}} = \sqrt{\frac{(.50)(.50)}{n}} = \sqrt{\frac{.25}{20}} = \sqrt{.0125} = .1118$$

Second, find the z score, where P_s is the proportion occurring in the sample, and P_t is the true proportion known to occur in the population.

$$z = \frac{P_s - P_t}{\sigma_p} = \frac{.80 - .50}{.1118} = \frac{.30}{.1118} = 2.6832$$

1. Null hypothesis: $H_0: P_s = P_t$ (where P_s is the hypothesized proportion found in the sample and P_t is the known population proportion)
2. Alternative hypothesis $H_a: P_s \neq P_t$
3. Critical value of z at $.01 = \pm 2.58$
4. Calculated value of z test:

$z = 2.6832$ Reject H_0.

We thus reject the hypothesis that our friend's coin is representative of the population of unbiased coins, and conclude instead that the coin is indeed probably "loaded."

Remember, as we pointed out earlier, that the binomial approximation of the z distribution is valid only when the number of events is sufficiently large (at least 8) and the P value is close to .50.

SECTION TWO PROBLEMS

1. Assume that the proportion of women and men on the faculty of a large university is equal, $p = .50$ and $q = .50$. However, among the 30 departmental chairpersons, there are only 9 women. Test the hypothesis that bias exists in the selection of chairpersons.
Answer: $z = -2.1905$, z at $.05 = \pm 1.96$. Reject the null hypothesis and conclude that bias probably does exist.

2. Assume that the proportion of true answers is equal to the proportion of false answers on a 50-item true-false exam in botany. One student, who claims that he hasn't studied for the test for even 1 minute and in general says he knows nothing about botany, answers 30 of the items correctly. Test the hypothesis that he probably did sneak a look at the text and does know more about botany than he's willing to admit.

    Answer: $z = 1.43$, z at .05 $= \pm 1.96$. Accept H_0: the student is apparently telling the truth. He didn't study, and he sure doesn't know anything about botany.

3. Assume that the proportion of boys and girls in a large high school is equal, $p = .50$ and $q = .50$. Among a sample of 16 students elected to serve as class officers, however, there are only 4 girls. Test the hypothesis that bias exists in the election of class officers.

    Answer: $z = -2.00$, z at .05 $= \pm 1.96$. Reject H_0; significant at $p < .05$.

SECTION THREE

Testing the Difference Between Two Sample Proportions

Just as the binomial z test can assess whether differences exist between those proportions that have been observed in the sample and those that are known to exist in the population, z can also be used to test for differences between two independent *sample* proportions. To use this form of z, it is assumed that the sample sizes must be fairly large and the parameter proportions of p and q not too different from .50. One rule which can be used for testing this assumption is that if both samples have an N of at least 100, then the binomial z is appropriate. If, however, either of samples has an N of less than 100 and the p value is either greater than .90 or less than .10, then this test should not be used.

For example, a researcher wishes to find out if there is difference between Republicans and Democrats regarding their opinions on the statement, "More tax money should be used to support welfare programs." A random sample of 100 Democrats was selected ($N = 100$) and a second random sample of 80 Republicans was also selected ($N = 80$). Of the Democrats, 60 agreed with the statement, and 40 disagreed. Of the Republicans, 30 agreed and 50 disagreed.

First, we calculate the two proportions, p_1 and p_2. To obtain these values, simply calculate the proportion in each group who "agree," or $60/100 = .60$ for p_1 (the Democrats) and $30/80 = .375$ for p_2 (the Republicans).

Next, calculate the standard error for each proportion, using the equation:

$$\sigma_p = \sqrt{\frac{pq}{n}}$$

For sample 1, the Democrats, this would be

$$\sigma_{p_1} = \sqrt{\frac{pq}{n}} = \sqrt{\frac{(.60)(.40)}{100}} = \sqrt{\frac{.24}{100}} = \sqrt{.0024} = .0490$$

And for sample 2, the Republicans, it would be

$$\sigma_{p_2} = \sqrt{\frac{pq}{n}} = \sqrt{\frac{(.375)(.625)}{80}} = \sqrt{\frac{.2344}{80}} = \sqrt{.0029} = .0539$$

Both these values are then used to obtain the standard error of the difference between two proportions, or σ_{DP}

$$\sigma_{DP} = \sqrt{\sigma_{p_1}{}^2 + \sigma_{p_2}{}^2} = \sqrt{.0490^2 + .0539^2} = \sqrt{.0024 + .0029}$$
$$\sigma_{DP} = \sqrt{.0053} = .0748$$

Finally, use p_1, the proportion of Democrats who "agreed," .60, and p_2, the proportion of Republicans who "agreed," .375, and calculate z as follows:

$$z = \frac{p_1 - p_2}{\sigma_{DP}} = \frac{.60 - .375}{.0748} = 3.01$$
$$z = 3.01$$

1. Null hypothesis H_0: $p_1 = p_2$ (the sample proportions are equal)
2. Alternative hypothesis H_a: $p_1 \neq p_2$ (the sample proportions are not equal)
3. Critical value of z at .01 $= \pm 2.58$
4. Calculated value of z test $= 3.01$ Reject H_0. $p < .01$, and conclude that the samples are significantly different.

Chi Square and Proportions

For those of you who have already completed Chapter 13, you may prefer a chi square analysis of these types of proportion problems. For example, since with one degree of freedom, chi square is equal to z^2, the proportion problems just shown could be done via the chi square.

First, to test whether a sample proportion is consistent with a known population proportion, the problem shown on page 431 produced a significant z of 2.68. Setting these data in a chi square table would result in

	HEADS	TAILS	
f_o	16	4	
f_e	10	10	← (They should split evenly in the population)
$f_o - f_e$	6	−6	
$(f_o - f_e)^2$	36	36	
$\dfrac{(f_o - f_e)^2}{f_e}$	3.60 +	3.60 $= x^2 = 7.20$	

$$x^2_{.01(1)} = 6.64$$
Reject H_0; significant at $P < .01$.

The z value of 2.68, shown on page 431, when squared, will then equal chi square. ($2.6832^2 = 7.20$.)

Or, to test the difference between two independent samples, the problem shown on page 432 would be solved by chi square as follows:

	AGREE	DISAGREE	
Democrats	a 60	b 40	100
Republicans	c 30	d 50	80
	90	90	180

	a	b	c	d
f_o	60	40	30	50
f_e	50	50	40	40
$f_o - f_e$	10	−10	−10	10
$(f_o - f_e)^2$	100	100	100	100

$$\frac{(f_o - f_e)^2}{f_e} \qquad 2.00 \; + \; 2.00 \; + \; 2.50 \; + \; 2.50 = x^2 = 9.00$$

$$x^2_{.01(1)} = 6.64$$

Reject H_0; significant at $P < .01$.

The z value for this problem, shown on page 433, was 3.01, which when squared, equals, within rounding errors, the chi square value of 9.00.

SECTION THREE PROBLEMS

1. A political scientist wishes to assess the possible difference between white and blue-collar workers regarding their preferences for a certain political candidate. A sample of 50 blue-collar workers was randomly selected, and 30 of those said they would vote for the candidate. A random sample of 50 white-collar workers was selected and 20 said they would vote for the candidate. Do the voting proportions in the two sample groups differ significantly?
   Answer: (For all the answers in this section, the z^2 values will equal chi square *within rounding errors*.) The difference is significant: $z = 2.00$, $z.05 = \pm 1.96$. Reject H_0; $p < .05$. Or, as a chi square, $x^2 = 4.00$, $x^2.05(1) = 3.84$. Reject H_0; significant at $p < .05$.

2. A sociologist is interested in discovering whether a difference exists between the sample proportions of men and women autoworkers regarding attitudes toward labor unions. Random samples of 100 men and 100 women were selected from the assembly-line workers at a large auto plant. The workers were asked whether they agreed with the statement "Only through collective bargaining can a worker expect a fair wage." Of the men, 70 agreed, but of the women, only 60 agreed. Test the hypothesis that men and women differ regarding attitudes toward unions.

Answer: The difference is not significant: $z = 1.49$, $z.05 = \pm 1.96$. Accept H_0. Or, as a chi square, $x^2 = 2.18$, $x^2.05(1) = 3.84$. Accept H_0.

3. Two groups of rats were randomly selected. Group A, consisting of 100 rats, was fed a diet containing glutamic acid. Group B, also consisting of 100 rats, was fed an identical diet but without containing the glutamic acid supplement. Both groups were then given 25 reinforced trials on a certain maze. The results of this experiment are as follows:

	ERROR FREE	NOT ERROR FREE
GROUP A	55	45
GROUP B	40	60

Test the hypothesis that glutamic acid affected maze learning among these rats.

Answer: The difference is significant: $z = 2.14$, $z.05 = \pm 1.96$. Reject H_0. Or, as a chi square, $x^2 = 4.51$, $x^2.05(1) = 3.84$. Reject H_0.

4. A market researcher sent a free sample-size box of a certain dishwashing detergent to 200 subjects chosen at random. A separate random sample of 200 subjects was also chosen and received a discount coupon toward the purchase of the product. Six weeks later, all subjects were contacted and data were gathered on the number of subjects from each group who had actually gone out and purchased the product.

	BOUGHT PRODUCT	DID NOT BUY PRODUCT
SAMPLED GROUP	40	160
COUPON GROUP	50	150

Determine whether a significant difference exists between the two groups regarding their purchase of the dishwashing product.

Answer: The difference is not significant: $z = -1.1995$, $z.05 = \pm 1.96$. Accept H_0. Or, as a chi square, $x^2 = 1.43$, $x^2.05(1) = 3.84$. Accept H_0.

5. A psychologist is interested in whether there is a gender difference regarding the manner in which children play in the schoolyard. To investigate this question, data were gathered for the previous year on the number of girls versus boys who had broken at least one bone while playing during recess. By checking the records of a small school district, information was provided on a total of 150 girls and 200 boys.

	BROKEN BONE	NO BROKEN BONE
GIRLS	30	120
BOYS	60	140

Answer: The difference is significant: $z = -2.1186$, $z.05 = \pm 1.96$. Reject H_0. Or, as a chi square, $x^2 = 4.48$, $x^2.05(1) = 3.84$. Reject H_0.

Answer Key

The answers given here are those for all the odd-numbered problems. For all problems requiring mathematical calculations, the rounding method used in this book is to round each result to two digits to the right of the decimal point (the hundredths place). The convention of rounding up one digit when the last dropped number is five or greater has been adopted. This is consistent with the method built into hand calculators (on those models that can be set automati-cally to round off). Although there may well be some minor discrepancies between your answers and those that follow, these will most probably be due to rounding differ-ences and, therefore, should be small. Ex-cept for a few problems in the later chapters where you have been instructed to chain multiply, all calculations are to be rounded, that is, round each and every time you add, subtract, multiply, divide, square, or take a square root.

Chapter 1

1. The difference between the two popula-tion sizes makes it impossible to com-pare them on the basis of the number of fatalities. Although the population of Americans in Vietnam during the war may have exceeded 500,000, the number of Americans in the United States at the time was well over 200 million.

3. On the sole basis of the "evidence" from the commercial, the acid can do nothing at all to your stomach—unless your stomach is made out of a napkin.

5. The assumption may be criticized on a number of points. First, no comparison or control group is mentioned. One could therefore ask, "Twenty-seven per-cent fewer than what?" If it is assumed that 27% fewer means fewer than before the toothpaste was used, then some mention of other variables should be made, for example, age, sex, and brush-ing habits. It is known, for one thing, that teenagers have more dental caries than do adults. Was the before measure done on teenagers, and the after measure taken on these same individuals when

they reached their twenties? Or does the 27% fewer comparison refer to another group of individuals who used a differ-ent kind of toothpaste? If so, again, were the groups equivalent before the study began, that is, same age, sex, brushing habits, and so on?

7. The consumer should first learn the an-swer to the question: "35% more than what?" Without any basis for compari-son the advertiser's claim is meaning-less. Perhaps the new detergent is 35% more effective than seawater, bacon fat, or beer.

9. The conclusions can be criticized on at least three counts. First, antisocial be-havior is not always followed by a court appearance. Perhaps the seemingly high number of these particular juve-niles appearing in court can be attrib-uted to the fact that they did not pos-sess the skills needed to avoid arrest. Second, we would have to know the per-centage of juveniles in that particular area who are assumed to be afflicted with learning disabilities. That is, if 40%

of the juveniles in that area are considered to be learning disabled, then the arrest statistics are simply reflecting the population parameters. In other words, we should be asking the question, "compared to what?" What percentage of the total population of learning-disabled juveniles in that area do not appear in court. Third, we should have some information regarding how the researchers arrived at the diagnosis of "learning disability."

11. At least interval.
13. Ordinal.

Chapter 2

1. $\bar{X}$ = 11.50, Mdn = 12.00, Mo = 12.
3. a. Mean = 10.62.
   b. Median = 12.00.
   c. Mode = 12.
5. a. $\bar{X}$ = 21.25, Mdn = 12.00, Mo = 12.
   b. The median, since the distribution is skewed (to the right or Sk +).
7. a. Not skewed.
   b. Skewed to the right (Sk +).
   c. Skewed to the left (Sk −).

9. a. Mo.
   b. $\bar{X}$.
11. To the left (Sk −).
13. Mode.
15. T.
17. F.
19. F (even a bimodal distribution can be perfectly symmetrical).
21. T.

Chapter 3

1. $\bar{X}$ = 10.38, R = 19, standard deviation = 4.81.
3. At Company X more of the scores are centrally located. Group X is, therefore, somewhat more homogeneous.
5. a. Mean = 2.71.
   b. Range = 2.00.
   c. Standard deviation = .69.
7. R = 0, standard deviation = 0.
9. For a given range, the larger the standard deviation, the more platykurtic the distribution.
11. The standard deviation equals approximately 20.
13. The range has an approximate value of 15 × 6 = 90.
15. T.
17. T.
19. T.
21. T.

Chapter 4

Within Chapter
4a. 1. 39.44%.
    2. 24.86%.
    3. 49.51%.
    4. 3.98%.
    5. 44.74%.
4b. 1. 99.49%.
    2. 14.69%.
    3. 21.48%.
    4. 94.63%.
    5. 55.57%.
4c. 1. 82.90% (83rd percentile).

2. 97.50% (98th percentile).
3. 25.14% (25th percentile).
4. 4.95% (5th percentile).
5. 69.15% (69th percentile).
4d. 1. 55.96%.
    2. 2.28%.
    3. 15.15%.
    4. 99.56%.
    5. 46.41%.
4e. 1. 78.19%.
    2. 53.47%.
    3. 15.34%.

4. 37.99%.
5. 2.69%.
4f. 1. 23.89%.
    2. 33.15%.
    3. 36.86%.
    4. 46.16%.
    5. 39.97%.
4g. 1. 76.11%.
    2. 11.90%.
    3. 92.07%.
    4. 54.78%.
    5. 5.59%.

4h. 1. 7.78%.
2. 46.81%.
3. 73.57%.
4. 56.75%.
5. 22.66%.
4i. 1. 90.04%.
2. 81.15%.
3. 20.62%.
4. 23.28%.
5. 36.91%.

End of Chapter
1. a. 25.80%.

b. 39.97%.
c. 16.28%.
d. 3.59%.
3. a. 10.56%.
b. 95.05%.
c. 9.18%.
d. 57.53%.
5. 62.55%.
7. a. 79.67%.
b. .62%.
c. 44.01%.
d. 15.58%.

9. a. .62%.
b. 6.68%.
c. 30.23%.
d. 13.59%.
11. $\bar{X}$ (the mean).
13. F.
15. T.
17. T.
19. F.
21. F.

Chapter 5

Within Chapter
5a. 1. 108.40.
2. 140.95.
3. 75.40.
4. 98.50.
5. 138.70.
5b. 1. 1.65.
2. − 1.28.
3. − .28.
4. .13.
5. .67.
5c. 1. 1488.30.
2. 1511.70.
3. 1406.40.
4. 1572.90.
5. 1351.50.
5d. 1. 20.00.
2. 28.85.
3. 62.50.
4. 20.27.
5. 26.32.

5e. 1. 53.90.
2. 264.50.
3. 10.05.
4. 83.43.
5. 1496.00.
5f. 1. 39.50.
2. 61.60.
3. 32.10.
4. 78.90.
5. 55.30.
5g. 1. 31.85.
2. 42.65.
3. 23.75.
4. 26.45.
5. 47.60.

End of Chapter
1. a. 15.25.
b. 13.28.
c. 7.96.
d. 7.08.

3. 75.20.
5. a. Standard deviation = 3.60.
b. Range = (6)(3.60) = 21.60.
c. 57.20.
d. 78.80.
7. a. Mean = 2.92.
b. Median = 2.92.
c. Mode = 2.92.
9. a. 51.00.
b. 35.80.
c. 70.30.
d. 45.00.
11. a. 531.50.
b. 340.60.
c. 614.50.
d. 465.10.
13. a. − .67.
b. 43.30.
15. F.
17. T.
19. T.

Chapter 6

Within Chapter
6a. 1. .20.
2. .98.
3. .87.
4. .29.
5. .05.
6b. 1. 41.18 and 42.82.
2. 39.55 and 44.45.
3. 37.80 and 46.20.

4. 31.65 and 52.35.
5. 25.82 and 58.18.
6c. 1. 83.30 and 208.70.
2. 71.52 and 220.48.
3. 91.28 and 200.72.
4. 99.26 and 192.74.
5. 47.96 and 244.04.

End of Chapter

1. a. .17.
b. 5 to 1.
3. a. $P = .02$.
b. 51 to 1.
c. $P = .02$.
5. a. $P = .28$.
b. $P = .13$.
7. $P = .03$.
9. z's of − 1.75 and + 1.75.

11. a. 22.13 and 29.11.
    b. 18.03 and 33.21.
    c. 14.53 and 36.71.
    d. 12.45 and 38.79.
13. $P = .24$.
15. a. $P = .45$.

b. $P = .55$.
c. $P = 1.00$.
17. a. $P = .08$.
    b. $P = .67$.
    c. $P = .17$.
19. 1.00.

21. 50.
23. T.
25. F.
27. T.
29. T.

Chapter 7

1. $P = .31$.
3. a. 2.37.
   b. $z = 2.11$; reject H_0 at $P < .05$.
5. Parameter.
7. Inferential or predictive.
9. Statistic.

11. Random sampling.
13. Bias.
15. Normal.
17. a. $P = .50$.
    b. $P = .50$.
19. T.

21. F.
23. Statistic.
25. Parameter.
27. Statistic.

Chapter 8

1. $t = 1.39$; accept H_0. Population and sample mean do *not* differ significantly at .05.
3. $t = -2.89$; reject H_0 at the .05 level. The sample is probably *not* representative of a population whose mean could be 50.
5. a. The point estimate of parameter mean = 470.
   b. The .95 confidence interval is between 452.46 and 487.54.
   c. No.
   d. In that case, the national mean could have been included within the interval.

7. a. 1.69.
   b. 8.18 to 15.82.
9. a. 1.20.
   b. 6.79 to 12.33.
   c. 5.53 to 13.59.
11. The z, or normal, distribution.
13. Null.
15. Alpha (Type 1) error.
17. 3.11.
19. Confidence interval.
21. Sample.
23. T.
25. T.
27. T.

Chapter 9

1. $t = 1.80$.
   a. Accept H_0 as a two-tail test.
   b. Did not reject.
   c. Reject H_0.
   d. $P = .05$.
3. $t = 2.77$.
   a. No.
   b. The probability of alpha error is .01 as a one-tail test, or .05 as a two-tail.
   c. Since the direction of the difference is specified in the alternative hypothesis, this can be a one-tail test. In either case, however, it's a reject of H_0.

5. Yes, since the t ratio is -2.00, H_0 is rejected. Significant at $P < .05$.
7. Less likely.
9. At least interval.
11. Estimated standard error of difference.
13. Estimated standard error of difference.
15. Null hypothesis (H_0).
17. Alternative hypothesis (H_a).
19. F.
21. T.
23. F.

Chapter 10

1. $r = .91$; reject H_0, $P < .01$. Shows a marked, positive relationship between "party loyalty" and hours worked. Cannot, on the face of it, determine the reason for this relationship. Perhaps party loyalty produces work. Or perhaps the more one works the more one develops a sense of loyalty. Perhaps other factor(s) (age, personality, etc.) produce both.

3. a. $r = -.87$.
   b. Yes, reject H_0. Correlation is significant at $P < .01$.
   c. Both of these hypotheses *may* be valid, although neither is proven by the analysis.

5. $r = .92$; reject H_0, $P < .01$. The evidence from this sample suggests that the two variables are related.

7. $r_s = -.91$; reject H_0, $P < .01$. Sample scores indicate an inverse relationship between reading scores and musical ability. The correlation itself, however, cannot produce the reason for this relationship.

9. a. Not necessarily.
   b. It could mean that higher-IQ parents are more willing to have their taxes increased or can afford to move to areas where taxes are already higher.

11. Pearson r.
13. Spearman r_s.
15. T.
17. F.
19. F.
21. T.
23. F.
25. F.

Chapter 11

1. Hypothesis of difference: post-facto.
3. Hypothesis of difference: post-facto.
5. Hypothesis of association: post-facto.
7. I.V. is gender (assigned-subject).
9. I.V. is level of income (assigned-subject).
11. I.V. is made up of scores on the anxiety test. The prejudice measures comprise the D.V. The I.V. is an assigned-subject variable.
13. This study, since it is before–after with no separate control, is open to several possible confounding variables, such as (1) the Hawthorne effect—perhaps the flattery and attention paid by the researchers produced some gains; (2) growth and development—the children are, after all, six months more mature when the second test was given; and (3) reading experience outside the classroom might produce gains, with or without the special training.
15. The time span between the pre- and post-testing. Subjects may change as a result of a host of other variables, perhaps none of which is related to the alleged independent variable.

17. a. Experimental.
    b. I.V. is whether or not subjects received the psychotherapy, and the D.V. is the judgment of the subject's symptoms.
    c. It would be best to keep the judges on a "blind" basis to help avert any unconscious bias in symptom evaluation.
    d. Perhaps the subjects should also be equated on the basis of gender. The difficulty in this study is not that many matching variables have been overlooked, but in finding subjects who indeed match up on that rather long list of variables.

19. F.
21. T.
23. T.
25. F.

Chapter 12

1. $F = 19.72$; reject H_0, $P < .01$.
3. Accept H_0; $F = 4.00$, and, within rounding error, $t = \sqrt{4.00} = 2.00$.
5. a. $F = 1.34$.
   b. Accept H_0; drug had no effect on retention.
   c. Not appropriate, since F is not significant.
   d. Experimental.
   e. Independent variable—amount of magnesium pemoline. Dependent variable—retention scores.
   f. That the independent variable (magnesium pemoline) has no significant effect on the dependent variable (retention).
7. F for rows (exercise) is 119.10, and is significant. Reject H_0, $P < .01$.

F for columns (temperature) is 41.71, and is significant. Reject H_0, $P < .01$. F for interaction is 2.45, and is not significant. Accept H_0. Whereas each of the main effects had a significant influence on the dependent variable, the interaction had no effect.
9. The total mean.
11. The mean of its own sample.
13. The appropriate degrees of freedom.
15. Between, within.
17. 4 and 25.
19. T.
21. F.
23. T.
25. F.
27. F.

Chapter 13

1. Chi square $= 30.00$; Reject H_0; $P < .01$. The results differ significantly from chance.
3. Chi square $= 1.66$; accept H_0. The results do *not* differ from chance.
5. Chi square $= 33.58$. Reject H_0; significant at $P < .01$.
7. Chi square $= 7.51$.
   a. Reject H_0, $P < .01$.
   b. Post-facto.
   c. Independent variable, sex (assigned). Dependent variable, union membership.

9. Chi square $= 21.39$; reject H_0, $P < .01$. The results differ significantly from chance.
11. Chi square $= 4.95$; accept H_0. Differences are *not* significant.
13. Chi square $= 22.27$; reject H_0, $P < .01$. Significantly more subjects were judged assertive in the posttest.
15. Chi square $= 6.62$; reject H_0, $P < .05$. $C = .24$.
17. H_0: $f_o = f_e$.
19. Matched subjects or before-after.
21. T.
23. T.

Chapter 14

1. a. 545.
   b. 590.
   c. 680.
3. a. $r = .86$; reject H_0, $P < .01$.
   b. 11.55.
5. Predicted IQ $= 125$, the mean of the Y distribution.
7. a. 66.78 to 72.22 (for 69.50).

   b. 68.78 to 74.22 (for 71.50).
   c. 73.28 to 78.72 (for 76.00).
9. Partial correlation $= .14$.
11. 112.85.
13. $r_{y1 \cdot 2} = .08$.
15. Slope.
17. Intercept.
19. Multiple R.

21. .50.
23. Negative.
25. *a* (the intercept).

27. T.
29. F.
31. F.

Chapter 15

1. $r = .79$.
   $t = 2.74$.
   Reject H_0, $P < .05$.
3. $F = 107.35$; reject H_0, $P < .01$.
5. $F = 215.36$; reject H_0, $P < .01$. Higher anxiety is presumably generated by warmer room colors.
7. $t = 3.87$. Reject H_0; difference is significant at $P < .01$.
9. B/A and M/S.
11. Paired t has half the independent t's degrees of freedom for equal numbers of observations.
13. Higher (assuming that the matching process produces a correlation).
15. Dependent.
17. Interval.
19. T.
21. T.
23. T.
25. F.
27. F.

Chapter 16

1. $z_u = 1.81$; accept H_0. Difference is *not* significant.
3. $T = 8.50$; accept H_0. Difference is *not* significant.
5. $H = 6.74$; reject H_0, $P < .05$. Background music increases judge's ranking, especially classical music.
7. $T = 18.50$; accept H_0. Difference is not significant. Inservice workshop did not produce higher scores.
9. Mann–Whitney U test.
11. Friedman ANOVA by ranks.
13. T.
15. F.
17. F.
19. Null.

Appendix

MOMENTS OF THE CURVE

1. In any unimodal distribution, the first moment defines the mean, where the average deviations equal zero. A deviation score equals the raw score minus the mean, $x = X - \bar{X}$.

$$m_1 = \frac{\Sigma x}{N} = 0$$

2. The second moment defines the variance, so that

$$m_2 = \frac{\Sigma x^2}{N}$$

The square root of the variance equals the standard deviation.

$$SD = \sqrt{\frac{\Sigma x^2}{N}}$$

3. The third moment defines skewness, so that

$$m_3 = \frac{\Sigma x^3}{N}$$

As a standard value,

$$Sk = \frac{m_3}{SD^3}$$

Any standard value greater than ± 1.00 indicates marked skewness. (Note, cubing the deviations allows for Sk to take on negative values.)

4. The fourth moment defines kurtosis, such that

$$m_4 = \frac{\Sigma x^4}{N}$$

As a standard value, which will assess the amount of deviation from mesokurtosis,

$$Ku = \frac{m_4}{SD^4} - 3.00.$$

(Note, this is because $m_4/SD^4 = 3.00$ defines mesokurtosis.)

Leptokurtic distributions will yield a positive standard value, whereas platykurtic distributions produce a negative standard value. Extreme leptokurtic curves produce a positive value of greater than +1.00, such as +1.50 and extreme platykurtic curves a negative value of greater than −1.00 such as −1.50.

Table A Percent of area under the normal curve between $\bar{X}$ and z.

z	.00	.01	.02	.03	.04	.05	.06	.07	.08	.09
0.0	00.00	00.40	00.80	01.20	01.60	01.99	02.39	02.79	03.19	03.59
0.1	03.98	04.38	04.78	05.17	05.57	05.96	06.36	06.75	07.14	07.53
0.2	07.93	08.32	08.71	09.10	09.48	09.87	10.26	10.64	11.03	11.41
0.3	11.79	12.17	12.55	12.93	13.31	13.68	14.06	14.43	14.80	15.17
0.4	15.54	15.91	16.28	16.64	17.00	17.36	17.72	18.08	18.44	18.79
0.5	19.15	19.50	19.85	20.19	20.54	20.88	21.23	21.57	21.90	22.24
0.6	22.57	22.91	23.24	23.57	23.89	24.22	24.54	24.86	25.17	25.49
0.7	25.80	26.11	26.42	26.73	27.04	27.34	27.64	27.94	28.23	28.52
0.8	28.81	29.10	29.39	29.67	29.95	30.23	30.51	30.78	31.06	31.33
0.9	31.59	31.86	32.12	32.38	32.64	32.90	33.15	33.40	33.65	33.89
1.0	34.13	34.38	34.61	34.85	35.08	35.31	35.54	35.77	35.99	36.21
1.1	36.43	36.65	36.86	37.08	37.29	37.49	37.70	37.90	38.10	38.30
1.2	38.49	38.69	38.88	39.07	39.25	39.44	39.62	39.80	39.97	40.15
1.3	40.32	40.49	40.66	40.82	40.99	41.15	41.31	41.47	41.62	41.77
1.4	41.92	42.07	42.22	42.36	42.51	42.65	42.79	42.92	43.06	43.19
1.5	43.32	43.45	43.57	43.70	43.83	43.94	44.06	44.18	44.29	44.41
1.6	44.52	44.63	44.74	44.84	44.95	45.05	45.15	45.25	45.35	45.45
1.7	45.54	45.64	45.73	45.82	45.91	45.99	46.08	46.16	46.25	46.33
1.8	46.41	46.49	46.56	46.64	46.71	46.78	46.86	46.93	46.99	47.06
1.9	47.13	47.19	47.26	47.32	47.38	47.44	47.50	47.56	47.61	47.67
2.0	47.72	47.78	47.83	47.88	47.93	47.98	48.03	48.08	48.12	48.17
2.1	48.21	48.26	48.30	48.34	48.38	48.42	48.46	48.50	48.54	48.57
2.2	48.61	48.64	48.68	48.71	48.75	48.78	48.81	48.84	48.87	48.90
2.3	48.93	48.96	48.98	49.01	49.04	49.06	49.09	49.11	49.13	49.16
2.4	49.18	49.20	49.22	49.25	49.27	49.29	49.31	49.32	49.34	49.36
2.5	49.38	49.40	49.41	49.43	49.45	49.46	49.48	49.49	49.51	49.52
2.6	49.53	49.55	49.56	49.57	49.59	49.60	49.61	49.62	49.63	49.64
2.7	49.65	49.66	49.67	49.68	49.69	49.70	49.71	49.72	49.73	49.74
2.8	49.74	49.75	49.76	49.77	49.77	49.78	49.79	49.79	49.80	49.81
2.9	49.81	49.82	49.82	49.83	49.84	49.84	49.85	49.85	49.86	49.86
3.0	49.87									
4.0	49.997									

Source: Karl Pearson, *Tables for Statisticians and Biometricians,* Cambridge University Press, London, pp. 98–101, by permission of the Biometrika Trustees.

Table B Conversion table—percentiles to z scores.

Percentile	z Score	Percentile	z Score	Percentile	z Score	Percentile	z Score
1st	−2.41	26th	−0.64	51st	0.03	76th	0.71
2nd	−2.05	27th	−0.61	52nd	0.05	77th	0.74
3rd	−1.88	28th	−0.58	53rd	0.08	78th	0.77
4th	−1.75	29th	−0.55	54th	0.10	79th	0.81
5th	−1.65	30th	−0.52	55th	0.13	80th	0.84
6th	−1.56	31st	−0.50	56th	0.15	81st	0.88
7th	−1.48	32nd	−0.47	57th	0.18	82nd	0.92
8th	−1.41	33rd	−0.44	58th	0.20	83rd	0.95
9th	−1.34	34th	−0.41	59th	0.23	84th	1.00
10th	−1.28	35th	−0.39	60th	0.25	85th	1.04
11th	−1.23	36th	−0.36	61st	0.28	86th	1.08
12th	−1.18	37th	−0.33	62nd	0.31	87th	1.13
13th	−1.13	38th	−0.31	63rd	0.33	88th	1.18
14th	−1.08	39th	−0.28	64th	0.36	89th	1.23
15th	−1.04	40th	−0.25	65th	0.39	90th	1.28
16th	−1.00	41st	−0.23	66th	0.41	91st	1.34
17th	−0.95	42nd	−0.20	67th	0.44	92nd	1.41
18th	−0.92	43rd	−0.18	68th	0.47	93rd	1.48
19th	−0.88	44th	−0.15	69th	0.50	94th	1.56
20th	−0.84	45th	−0.13	70th	0.52	95th	1.65
21st	−0.81	46th	−0.10	71st	0.55	96th	1.75
22nd	−0.77	47th	−0.08	72nd	0.58	97th	1.88
23rd	−0.74	48th	−0.05	73rd	0.61	98th	2.05
24th	−0.71	49th	−0.03	74th	0.64	99th	2.41
25th	−0.67	50th	0.00	75th	0.67	100th	∞

Table C Critical values of *t*

df	LEVEL OF SIGNIFICANCE FOR TWO-TAIL TEST	
	.05	.01
1	12.706	63.657
2	4.303	9.925
3	3.182	5.841
4	2.776	4.604
5	2.571	4.032
6	2.447	3.707
7	2.365	3.499
8	2.306	3.355
9	2.262	3.250
10	2.228	3.169
11	2.201	3.106
12	2.179	3.055
13	2.160	3.012
14	2.145	2.977
15	2.131	2.947
16	2.120	2.921
17	2.110	2.898
18	2.101	2.878
19	2.093	2.861
20	2.086	2.845
21	2.080	2.831
22	2.074	2.819
23	2.069	2.807
24	2.064	2.797
25	2.060	2.787
26	2.056	2.779
27	2.052	2.771
28	2.048	2.763
29	2.045	2.756
30	2.042	2.750
40	2.021	2.704
60	2.000	2.660
120	1.980	2.617
∞	1.960	2.576

Source: Abridged from Table III of Fisher and Yates, *Statistical Tables for Biological, Agricultural, and Medical Research.* Published by Longman Group Ltd., London (previously published by Oliver and Boyd Ltd., Edinburgh), and by permission of the authors and publishers.

Table D Critical values of *t*

df	LEVEL OF SIGNIFICANCE FOR ONE-TAIL TEST	
	.05	.01
1	6.314	31.821
2	2.920	6.965
3	2.353	4.541
4	2.132	3.747
5	2.015	3.365
6	1.943	3.143
7	1.895	2.998
8	1.860	2.896
9	1.833	2.821
10	1.812	2.764
11	1.796	2.718
12	1.782	2.681
13	1.771	2.650
14	1.761	2.624
15	1.753	2.602
16	1.746	2.583
17	1.740	2.567
18	1.734	2.552
19	1.729	2.539
20	1.725	2.528
21	1.721	2.518
22	1.717	2.508
23	1.714	2.500
24	1.711	2.492
25	1.708	2.485
26	1.706	2.479
27	1.703	2.473
28	1.701	2.467
29	1.699	2.462
30	1.697	2.457
40	1.684	2.423
60	1.671	2.390
120	1.658	2.358
∞	1.645	2.326

Source: Abridged from Table III of Fisher and Yates, *Statistical Tables for Biological, Agricultural, and Medical Research.* Published by Longman Group Ltd., London (previously published by Oliver and Boyd Ltd., Edinburgh), and by permission of the authors and publishers.

Table E Critical values of r for the Pearson correlation coefficient (degrees of freedom = number of pairs of scores − 2).

df	$\alpha = .05$	$\alpha = .01$	df	$\alpha = .05$	$\alpha = .01$
1	.997	.9999	21	.413	.526
2	.950	.990	22	.404	.515
3	.878	.959	23	.396	.505
4	.811	.917	24	.388	.496
5	.754	.874	25	.381	.487
6	.707	.834	26	.374	.479
7	.666	.798	27	.367	.471
8	.632	.765	28	.361	.463
9	.602	.735	29	.355	.456
10	.576	.708	30	.349	.449
11	.553	.684	35	.325	.418
12	.532	.661	40	.304	.393
13	.514	.641	45	.288	.372
14	.497	.623	50	.273	.354
15	.482	.606	60	.250	.325
16	.468	.590	70	.232	.302
17	.456	.575	80	.217	.283
18	.444	.561	90	.205	.267
19	.433	.549	100	.195	.254
20	.423	.537			

If your calculated r is greater than table r, reject H_0. If your value of degrees of freedom is not listed, use table r for the next smaller value of degrees of freedom.

Source: Table VI of Fisher and Yates: *Statistical Tables for Biological, Agricultural and Medical Research,* 6th edition, 1974, published by Longman Group Ltd., London; (previously published by Oliver and Boyd Ltd., Edinburgh), and by permission of the authors and publishers.

Table F Critical values for the Spearman rank-order correlation coefficient (N = number of pairs of scores).

N	.05	.01	N	.05	.01
5	1.000		18	.474	.600
6	.886	1.000	19	.460	.585
7	.786	.929	20	.447	.570
8	.715	.881	21	.437	.556
9	.700	.834	22	.426	.544
10	.649	.794	23	.417	.532
11	.619	.764	24	.407	.521
12	.588	.735	25	.399	.511
13	.561	.704	26	.391	.501
14	.539	.680	27	.383	.493
15	.522	.658	28	.376	.484
16	.503	.636	29	.369	.475
17	.488	.618	30	.363	.467

Source: Glasser, G. J., and R. F. Winter, "Critical Values of the Coefficient of Rank Correlation for Testing the Hypothesis of Independence," *Biometrika,* 48, 444 (1961).

Table G Critical values of F for the analysis of variance.

		1	2	3	4	5	6	7	8	9	10	11	12
					DEGREES OF FREEDOM FOR NUMERATOR								
1		161	200	216	225	230	234	237	239	241	242	243	244
		4052	**4999**	**5403**	**5625**	**5764**	**5859**	**5928**	**5981**	**6022**	**6056**	**6082**	**6106**
2		18.51	19.00	19.16	19.25	19.30	19.33	19.36	19.37	19.38	19.39	19.40	19.41
		98.49	**99.01**	**99.17**	**99.25**	**99.30**	**99.33**	**99.34**	**99.36**	**99.38**	**99.40**	**99.41**	**99.42**
3		10.13	9.55	9.28	9.12	9.01	8.94	8.88	8.84	8.81	8.78	8.76	8.74
		34.12	**30.81**	**29.46**	**28.71**	**28.24**	**27.91**	**27.67**	**27.49**	**27.34**	**27.23**	**27.13**	**27.05**
4		7.71	6.94	6.59	6.39	6.26	6.16	6.09	6.04	6.00	5.96	5.93	5.91
		21.20	**18.00**	**16.69**	**15.98**	**15.52**	**15.21**	**14.98**	**14.80**	**14.66**	**14.54**	**14.45**	**14.37**
5		6.61	5.79	5.41	5.19	5.05	4.95	4.88	4.82	4.78	4.74	4.70	4.68
		16.26	**13.27**	**12.06**	**11.39**	**10.97**	**10.67**	**10.45**	**10.27**	**10.15**	**10.05**	**9.96**	**9.89**
6		5.99	5.14	4.76	4.53	4.39	4.28	4.21	4.15	4.10	4.06	4.03	4.00
		13.74	**10.92**	**9.78**	**9.15**	**8.75**	**8.47**	**8.26**	**8.10**	**7.98**	**7.87**	**7.79**	**7.72**
7		5.59	4.74	4.35	4.12	3.97	3.87	3.79	3.73	3.68	3.63	3.60	3.57
		12.25	**9.55**	**8.45**	**7.85**	**7.46**	**7.19**	**7.00**	**6.84**	**6.71**	**6.62**	**6.54**	**6.47**
8		5.32	4.46	4.07	3.84	3.69	3.58	3.50	3.44	3.39	3.34	3.31	3.28
		11.26	**8.65**	**7.59**	**7.01**	**6.63**	**6.37**	**6.19**	**6.03**	**5.91**	**5.82**	**5.74**	**5.67**
9		5.12	4.26	3.86	3.63	3.48	3.37	3.29	3.23	3.18	3.13	3.10	3.07
		10.56	**8.02**	**6.99**	**6.42**	**6.06**	**5.80**	**5.62**	**5.47**	**5.35**	**5.26**	**5.18**	**5.11**
10		4.96	4.10	3.71	3.48	3.33	3.22	3.14	3.07	3.02	2.97	2.94	2.91
		10.04	**7.56**	**6.55**	**5.99**	**5.64**	**5.39**	**5.21**	**5.06**	**4.95**	**4.85**	**4.78**	**4.71**
11		4.84	3.98	3.59	3.36	3.20	3.09	3.01	2.95	2.90	2.86	2.82	2.79
		9.65	**7.20**	**6.22**	**5.67**	**5.32**	**5.07**	**4.88**	**4.74**	**4.63**	**4.54**	**4.46**	**4.40**
12		4.75	3.88	3.49	3.26	3.11	3.00	2.92	2.85	2.80	2.76	2.72	2.69
		9.33	**6.93**	**5.95**	**5.41**	**5.06**	**4.82**	**4.65**	**4.50**	**4.39**	**4.30**	**4.22**	**4.16**
13		4.67	3.80	3.41	3.18	3.02	2.92	2.84	2.77	2.72	2.67	2.63	2.60
		9.07	**6.70**	**5.74**	**5.20**	**4.86**	**4.62**	**4.44**	**4.30**	**4.19**	**4.10**	**4.02**	**3.96**
14		4.60	3.74	3.34	3.11	2.96	2.85	2.77	2.70	2.65	2.60	2.56	2.53
		8.86	**6.51**	**5.56**	**5.03**	**4.69**	**4.46**	**4.28**	**4.14**	**4.03**	**3.94**	**3.86**	**3.80**
15		4.54	3.68	3.29	3.06	2.90	2.79	2.70	2.64	2.59	2.55	2.51	2.48
		8.68	**6.36**	**5.42**	**4.89**	**4.56**	**4.32**	**4.14**	**4.00**	**3.89**	**3.80**	**3.73**	**3.67**
16		4.49	3.63	3.24	3.01	2.85	2.74	2.66	2.59	2.54	2.49	2.45	2.42
		8.53	**6.23**	**5.29**	**4.77**	**4.44**	**4.20**	**4.03**	**3.89**	**3.78**	**3.69**	**3.61**	**3.55**
17		4.45	3.59	3.20	2.96	2.81	2.70	2.62	2.55	2.50	2.45	2.41	2.38
		8.40	**6.11**	**5.18**	**4.67**	**4.34**	**4.10**	**3.93**	**3.79**	**3.68**	**3.59**	**3.52**	**3.45**
18		4.41	3.55	3.16	2.93	2.77	2.66	2.58	2.51	2.46	2.41	2.37	2.34
		8.28	**6.01**	**5.09**	**4.58**	**4.25**	**4.01**	**3.85**	**3.71**	**3.60**	**3.51**	**3.44**	**3.37**

Table G   (con't)

	DEGREES OF FREEDOM FOR NUMERATOR											
	1	*2*	*3*	*4*	*5*	*6*	*7*	*8*	*9*	*10*	*11*	*12*
19	4.38	3.52	3.13	2.90	2.74	2.63	2.55	2.48	2.43	2.38	2.34	2.31
	8.18	**5.93**	**5.01**	**4.50**	**4.17**	**3.94**	**3.77**	**3.63**	**3.52**	**3.43**	**3.36**	**3.30**
20	4.35	3.49	3.10	2.87	2.71	2.60	2.52	2.45	2.40	2.35	2.31	2.28
	8.10	**5.85**	**4.94**	**4.43**	**4.10**	**3.87**	**3.71**	**3.56**	**3.45**	**3.37**	**3.30**	**3.23**
21	4.32	3.47	3.07	2.84	2.68	2.57	2.49	2.42	2.37	2.32	2.28	2.25
	8.02	**5.78**	**4.87**	**4.37**	**4.04**	**3.81**	**3.65**	**3.51**	**3.40**	**3.31**	**3.24**	**3.17**
22	4.30	3.44	3.05	2.82	2.66	2.55	2.47	2.40	2.35	2.30	2.26	2.23
	7.94	**5.72**	**4.82**	**4.31**	**3.99**	**3.76**	**3.59**	**3.45**	**3.35**	**3.26**	**3.18**	**3.12**
23	4.28	3.42	3.03	2.80	2.64	2.53	2.45	2.38	2.32	2.28	2.24	2.20
	7.88	**5.66**	**4.76**	**4.26**	**3.94**	**3.71**	**3.54**	**3.41**	**3.30**	**3.21**	**3.14**	**3.07**
24	4.26	3.40	3.01	2.78	2.62	2.51	2.43	2.36	2.30	2.26	2.22	2.18
	7.82	**5.61**	**4.72**	**4.22**	**3.90**	**3.67**	**3.50**	**3.36**	**3.25**	**3.17**	**3.09**	**3.03**
25	4.24	3.38	2.99	2.76	2.60	2.49	2.41	2.34	2.28	2.24	2.20	2.16
	7.77	**5.57**	**4.68**	**4.18**	**3.86**	**3.63**	**3.46**	**3.32**	**3.21**	**3.13**	**3.05**	**2.99**
26	4.22	3.37	2.89	2.74	2.59	2.47	2.39	2.32	2.27	2.22	2.18	2.15
	7.72	**5.53**	**4.64**	**4.14**	**3.82**	**3.59**	**3.42**	**3.29**	**3.17**	**3.09**	**3.02**	**2.96**
27	4.21	3.35	2.96	2.73	2.57	2.46	2.37	2.30	2.25	2.20	2.16	2.13
	7.68	**5.49**	**4.60**	**4.11**	**3.79**	**3.56**	**3.39**	**3.26**	**3.14**	**3.06**	**2.98**	**2.93**
28	4.20	3.34	2.95	2.71	2.56	2.44	2.36	2.29	2.24	2.19	2.15	2.12
	7.64	**5.45**	**4.57**	**4.07**	**3.76**	**3.53**	**3.36**	**3.23**	**3.11**	**3.03**	**2.95**	**2.90**
29	4.18	3.33	2.93	2.70	2.54	2.43	2.35	2.28	2.22	2.18	2.14	2.10
	7.60	**5.52**	**4.54**	**4.04**	**3.73**	**3.50**	**3.32**	**3.20**	**3.08**	**3.00**	**2.92**	**2.87**
30	4.17	3.32	2.92	2.69	2.53	2.42	2.34	2.27	2.21	2.16	2.12	2.09
	7.56	**5.39**	**4.51**	**4.02**	**3.70**	**3.47**	**3.30**	**3.17**	**3.06**	**2.98**	**2.90**	**2.84**
32	4.15	3.30	2.90	2.67	2.51	2.40	2.32	2.25	2.19	2.14	2.10	2.07
	7.50	**5.34**	**4.46**	**3.97**	**3.66**	**3.42**	**3.25**	**3.12**	**3.01**	**2.94**	**2.86**	**2.80**
34	4.13	3.28	2.88	2.65	2.49	2.38	2.30	2.23	2.17	2.12	2.08	2.05
	7.44	**5.29**	**4.42**	**3.93**	**3.61**	**3.38**	**3.21**	**3.08**	**2.97**	**2.89**	**2.82**	**2.76**
36	4.11	3.26	2.86	2.63	2.48	2.36	2.28	2.21	2.15	2.10	2.06	2.03
	7.39	**5.25**	**4.38**	**3.89**	**3.58**	**3.35**	**3.18**	**3.04**	**2.94**	**2.86**	**2.78**	**2.72**
38	4.10	3.25	2.85	2.62	2.46	2.35	2.26	2.19	2.14	2.09	2.05	2.02
	7.35	**5.21**	**4.34**	**3.86**	**3.54**	**3.32**	**3.15**	**3.02**	**2.91**	**2.82**	**2.75**	**2.69**
40	4.08	3.23	2.84	2.61	2.45	2.34	2.25	2.18	2.12	2.07	2.04	2.00
	7.31	**5.18**	**4.31**	**3.83**	**3.51**	**3.29**	**3.12**	**2.99**	**2.88**	**2.80**	**2.73**	**2.66**
42	4.07	3.22	2.83	2.59	2.44	2.32	2.24	2.17	2.11	2.06	2.02	1.90
	7.27	**5.15**	**4.29**	**3.80**	**3.49**	**3.26**	**3.10**	**2.96**	**2.86**	**2.77**	**2.70**	**2.64**

Table G (con't)

	DEGREES OF FREEDOM FOR NUMERATOR											
	1	2	3	4	5	6	7	8	9	10	11	12
44	4.06	3.21	2.82	2.58	2.43	2.31	2.23	2.16	2.10	2.05	2.01	1.98
	7.24	**5.12**	**4.26**	**3.78**	**3.46**	**3.24**	**3.07**	**2.94**	**2.84**	**2.75**	**2.68**	**2.62**
46	4.05	3.20	2.81	2.57	2.42	2.30	2.22	2.14	2.09	2.04	2.00	1.97
	7.21	**5.10**	**4.24**	**3.76**	**3.44**	**3.22**	**3.05**	**2.92**	**2.82**	**2.73**	**2.66**	**2.60**
48	4.04	3.19	2.80	2.56	2.41	2.30	2.21	2.14	2.08	2.03	1.99	1.96
	7.19	**5.08**	**4.22**	**3.74**	**3.42**	**3.20**	**3.04**	**2.90**	**2.80**	**2.71**	**2.64**	**2.58**
50	4.03	3.18	2.79	2.56	2.40	2.29	2.20	2.13	2.07	2.02	1.98	1.95
	7.17	**5.06**	**4.20**	**3.72**	**3.41**	**3.18**	**3.02**	**2.88**	**2.78**	**2.70**	**2.62**	**2.56**
55	4.02	3.17	2.78	2.54	2.38	2.27	2.18	2.11	2.05	2.00	1.97	1.93
	7.12	**5.01**	**4.16**	**3.68**	**3.37**	**3.15**	**2.98**	**2.85**	**2.75**	**2.66**	**2.59**	**2.53**
60	4.00	3.15	2.76	2.52	2.37	2.25	2.17	2.10	2.04	1.99	1.95	1.92
	7.08	**4.98**	**4.13**	**3.65**	**3.34**	**3.12**	**2.95**	**2.82**	**2.72**	**2.63**	**2.56**	**2.50**
65	3.99	3.14	2.75	2.51	2.36	2.24	2.15	2.08	2.02	1.98	1.94	1.90
	7.04	**4.95**	**4.10**	**3.62**	**3.31**	**3.09**	**2.93**	**2.79**	**2.70**	**2.61**	**2.54**	**2.47**
70	3.98	3.13	2.74	2.50	2.35	2.22	2.14	2.07	2.01	1.97	1.93	1.89
	7.01	**4.92**	**4.08**	**3.60**	**3.29**	**3.07**	**2.91**	**2.77**	**2.67**	**2.59**	**2.51**	**2.45**
80	3.96	3.11	2.72	2.48	2.33	2.21	2.12	2.05	1.99	1.95	1.91	1.88
	6.96	**4.88**	**4.04**	**3.56**	**3.25**	**3.04**	**2.87**	**2.74**	**2.64**	**2.55**	**2.48**	**2.41**
100	3.94	3.09	2.70	2.46	2.30	2.19	2.10	2.03	1.97	1.92	1.88	1.85
	6.90	**4.82**	**3.98**	**3.51**	**3.20**	**2.99**	**2.82**	**2.69**	**2.59**	**2.51**	**2.43**	**2.36**
125	3.92	3.07	2.68	2.44	2.29	2.17	2.08	2.01	1.95	1.90	1.86	1.83
	6.84	**4.78**	**3.94**	**3.47**	**3.17**	**2.95**	**2.79**	**2.65**	**2.56**	**2.47**	**2.40**	**2.33**
150	3.91	3.06	2.67	2.43	2.27	2.16	2.07	2.00	1.94	1.89	1.85	1.82
	6.81	**4.75**	**3.91**	**3.44**	**3.13**	**2.92**	**2.76**	**2.62**	**2.53**	**2.44**	**2.37**	**2.30**
200	3.89	3.04	2.65	2.41	2.26	2.14	2.05	1.98	1.92	1.87	1.83	1.80
	6.76	**4.71**	**3.88**	**3.41**	**3.11**	**2.90**	**2.73**	**2.60**	**2.50**	**2.41**	**2.34**	**2.28**
400	3.86	3.02	2.62	2.39	2.23	2.12	2.03	1.96	1.90	1.85	1.81	1.78
	6.70	**4.66**	**3.83**	**3.36**	**3.06**	**2.85**	**2.69**	**2.55**	**2.46**	**2.37**	**2.29**	**2.23**
1000	3.85	3.00	2.61	2.38	2.22	2.10	2.02	1.95	1.89	1.84	1.80	1.76
	6.66	**4.62**	**3.80**	**3.34**	**3.04**	**2.82**	**2.66**	**2.53**	**2.43**	**2.34**	**2.26**	**2.20**
∞	3.84	2.99	2.60	2.37	2.21	2.09	2.01	1.94	1.88	1.83	1.79	1.75
	6.64	**4.60**	**3.78**	**3.32**	**3.02**	**2.80**	**2.64**	**2.51**	**2.41**	**2.32**	**2.24**	**2.18**

Values of F for $\alpha = .05$ are given in lightface type, and values of F for $\alpha = .01$ are given in boldface type.

Source: Reprinted by permission from *Statistical Methods* by George W. Snedecor and William G. Cochran © 1980 by The Iowa State University Press, Ames, Iowa 50010.

Table H Percentage points of the studentized range (Critical values for Tukey's HSD).

MS_w df	α	2	3	4	5	6	7	8	9	10	11
					$k = $	NUMBER	OF MEANS				
5	.05	3.64	4.60	5.22	5.67	6.03	6.33	6.58	6.80	6.99	7.17
	.01	5.70	6.98	7.80	8.42	8.91	9.32	9.67	9.97	10.24	10.48
6	.05	3.46	4.34	4.90	5.30	5.63	5.90	6.12	6.32	6.49	6.65
	.01	5.24	6.33	7.03	7.56	7.97	8.32	8.61	8.87	9.10	9.30
7	.05	3.34	4.16	4.68	5.06	5.36	5.61	5.82	6.00	6.16	6.30
	.01	4.95	5.92	6.54	7.01	7.37	7.68	7.94	8.17	8.37	8.55
8	.05	3.26	4.04	4.53	4.89	5.17	5.40	5.60	5.77	5.92	6.05
	.01	4.75	5.64	6.20	6.62	6.96	7.24	7.47	7.68	7.86	8.03
9	.05	3.20	3.95	4.41	4.76	5.02	5.24	5.43	5.59	5.74	5.87
	.01	4.60	5.43	5.96	6.35	6.66	6.91	7.13	7.33	7.49	7.65
10	.05	3.15	3.88	4.33	4.65	4.91	5.12	5.30	5.46	5.60	5.72
	.01	4.48	5.27	5.77	6.14	6.43	6.67	6.87	7.05	7.21	7.36
11	.05	3.11	3.82	4.26	4.57	4.82	5.03	5.20	5.35	5.49	5.61
	.01	4.39	5.15	5.62	5.97	6.25	6.48	6.67	6.84	6.99	7.13
12	.05	3.08	3.77	4.20	4.51	4.75	4.95	5.12	5.27	5.39	5.51
	.01	4.32	5.05	5.50	5.84	6.10	6.32	6.51	6.67	6.81	6.94
13	.05	3.06	3.73	4.15	4.45	4.69	4.88	5.05	5.19	5.32	5.43
	.01	4.26	4.96	5.40	5.73	5.98	6.19	6.37	6.53	6.67	6.79
14	.05	3.03	3.70	4.11	4.41	4.64	4.83	4.99	5.13	5.25	5.36
	.01	4.21	4.89	5.32	5.63	5.88	6.08	6.26	6.41	6.54	6.66
15	.05	3.01	3.67	4.08	4.37	4.59	4.78	4.94	5.08	5.20	5.31
	.01	4.17	4.84	5.25	5.56	5.80	5.99	6.16	6.31	6.44	6.55
16	.05	3.00	3.65	4.05	4.33	4.56	4.74	4.90	5.03	5.15	5.26
	.01	4.13	4.79	5.19	5.49	5.72	5.92	6.08	6.22	6.35	6.46
17	.05	2.98	3.63	4.02	4.30	4.52	4.70	4.86	4.99	5.11	5.21
	.01	4.10	4.74	5.14	5.43	5.66	5.85	6.01	6.15	6.27	6.38
18	.05	2.97	3.61	4.00	4.28	4.49	4.67	4.82	4.96	5.07	5.17
	.01	4.07	4.70	5.09	5.38	5.60	5.79	5.94	6.08	6.20	6.31
19	.05	2.96	3.59	3.98	4.25	4.47	4.65	4.79	4.92	5.04	5.14
	.01	4.05	4.67	5.05	5.33	5.55	5.73	5.89	6.02	6.14	6.25
20	.05	2.95	3.58	3.96	4.23	4.45	4.62	4.77	4.90	5.01	5.11
	.01	4.02	4.64	5.02	5.29	5.51	5.69	5.84	5.97	6.09	6.19
24	.05	2.92	3.53	3.90	4.17	4.37	4.54	4.68	4.81	4.92	5.01
	.01	3.96	4.55	4.91	5.17	5.37	5.54	5.69	5.81	5.92	6.02
30	.05	2.89	3.49	3.85	4.10	4.30	4.46	4.60	4.72	4.82	4.92
	.01	3.89	4.45	4.80	5.05	5.24	5.40	5.54	5.65	5.76	5.85
40	.05	2.86	3.44	3.79	4.04	4.23	4.39	4.52	4.63	4.73	4.82
	.01	3.82	4.37	4.70	4.93	5.11	5.26	5.39	5.50	5.60	5.69

Table H (con't)

MS$_w$ df	α	2	3	4	5	6	7	8	9	10	11
					k = NUMBER OF MEANS						
60	.05	2.83	3.40	3.74	3.98	4.16	4.31	4.44	4.55	4.65	4.73
	.01	3.76	4.28	4.59	4.82	4.99	5.13	5.25	5.36	5.45	5.53
120	.05	2.80	3.36	3.68	3.92	4.10	4.24	4.36	4.47	4.56	4.64
	.01	3.70	4.20	4.50	4.71	4.87	5.01	5.12	5.21	5.30	5.37
∞	.05	2.77	3.31	3.63	3.86	4.03	4.17	4.29	4.39	4.47	4.55
	.01	3.64	4.12	4.40	4.60	4.76	4.88	4.99	5.08	5.16	5.23

Source: E. S. Pearson and H. O. Hartley, *Biometrika Tables for Statisticians,* Vol. 1, 3rd ed., Cambridge Press, New York, 1966, by permission of the Biometrika Trustees.

Table I Critical values of chi square.

df	.05	.01	df	.05	.01
1	3.84	6.64	16	26.30	32.00
2	5.99	9.21	17	27.59	33.41
3	7.82	11.34	18	28.87	34.80
4	9.49	13.28	19	30.14	36.19
5	11.07	15.09	20	31.41	37.57
6	12.59	16.81	21	32.67	38.93
7	14.07	18.48	22	33.92	40.29
8	15.51	20.09	23	35.17	41.64
9	16.92	21.67	24	36.42	42.98
10	18.31	23.21	25	37.65	44.31
11	19.68	24.72	26	38.88	45.64
12	21.03	26.22	27	40.11	46.96
13	22.36	27.69	28	41.34	48.28
14	23.68	29.14	29	42.56	49.59
15	25.00	30.58	30	43.77	50.89

Source: Table I is abridged from Table IV of Fisher and Yates: *Statistical Tables for Biological, Agricultural, and Medical Research,* Published by Longman Group Ltd., London (previously published by Oliver and Boyd Ltd., Edinburgh), and by permission of the authors and publishers.

Table J   Values of T at the .05 and .01 levels of significance in the Wilcoxon signed-ranks test.

N	.05	.01	N	.05	.01
6	0	—	16	30	20
7	2	—	17	35	23
8	4	0	18	40	28
9	6	2	19	46	32
10	8	3	20	52	38
11	11	5	21	59	43
12	14	7	22	66	49
13	17	10	23	73	55
14	21	13	24	81	61
15	25	16	25	89	58

Source: Table J is abridged from Table I of G. Wilcoxon, *Some rapid approximate statistical procedures,* New York, American Cyanamid Co., 1949, with the permission of the publisher.

Glossary

Abscissa The horizontal or X axis of the co-ordinate system. On a frequency distribution, the abscissa typically measures the variable in question (performance measures), whereas the Y axis (ordinate) represents the frequency of occurrence.

Alpha Error (or Type 1 Error) The probability of being wrong whenever the null hypothesis is rejected, or the probability of rejecting the null hypothesis when it should have been accepted. By definition, then, the alpha error can only occur when H_0 has been rejected.

Alternative Hypothesis The opposite of the null hypothesis. The alternative hypothesis states that chance has been ruled out—that there are population differences (when testing the hypothesis of difference) or that a correlation does exist in the population (when testing the hypothesis of association).

Analysis of Covariance (ANCOVA) A statistical procedure designed to control the effects of any variable(s) which is known to be correlated with the dependent variable. ANCOVA is typically used as an after-the-fact adjustment for controlling any possible differences which may have already existed between comparison groups.

Analysis of Variance Statistical test of significance developed by Sir Ronald Fisher. It is also called the F ratio, or ANOVA, for ANalysis Of VAriance. The test is designed to establish whether or not a significant (nonchance) difference exists among several sample means. Statistically, it is the ratio of the variance occurring between the sample means to the variance occurring within the sample groups. A large F ratio, that is when the variance between is larger than the variance within, usually indicates a nonchance or significant difference.

Beta Coefficient (b) or Slope In a scatter plot, the slope of the regression line indicates how much change on the Y variable accompanies a one-unit change in the X variable. When the slope is positive (lower left to upper right), Y will show an increase as X increases, whereas a negative slope (upper left to lower right) indicates a decrease in Y is accompanying an increase in X. In the regression equation, $Y = bX + a$, the slope is symbolized by the b term.

$$b = \frac{rSD_y}{SD_x} \quad \text{or} \quad \frac{rs_y}{s_x}$$

Beta Error (or Type 2 Error) The probability of being wrong whenever the null hypothesis is accepted, or the probability of accepting the null hypothesis when it should have been rejected.

Bias Systematic or nonrandom sampling error. Occurs when the difference between $\bar{X}$ and μ is consistently in *one direction*. Bias results when samples *are not* representative of the population.

Central Limit Theorem The theoretical statement that when sample means are selected randomly from a single population, the means will distribute as an approximation of the normal distribution, even if the population distribution deviates from normality. The theorem assumes that sample sizes are relatively large (at least 30) and that they are all selected from *one* population.

Central Tendency (Measures of) A statistical term used for describing the typical, middle, or central scores in a distribution of scores. Measures of central tendency are used when the researcher wants to describe a group as a whole with a view toward characterizing that group on the basis of its most common measurement. The researcher wishes to know what score best represents a group of differing scores. The three measures of central tendency are the mean (or arithmetic average), the median (or the midpoint of the distribution), and the mode (the most frequently occurring score in the distribution).

Chi Square (χ^2)   A statistical test of significance used to determine whether or not frequency differences have occurred on the basis of chance. Chi square requires that the data be in nominal form, or the actual number of cases (frequency of occurrence) that fall into two or more discrete categories. It is considered to be a nonparametric test (no population assumptions are required for its use). The basic equation is as follows:

$$\chi^2 = \Sigma \, \frac{(f_0 - f_e)^2}{f_e}$$

where f_0 denotes the frequencies actually observed and f_e the frequencies expected on the basis of chance.

Coefficient of Contingency (C)   A test of correlation on nominal data sorted into any number of independent cells.

$$C = \sqrt{\frac{\chi^2}{N + \chi^2}}$$

Coefficient of Determination (r^2)   A method for determining what proportion of the information about Y is contained in X; found by squaring the Pearson r.

Confidence Interval   The range of predicted values within which one presumes with a stated degree of confidence that the true parameter will fall. Usually, confidence intervals are determined on the basis of a probability value of .95 (95% certainty) or .99 (99% certainty). Since sample values differ, even when drawn from the same population, one can never be sure that any particular interval contains the parameter.

Control Group   In experimental research, the control group is the comparison group, or the group that ideally receives zero magnitude of the independent variable. The use of a control group is critical in evaluating the pure effects of the independent variable on the measured responses of the subjects.

Correlated (or Dependent) Samples   In experimental research, two or more samples that are *not selected independently*. The selection of one sample determines who will be selected for the other sample(s), as in a matched-subjects design.

Correlation Coefficient   A quantitative formulation of the relationship existing among two or more variables. A correlation is said to be positive when high scores on one variable associate with high scores on another variable, and low scores on the first variable associate with low scores on the second. A correlation is said to be negative when high scores on the first variable associate with low scores on the second, and vice versa. Correlation coefficients range in value from +1.00 to −1.00. Correlation coefficients falling near the zero point indicate no consistent relationship among the measured variables. In social research, the correlation coefficient is usually based on taking several response measures of *one group of subjects.*

Counterbalancing   A technique used in repeated-measures designs to help prevent confounding of the independent variable by evenly distributing sequencing effects across all treatment conditions.

Cross-Sectional Research   Type of nonexperimental research, sometimes used to obtain data on possible growth trends in a population. The researcher selects a sample (cross section) at one age level, say, 20-year-olds, and compares these measurements with those taken on a sample of older subjects, say, 65-year-olds. Comparisons of this type are often misleading (today's 20-year-olds may have very different environmental backgrounds, educational experience, for example, than the 65-year-old subjects have).

Deciles   Divisions of a distribution representing tenths, the first decile representing the 10th percentile, and so on. The 5th decile, therefore, equals the 50th percentile, the 2nd quartile, and the median.

Degrees of Freedom (df)   With interval (or ratio) data, degrees of freedom refer to the number of scores free to vary after certain restrictions have been placed on the data. With six scores and a restriction that the sum of these scores must equal a specified value, then five of these scores are free to take on any value whereas the sixth score is fixed (not free to vary). In inferential statistics, the larger the size of the sample, the larger the number of degrees of freedom. With nominal data, degrees of freedom depend, *not on the size of the sample*, but on the number of categories in which the observations are allocated. Degrees of freedom are here based on the number of frequency values free to vary after the sum of the frequencies from all the cells has been fixed.

Dependent Variable In any antecedent-consequent relationship, the consequent variable is called the dependent variable. The dependent variable is a measure of the output side of the input–output relationship. In experimental research, the dependent variable is the possible effect half of the cause-and-effect relationship, whereas in correlational research, the dependent variable is the measure being predicted and is called the *criterion variable*. In the social sciences, the dependent variable is usually a response measure.

Descriptive Statistics Techniques for describing and summarizing data in abbreviated, symbolic form; shorthand symbols for describing large amounts of data.

Deviation Score (x) The difference between a single score and the mean of the distribution. It is found by subtracting the mean, $\bar{X}$, from the score X. The deviation score $(X - \bar{X})$ is symbolized as x. Thus, $x = X - \bar{X}$.

Distribution The arrangement of measured scores in order of magnitude. Listing scores in distribution form allows the researcher to notice general trends more readily than with an unordered set of raw scores. A *frequency distribution* is a listing of each score achieved, together with the number of individuals receiving that score. When graphing frequency distributions, one usually indicates the scores on the horizontal axis (abscissa) and the frequency of occurrence on the vertical axis (ordinate).

Double-Blind Study A method used by researchers in an attempt to reduce one form of experimental error. In a double-blind study neither the person conducting the study nor the subjects are aware of which group is the experimental group and which the control. This helps to prevent any unconscious bias on the part of the experimenter or any contaminating motivational sets on the part of the subjects.

Exclusion Area The extreme areas under the normal curve. Because of the curve's symmetry, the extreme areas at both the top and bottom of the curve are excluded by two z scores that are equidistant from the mean.

Experimental Design Techniques used in experimental research for creating equivalent

groups of subjects. There are three basic experimental designs: (1) after-only, where subjects are randomly assigned to control and experimental groups and the dependent variable is measured only after the introduction of the independent variable; (2) before–after (repeated measures), where a group of subjects is used as its own control and the dependent variable is measured both pre and post the introduction of the independent variable; and (3) matched-subjects, where subjects are equated person for person on some variable deemed relevant to the dependent variable.

Experimental Research Research conducted using the experimental method, where an independent variable is manipulated (stimulus) in order to bring about a change in the dependent variable (response). Using this method, the experimenter is allowed the opportunity for making cause-and-effect inferences. Experimental research requires careful controls in order to establish the pure effects of the independent variable. Equivalent groups of subjects are formed, then exposed to different stimulus conditions, and then measured to see if differences can be observed.

External Validity The extent to which an experimental finding can be projected to the population at large. An experiment is high in external validity when the sample is representative of the population and when it simulates real-life conditions.

Factorial ANOVA As opposed to a one-way ANOVA, the factorial ANOVA allows for the analysis of data when more than one independent variable is involved. Results can be analyzed on the basis of the main effects of each independent variable or on the basis of the possible interaction among the independent variables. Data to be analyzed should be in at least interval form.

Fisher, Sir Ronald (1890–1962) English mathematician and statistician who developed the analysis of variance technique, or F (for Fisher) ratio.

Frequency Distribution Curve Graphing procedure where measures, such as raw scores or z scores, are plotted on the X axis (abscissa) and frequency of occurrence on the Y axis (ordinate).

Frequency Polygon A graphic display of data

where single points are plotted above the measures of performance. The height where the point is placed indicates the frequency of occurrence. The points are connected by a series of straight lines.

Friedman ANOVA by Ranks (χ_r^2) A test of the hypothesis of difference on ordinal data when the sample groups have either been matched or a single sample has been repeatedly measured. The Friedman ANOVA is the ordinal counterpart of the within-subjects F.

$$\chi_r^2 = \frac{12}{Nk(k+1)}(\Sigma R_1^2 + \Sigma R_2^2 + \Sigma R_3^2 + \cdots) - 3N(k+1)$$

Galton, Sir Francis (1822–1911) The "father of intelligence testing" and the creator of the concept of individual differences. Galton also introduced the concepts of regression and correlation (although it was left to his friend and colleague Karl Pearson to work out the mathematical equations).

Gambler's Fallacy An erroneous assumption that independent events are somehow related. If a coin is flipped 10 times and comes up heads each of those times, the gambler's fallacy predicts that it is virtually certain for the coin to come up tails on the next flip. Since each coin flip is independent of the preceding one, the probability remains the same (.50) for each and every coin flip, regardless of what has happened in the past. The gambler remembers the past, but the coin does not.

Gauss, Karl Friedrich (1777–1855) German mathematician credited with having originated the normal curve. For this reason the normal curve is often called the Gaussian curve.

Gossett, William Sealy (1876–1937) Using the pen name "Student," Gossett, while working for the Guinness Brewing Company in Ireland, developed the technique of using sample data to predict population parameters, which led to the development of the t test.

Halo Effect A research error arising from the fact that people who are viewed positively on one trait are often also thought to have many other positive traits. Advertisers depend on this effect when they use famous personalities to endorse various products—anyone

who can throw touchdown passes *must be* an expert in evaluating razor blades. Researchers must guard against the halo effect, as it will contaminate the independent variable.

Hawthorne Effect A major research error due to response differences resulting not from the action of the independent variable, but from the flattery or attention paid to the subjects by the experimenter. Typically, the potential for this error is inherent in any study using the before–after experimental design without an adequate control group. Any research, for example, where subjects are measured, then subjected to some form of training, then measured again, should be viewed with suspicion unless an appropriate control group is used—that is, an equivalent group that is measured, *then not subjected to the training,* and then measured again. Only then can the researcher be reasonably confident that the response differences are due to the pure effects of the independent variable.

Histogram (Bar Graph) A graphic representation of data in which a series of rectangles (bars) is drawn above the measure of performance. The height of each bar indicates the frequency of occurrence.

Homogeneity of Variance An assumption of both the t and F ratios, which demands that the variability within each of the sample groups being compared should be fairly similar.

Homoscedasticity The fact that the standard deviations of the Y scores along the regression line should be fairly equal. Otherwise, the standard error of estimate is not a valid index of accuracy.

Inclusion Area The middlemost area of the normal curve, included between two z scores equidistant from the mean. Because of the symmetry of the normal curve, the middlemost area includes, in equal proportions, the area immediately to the left of the mean and the area immediately to the right of the mean.

Independent Variable In any antecedent-consequent relationship, the antecedent variable is called the independent variable. Independent variables may be manipulated or assigned. A manipulated independent variable occurs when the researcher deliberately alters the environmental conditions to which subjects are being exposed. An assigned independent variable occurs when the researcher categorizes subjects on the basis of some preexisting trait.

Whether the independent variable is manipulated or assigned determines whether the research is experimental (manipulated independent variable) or post-facto (assigned independent variable). In experimental research, the independent variable may be the causal half of the cause-and-effect relationship. In correlational research, the independent variable is the measure from which the prediction will be made and is, thus, called the *predictor* variable.

Inductive Fallacy   An error in logic resulting from overgeneralizing on the basis of too few observations. The inductive fallacy occurs when one assumes that all members of a class have a certain characteristic because one member of that class has it. It would be fallacious to assume that all Mongolians are liars on the basis of having met one Mongolian who was a liar.

Inferential (Predictive) Statistics   Techniques for using the measurements taken on samples to predict the characteristics of the population—the use of descriptive statistics for inferring parameters.

Interaction Effect   When two or more independent variables are involved (as in an ANOVA), the interaction effect is produced by the combined influence of these independent variables working in concert.

Interdecile Range   Those scores that include the middlemost 80% of a distribution, or the difference between the first and ninth deciles.

Internal Validity   The extent to which the results of an experiment can unambiguously identify the cause-and-effect relationship. A high degree of internal validity indicates that the experiment is relatively free of the contaminating effects of confounding variables—thus allowing for a clear interpretation of the pure effects of the independent variable(s).

Interquartile Range   Those scores that include the middlemost 50% of a distribution, or the difference between the first and third quartiles.

Interval Data   Data (measurements) in which values are assigned such that both the order of the numbers and the *intervals* between

numbers are known. Thus, interval data not only provide information regarding greater than or less than status, but also information as to how much greater or less than.

Kruskal–Wallis *H* Test   A test of the hypothesis of difference on ordinal data among at least three independently selected random samples. The *H* test is the ordinal counterpart of the one-way ANOVA.

$$H = \frac{12}{N(N+1)} \left(\frac{\Sigma R_1^2}{n_1} + \frac{\Sigma R_2^2}{n_2} + \frac{\Sigma R_3^2}{n_3} \cdots \right) - 3(N+1)$$

Kurtosis (ku)   The state or degree of the curvature of a unimodal frequency distribution. Kurtosis refers to the peakedness or flatness of the curve.

Leptokurtic Distribution   A unimodal frequency distribution in which the curve is relatively peaked—most of the scores occur in the middle of the distribution—with very few scores occurring in the tails. A leptokurtic distribution yields a relatively small standard deviation.

Longitudinal Research   A type of post-facto research in which subjects are measured repeatedly throughout their lives in order to obtain data on possible trends in growth and development. Terman's* massive study of growth trends among intellectually gifted children is an example of this research technique. The study, begun in the early 1920s, is still in progress today and is still providing science with new data. Longitudinal research requires great patience on the part of the investigator, but the obtained data are considered to be more valid than those obtained using the cross-sectional approach.

Main Effects   When two or more independent variables are involved (as in an ANOVA), the main effects are produced by the action of each independent variable working separately.

Mann–Whitney *U* Test   A test on ordinal data of the hypothesis of difference between two independently selected random samples. The *U* test is the ordinal counterpart of the independent *t* test.

*L. M. Terman, *Genetic Studies of Genius* (Stanford, Calif.: Stanford University Press, 1925, 1926, 1930, 1947, 1959).

$$z_U = \frac{U - [(n_1)(n_2)]/2}{\sqrt{[n_1 n_2 (n_1 + n_2 + 1)]/12}}$$

McNemar Test Technique developed by the statistician Quinn McNemar that uses chi square for the analysis of nominal data from correlated samples.

$$\chi^2 = \frac{|a - d|^2}{a + d}$$

Mean ($\bar{X}$) A measure of central tendency specifying the arithmetic average. Scores are added and then divided by the number of cases.

$$\bar{X} = \frac{\Sigma X}{N}$$

The mean is best used when the distribution of scores is balanced and unimodal. In a normal distribution, the mean coincides with the median and the mode. When the entire population of scores is used, the mean is designated by the Greek letter μ (mu).

Measurement A method of quantifying observations by assigning numbers to them on the basis of specific rules. The rules chosen determine which scale of measurement is being used: nominal, ordinal, interval, or ratio.

Median (Mdn) A measure of central tendency that specifies the middlemost point in an ordered set of scores. The median always represents the 50th percentile. It is the most valid measure of central tendency whenever the distribution is skewed.

Mesokurtic A unimodal frequency distribution whose curve is normal. (*See* Normal Curve.)

Mode (Mo) A measure of central tendency that specifies the most frequently occurring score in a distribution of scores. When there are two most common points, the distribution is said to be bimodal.

Multiple R A single numerical value that quantifies the correlation among three or more variables. The equation for a three-variable multiple R is as follows:

$$R_{y \cdot 1,2} = \sqrt{\frac{r_{y,1}^2 + r_{y,2}^2 - 2 r_{y,1} r_{y,2} r_{1,2}}{1 - r_{1,2}^2}}$$

Multiple Regression Technique using the multiple R for making predictions of one variable given measures on two or more others. It requires the calculation of the intercept (a) and also at least two slopes (b_1 and b_2). For the three-variable situation, the multiple regression equation is as follows:

$$Y_{\mathrm{Mpred}} = b_1 X_1 + b_2 X_2 + a$$

Nominal (or Categorial) Data Data (measurements) in which numbers are used to label discrete, mutually exclusive categories; nose-counting data, which focuses on the frequency of occurrence within independent categories.

Nonparametrics Statistical tests that neither predict the population parameter, μ, nor make any assumptions regarding the normality of the underlying population distribution. These tests may be run on ordinal or nominal data, and typically have less power than do the parametric tests.

Normal Curve A frequency distribution curve resulting when scores are plotted on the horizontal axis (X) and frequency of occurrence is plotted on the vertical axis (Y). The normal curve is a theoretical curve shaped like a bell and fulfills the following conditions: (1) most of the scores cluster around the center, and as we move away from the center in either direction there are fewer and fewer scores; (2) the scores fall into a symmetrical shape— each half of the curve is a mirror image of the other; (3) the mean, median, and mode all fall at precisely the same point, the center; (4) there are constant area characteristics regarding the standard deviation; and (5) the curve is asymptotic to the abscissa.

Null or "Chance" Hypothesis The assumption that the results are simply due to chance. When testing the hypothesis of difference, the null hypothesis states that no real differences exist in the population from which the samples were drawn. When testing the hypothesis of association, the null hypothesis states that the correlation in the population is equal to zero (does not exist).

Odds The chances *against* a specific event occurring. When the odds are 5 to 1, for example, it means that the event will *not* occur five times for each single time that it will occur.

One-Tail (directional) Test The use of only

one tail of the distribution for testing the null hypothesis. For example, with the t test, if the direction of the difference has already been specified in the alternative hypothesis, then the critical value of the t statistic, for a given number of degrees of freedom, is taken from only one side of its distribution.

Ordinal Data   Rank-ordered data, that is, derived only from the order of the numbers, not the differences between them. Ordinal measures provide information regarding greater than or less than status, but *not* how much greater or less.

Ordinate   The vertical or Y axis in the coordinate system. On a frequency distribution, the ordinate indicates the frequency of occurrence.

Paired t Ratio   Statistical test of the hypothesis of difference between *two sample means* where the sample selection is not independent. The paired t (also called correlated t) requires interval data and is typically used when the design has been before–after or matched-subjects.

$$t = \frac{\bar{X}_1 - \bar{X}_2}{\sqrt{s_{\bar{x}_1}^2 + s_{\bar{x}_2}^2 - 2r_{1,2}s_{\bar{x}_1}s_{\bar{x}_2}}}$$

Parameter   Any measure obtained by having measured the entire population. Parameters are, therefore, usually inferred rather than directly measured.

Partial Correlation   Correlation technique that allows for the ruling out of the possible effects of one or more variables on the relationship among the remaining variables. In the three-variable situation, the partial correlation rules out the influence of the third variable on the correlation between the remaining two variables. The equation for partialing out the influence of the third variable is as follows:

$$r_{y,1\cdot2} = \frac{r_{y,1} - r_{y,2}r_{1,2}}{\sqrt{(1 - r_{y,2}^2)(1 - r_{1,2}^2)}}$$

Pascal, Blaise (1623–1662)   French mathematician and philosopher who introduced the concepts of probability and random events.

Path Analysis   Sophisticated correlational techniques have been used in a causal modeling method called path analysis. Although path analysis is a correlational procedure, it is used to test a set of hypothesized cause-and-effect relationships among a series of variables which are logically ordered on the basis of time. Since, as logic suggests, a causal variable must precede (in time) a variable it is supposed to influence, the correlational analysis is done on a set of variables, each presumed to show a causal ordering. The attempt is made to find out whether a given variable is being influenced by the variables that precede it and then, in turn, is influencing the variables that follow it. A "path" diagram is drawn that portrays the assumed direction of the various relationships. Although not as definitive a proof of causation as when the independent variable is experimentally manipulated, path analysis takes us a long step forward from the naive extrapolations of causation which at one time were taken from simple bivariate correlations.

Pearson, Karl (1857–1936)   English mathematician and colleague of Sir Francis Galton. It was Pearson who translated Galton's ideas on correlation into precise mathematical terms, creating the equation for the product-moment correlation coefficient, or the Pearson r.

Pearson r   Statistical technique introduced by Karl Pearson for showing the degree of relationship between two variables. Also called the product-moment correlation coefficient, it is used to test the hypothesis of association, that is, whether or not there is a relationship between two sets of measurements. The Pearson r can be calculated as follows:

$$r = \frac{(\Sigma XY)/N - (\bar{X})(\bar{Y})}{SD_x SD_y} \quad \text{or}$$

$$\frac{\Sigma XY - [(\Sigma X)(\Sigma Y)]/N}{\dfrac{N - 1}{s_x s_y}}$$

Computed correlation values range from $+1.00$ (perfect positive correlation) through zero to -1.00 (perfect negative correlation). The farther the Pearson r is from zero, whether in a positive or negative direction, the stronger is the relationship between the two variables. The Pearson r can be used for making better than chance predictions, but should not be used alone for isolating causal factors.

Percentiles (or Centiles) The percentage of cases falling below a given score. Thus, if an individual scores at the 95th percentile, that individual has exceeded 95 percent of all persons taking that particular test. If test scores are normally distributed, and if the standard deviation of the distribution is known, percentile scores can easily be converted to the resulting z scores.

Percentile Rank The value that indicates a given percentile. A point at the 75th percentile is said to have a percentile rank of 75.

Platykurtic Distribution A unimodal frequency distribution in which the curve is relatively flat. Large numbers of scores appear in both tails of the distribution. A platykurtic distribution of scores yields a relatively large standard deviation.

Point Estimate The use of a sample value for predicting a *single* population value. For example, the use of the sample mean for estimating μ is a point estimate.

Point of Intercept (a) In a scatter plot, the point of intercept is the location where the regression line crosses the ordinate, or the value of Y when X is equal to zero (or its minimum value). In the regression equation, $Y = bX + a$, the intercept is symbolized by the a term.

Population The entire number of persons, things, or events (observations) having at *least* one trait in common. Populations may be limited (finite) or unlimited (infinite).

Post-Facto Research A type of research that, while not allowing for direct cause-and-effect conclusions, does allow the researcher to make better than chance predictions. In such research, subjects are measured on one response dimension, and these measurements are compared with other trait or response measures.

Power (1 − β) A measure of the sensitivity of a statistical test. The more powerful a test is, the less is the likelihood of committing the beta error (accepting the null hypothesis when it should have been rejected). The higher a test's power, the higher is the probability of a small difference or a small correlation being found to be significant.

Primary Variance Variability of the dependent variable which is assumed to have been produced by the direct action of the independent variable.

Probability (P) The statement as to the number of times a specific event, s, can occur out of the total possible number of events, t.

$$P = \frac{s}{t}$$

Probability should be expressed in decimal form. Thus, a probability of 1/20 is written as .05.

Quartiles Divisions of a distribution representing quarters; the first quartile representing the 25th percentile, the second quartile the 50th percentile (or median), and the third quartile the 75th percentile.

Quota Sampling A method used for selecting a representative sample which is based on drawing subjects in proportion to their existing percentages in the population. If, for example, 55% of a certain population is composed of women, then 55% of the sample must also be made up of women.

Random Sample Sample selected in such a way that every element or individual in the entire population has an equal chance of being chosen. When samples are selected randomly, then sampling error should also be random and the samples representative of the population.

Range (R) A measure of variability that describes the entire width of the distribution. The range is the difference between the two most extreme scores in a distribution and is, thus, equal to the highest value minus the lowest value.

Ratio Data Data (measurements) that provide information regarding the order of numbers, the difference between numbers, and also an *absolute* zero point. It permits comparisons, such as A being three times B, or one-half of B.

Regression Line The single straight line that lies closest to all the points in a scatter plot. The regression line can be used for making correlational predictions when three important pieces of information are known: (1) the extent to which the scatter points deviate from the line, (2) the slope of the line, and (3) the point of intercept.

$$Y = bX + a$$

Representative Sample   A sample that reflects the characteristics of the entire population. Random sampling is assumed to result in representative samples.

Sample   A group of any number of observations selected from a population, as long as it is less than the total population.

Sampling Distributions   Distributions made up of measures taken on successive random samples. Such measures are called statistics, and when all samples in an entire population are measured, the resulting sampling distributions are assumed to approximate normality. (*See* Central Limit Theorem.) Two important sampling distributions are the distribution of means and the distribution of differences.

Sampling Error   The expected difference between a measure of the sample and a measure of the population ($\bar{X} - \mu$). Under conditions of random sampling, the probability of obtaining a sample mean greater than the population mean is identical to the probability of obtaining a sample mean less than the population mean ($P = .50$).

Scatter Plot   A graphic format in which each point represents a pair of scores, the score on X as well as the score on Y. The array of points in a scatter plot typically forms an elliptical shape (a result of the central tendency usually present in both the X and Y distributions).

Secular Trend Analysis   A method, using correlational techniques, for predicting linear trends across *time*. Historical data are used to forecast future results, based on the not-always-wise assumption that the past trend will continue.

Secondary Variance   Variability of the dependent variable that is not under experimental control but is instead a result of confounding variables. Holding secondary variance to a minimum helps to ensure a high degree of internal validity.

Significance   A statistical term used to indicate that the results of a study are not simply a matter of chance. Researchers talk about significant differences and significant correlations, the assumption being that chance has been ruled out (on a probability basis) as the explanation of these phenomena.

Skewed Distribution   An unbalanced distribution in which there are a few extreme scores in *one direction*. In a skewed distribution, the best measure of central tendency is the median.

Spearman, Charles E. (1863–1945)   English psychologist and test expert who worked in the area of measuring intelligence and identifying the factors that make up intelligence. In pursuing his correlational studies on intellectual factors, he produced a correlation technique for the analysis of ordinal data called the Spearman rho, or the r_S.

Spearman r_S, The   Correlation coefficient devised by Charles E. Spearman for use with rank-ordered (ordinal) data. Sometimes called the Spearman ρ (rho), the coefficient is found as follows:

$$r_S = 1 - \frac{6 \Sigma d^2}{N(N^2 - 1)}$$

Standard Deviation   A measure of variability that indicates how far *all* scores in a distribution vary from the mean. The standard deviation has a constant relationship with the area under the normal curve. (*See* Normal Curve.) The actual or true standard deviation of any distribution is calculated as follows:

$$SD = \sqrt{\frac{\Sigma X^2}{N} - \bar{X}^2}$$

The unbiased *estimate* of the standard deviation in the population is calculated with the following equation:

$$s = \sqrt{\frac{\Sigma X^2 - (\Sigma X)^2 / N}{N - 1}}$$

When the standard deviation is calculated on the basis of all scores in the entire population, it is designated as σ (lowercase Greek letter sigma).

Standard Error of Difference   The standard deviation of the entire distribution of differences between pairs of successively drawn random sample means. These pairs of sample means are taken from the population until that population is exhausted. An estimate of this value can be made on the basis of the information contained in just two samples.

$$s_{\bar{x}_1-\bar{x}_2} = \sqrt{s_{\bar{x}_1}^2 + s_{\bar{x}_2}^2 - 2r_{1,2}s_{\bar{x}_1}s_{\bar{x}_2}}$$

When sample selections are independent of each other, the correlation term $(2r_{1,2}s_{\bar{x}_1}s_{\bar{x}_2})$ is equal to zero and is, therefore, not used. The estimated standard error of difference for independent samples is, thus, as follows:

$$s_{\bar{x}_1-\bar{x}_2} = \sqrt{s_{\bar{x}_1}^2 + s_{\bar{x}_2}^2}$$

Standard Error of Estimate (SE$_{est}$) A technique for establishing the accuracy of a predicted Y value obtained by using the regression equation. It estimates the amount of variation that occurs between the predicted values and the regression line. The higher the correlation between X and Y, the lower is the resulting value of the standard error of estimate and the more accurate is the predicted Y value. When $r = 0$, the standard error of estimate is equal to the standard deviation of the Y distribution.

$$SE_{est} = SD_y\sqrt{1 - r^2}$$

Standard Error of the Mean The standard deviation of the entire distribution of random sample means successively selected from a single population until that population is exhausted. An estimate of the standard error of the mean can be made on the basis of the information contained in a single random sample, that is, the variability within the sample and the size of the sample. When using the estimated population standard deviation, the equation becomes

$$s_{\bar{x}} = \frac{s}{\sqrt{N}}$$

When using the actual standard deviation of the sample scores, the equation becomes

$$s_{\bar{x}} = \frac{SD}{\sqrt{N-1}}$$

Standard Error of Multiple Estimate A technique for assessing the accuracy of a prediction that has been generated from the multiple regression equation. For the three-variable situation, the standard error of multiple estimate is as follows:

$$SE_{M\ est} = SD_y\sqrt{1 - R_{y\cdot 1,2}^2}$$

Statistic Any measure that is obtained from a sample as opposed to the entire population. The range (or the standard deviation or the mean) of a set of sample scores is a statistic.

Stratified or Quota Sampling Selecting a sample that directly reflects the population characteristics. If it is known that 45% of the population is composed of males, and if it is assumed that gender is a relevant research variable, then the sample must contain 45% of male subjects.

Sum of Squares (SS) An important concept for ANOVA; the sum of squares equals the sum of the squared deviations of all scores around the mean.

$$SS = \Sigma x^2 = \Sigma X^2 - \frac{(\Sigma X)^2}{N}$$

When the sum of squares is divided by its appropriate degrees of freedom, the resulting value is called the mean square, or variance.

t **Ratio** Statistical test used to establish whether or not significant (nonchance) differences can be detected between *two* means. With two samples, it is the ratio of the difference between the sample means to an estimate of the standard error of difference. With one sample, it is the ratio of the difference between the sample mean and population mean to an estimate of the standard error of the mean. Two samples:

$$t = \frac{\bar{X}_1 - \bar{X}_2}{s_{\bar{x}_1-\bar{x}_2}}$$

One sample:

$$t = \frac{\bar{X} - \mu}{s_{\bar{x}}}$$

T **Score** A converted standard score such that the mean equals 50 and the standard deviation equals 10. *T* scores, thus, range from a low of 20 to a high of 80.

Tukey's HSD (Honestly Significant Difference) A multiple comparison technique developed by J. W. Tukey for establishing whether or not the differences among various sample means are significant. The test is performed *after* the ANOVA when the *F* ratio is significant. It is, thus, a post hoc test.

Two-Tail (nondirectional) Test   The use of both tails of the distribution for testing the null hypothesis. For example, with the *t* test, if the direction of the difference is not specified in the alternative hypothesis, then the critical value of the *t* statistic, for a given number of degrees of freedom, is taken from both tails of its distribution.

Unimodal Distribution   A distribution of scores in which only one mode (most frequently occurring score) is present.

Variability Measures   Measures that give information regarding individual differences, or how persons or events vary in their measured scores. The three most important measures of variability are the range, the standard deviation, and the variance (which is the standard deviation squared).

Variable   Anything that varies *and can be measured.* In experimental research, the two most important variables to be identified are the independent variable and the dependent variable. The independent variable is a stimulus, is actively manipulated by the experimenter, and is the causal half of the cause-and-effect relationship. The dependent variable is a measure of the subject's response and is the effect half of the cause-and-effect relationship.

Variance   A measure of variability that indicates how far all of the scores in a distribution vary from the *mean.* Variance is equal to the square of the standard deviation.

Wilcoxon *T* Test   A test on ordinal data of the hypothesis of difference between two sample groups when the selections are correlated (as in the matched-subjects design). The Wilcoxon *T* is the ordinal counterpart of the paired *t*.

Within-Subjects *F* Ratio   Statistical test of the hypothesis of difference among several sample means, where sample selection is not independent. It is used when samples are correlated, as in repeated-measure designs, and the data are in at least interval form.

Yates Correction   A correction for continuity usually applied to a 2×2 chi square analysis (df $= 1$) whenever any or all of the expected frequencies are less than 10. The absolute difference between f_o and f_e is reduced by .50, resulting in a slightly lower chi square value.

z Distribution   The standard normal distribution, or as it is sometimes called, the unit-normal distribution, where the mean is equal to zero and the standard deviation equal to 1.00. It is the distribution which is shown in Table A, page 438.

z Score (Standard Score)   A number that results from the transformation of a raw score into units of standard deviation. The *z* score specifies how far above or below the mean a given raw score is in these standard deviation units. Any raw score above the mean converts to a positive *z* score, while scores below the mean convert to negative *z* scores. The *z* score is also referred to as a standard score. The normal deviate, *z*, has a mean equal to 0 and a standard deviation equal to 1.00.

$$z = \frac{X - \bar{X}}{SD}$$

z Test   A method of hypothesis testing which can be used when the parameter values are normally distributed and the mean and standard deviation of the distribution are already known.

Index